# INSIDE ArcView® GIS

Second Edition

**Scott Hutchinson and Larry Daniel**

# INSIDE ArcView® GIS
## Second Edition

By Scott Hutchinson and Larry Daniel

Published by:
**OnWord Press**
**2530 Camino Entrada**
**Santa Fe, NM 87505-4835 USA**




SAN 694-0269

First Edition, 1995
Second Edition, 1997

10 9 8 7 6 5 4 3 2

Printed in the United States of America

**Library of Congress Cataloging-in-Publication Data**

Hutchinson, Scott, 1952–
Inside ArcView GIS / Scott Hutchinson and Larry Daniel -- 2nd ed. p. cm.
Includes index.
ISBN 1-56690-116-2
1. Geographic information systems. 2. ArcView. I. Daniel, Larry, 1961– . II. Title.
G70.212.H87 1996
910'.285--dc21 96-47248
CIP

# Trademark Acknowledgments



# Warning and Disclaimer



# About the Authors

Since 1983, **Scott Hutchinson** has worked with geographic information systems (GIS) in the U.S. Soil Conservation Service, the Arizona Department of Transportation, and the Arizona State Land Department. He has a B.S. degree in agriculture (soil science) from the University of Arizona. He recently completed a system, using ArcView, that linked digitized land parcel and other resource data to the Arizona State Land Department Business Systems database, to provide querying and mapping of the state's trust land. Scott is an ESRI-certified ArcView instructor, and is presently providing ArcView training in the Washington, D.C., metro area.

**Larry Daniel** is president and principal consultant of the Daniel Consulting Group, a GIS management consulting and software development organization out of Austin, Texas. In years prior to forming DCG, he was vice president of the Castillo Company in Phoenix, Arizona, and director of GIS at MPSI in Tulsa, Oklahoma. Larry holds an M.A. in geography from the University of Texas at Austin and a B.S. in computer engineering from Bucknell University. He has been involved with GIS for over ten years, has been an ongoing columnist for *Business Geographics* since 1993, has contributed to other professional journals such as *GIS World, GeoInfo Systems, Earth Observation Magazine, Computer-Aided Engineering,* and *Design Management,* and is also a co-author of *INSIDE MapInfo Professional.*

## Acknowledgments

First and foremost, I wish to acknowledge my debt to Larry Daniel, who initially approached me on this project, and without whom there would be nothing to acknowledge. His vision and expertise are evidenced on every page of this book. My thanks to Gary Irish and Lynn Larson at the Arizona State Land Department for providing the opportunity to participate in the ArcView Beta program, which gave me the expertise necessary to write this book. Thanks to my co-workers, Marleen Riggs and Rob Rutland, for their valuable perspective as ArcView end users. Special thanks to Joan Wakefield and Jim Cristea with the city of Tempe, for providing the image and CAD drawings used in the exercises in Chapter 9, and to Lisa Sivey with the city of Sedona for the insert on cartographic design, as well as for being a sounding board for seemingly half-baked ideas.

Thanks also to Environmental Systems Research Institute, Inc., especially Ali Fain, Jack McCarthy, Larry Batten, John Calkins, April Nichols, Kathleen Bertrand, and Rich Turner. Thanks also to my mother, for making it all possible, including the computer and software. And finally, thanks to my wife Beth, and Robert, and Ana, who watched as the second edition developed, first from a distance, and then face-to-face!

*Scott Hutchinson*

As second author, I want first to convey my continuing applause for Scott Hutchinson, for his efforts in staying abreast of the ArcView environment and for sharing his knowledge so capably with others. It was a pleasure to work with Scott and High Mountain Press, and to then witness success of the first edition—I hope that the second edition serves as well toward enriching the use of the ArcView product.

Thanks, too, to the many people who have been personally involved in fueling my success with GIS—Ken Foote and David Huff at the University of Texas, Rick Baumgartner at IBM, Mark Darling at American Isuzu, and the many clients I serve—with each day we seem to push, challenge, and advance each other's interests.

Lastly, a hearty and very special thanks to my wife and family: Mel, Melanie, Lianna, Dwight, and Meredith. You have endured with me many of the trials and long hours necessary to make these types of projects happen—thank you for being my support.

*Larry Daniel*

---

# Acknowledgment of Desktop Mapping Innovators

Many organizations and individuals in the private and public sectors have served as innovators in the use and development of GIS and desktop mapping. Several of these innovators provided data or ArcView applications for this book. Our special innovator list follows: Equifax/NDS; International House of Pancakes; Arizona State Land Department; American Isuzu Motors; Maine Department of Environmental Protection; Dr. James F. Campbell, School of Business Administration, University of Missouri-St. Louis; and Professor Ben Niemann, Land Resources Department, and Professor Steve Ventura, Environmental Studies Department, University of Wisconsin-Madison.

While compiling our second edition, we posted several calls for "real world" material and found many parties in the ArcView community willing to share their applications. Special thanks to Jayan Devasundarum (Maryland State Health Department), Gwen Ford (University of South Carolina), Robert Greene (Los Alamos National Laboratory), Leigh McNaught (ViGYAN), and Janko Tchoukanski

(South Africa, Department of Water Affairs and Forestry) for providing the basis of our case studies, and a general "thank you" to all who expressed interest.

## OnWord Press Credits

OnWord Press is dedicated to the fine art of professional documentation. In addition to the authors who developed the material for this book, other members of the OnWord Press team contributed their skills to make the book a reality. Thanks to the following people and other members of the OnWord Press team who contributed to the production and distribution of this book.

Dan Raker, President
Daniel Clavio, Market Development and Strategic Relations Director
David Talbott, Acquisitions Director
Dave Surette, Administration and Finance Director
Barbara Kohl, Editorial Manager, Acquisitions Manager
Lisa Levine, Senior Editor
Carol Leyba, Production Manager
Cynthia Welch, Production Artist
Lauri Hogan, Marketing Services Manager
Kristie Reilly, Assistant Editor
Lynne Egensteiner, Cover designer, Illustrator

# Contents

## Chapter 10: Editing Shapefiles 255



## Chapter 11: Beyond the Basics 283



## Chapter 12: Advanced Functionality 327



## Chapter 13: Optimizing Project Design 337



## Chapter 14: ArcView Customization 349

# Introduction

Congratulations on selecting ArcView GIS and for choosing this book. ArcView GIS is the premier PC-based geographic information system (GIS) software by Environmental Systems Research Institute, Inc. (ESRI). This book will help validate your organization's good judgment in choosing ArcView GIS.

*INSIDE ArcView GIS* is not just a rehash of the ArcView GIS manual and online help system created by writers who are not familiar with the software. This book contains the knowledge that comes from time spent on the front lines of public and private sector organizations, helping individuals, departments, corporations, and government agencies learn to work effectively using GIS and desktop mapping software. Our experience has ranged from helping business people and government agency professionals to get started with GIS to teaching developers how to customize GIS software to fit their unique environments. We have included tips and tricks that we have learned over time—many of which are either not in the manuals or are difficult to find.

We hope that *INSIDE ArcView GIS* enables you to get "up and running" with ArcView GIS. If you have any comments, questions, or recommendations about the book, we encourage you to contact us at OnWord Press, 2530 Camino Entrada, Santa Fe, NM 87505-4835, USA, or to e-mail us at *readers@hmp.com.*

# On the Current Status of Desktop Mapping and ArcView GIS

Recent developments by Microsoft and other companies incorporating mapping functionality into desktop applications support the observation that desktop mapping has joined the ranks of "mainstream" applications such as spreadsheets, databases, and presentation graphics. Advances in imaging, global positioning systems (GPS), and pen-based computing are further extending the functionality of desktop mapping. Given that an estimated 80 percent of all data has a spatial component, you might ask where desktop mapping does not apply!

All major functionality of geographic information systems (GIS) software is present in ArcView GIS. ArcView GIS has several other notable strong points: Avenue, the object-oriented programming language; direct access to ARC/INFO coverages; connectivity to CAD data; sophisticated ArcView extensions; and database connectivity through use of ESRI's Spatial Database Engine (SDE).

## Avenue

The advantages of using Avenue for customizing ArcView GIS are covered in Chapter 14, but they are worth summarizing here. ArcView GIS is comprised of objects and requests, and Avenue is an object-oriented programming language that allows you to directly access these objects. Because Avenue provides the ability to access the objects with which ArcView GIS is constructed, it allows a high level of control over all aspects of your environment, including data, applications, interface, and output. Even if you have no immediate intention of customizing ArcView GIS, you can think of Avenue as an insurance policy: when and if you require customization, Avenue's capabilities and power will be available.

## Direct Access to ARC/INFO

The ability to directly access ARC/INFO coverages is clearly an advantage to sites using ARC/INFO or with ready access to data in ARC/INFO format. Given that ARC/INFO is the leading UNIX-based GIS software, this is no small advantage. ArcView GIS provides the ability to leverage an existing ARC/INFO installation by adding ArcView GIS Windows or UNIX-based seats to a network. Existing ARC/INFO workspaces can be copied without translation from the UNIX host to stand-alone Windows ArcView GIS installations.

The advantage of being able to access spatial data in ARC/INFO format extends to users not directly associated with an ARC/INFO site. The widespread use of ARC/INFO among government agencies, coupled with the wide range of exchange formats supported by ARC/INFO, makes it highly probable that most third-party data providers will be running ARC/INFO and, as such, will be able to directly supply data in either ARC/INFO coverage or ArcView shapefile format. If spatial data in whatever format is available for your project, it can likely be converted to ArcView format via ARC/INFO.

## Connectivity to CAD Data

With ArcView GIS 3.*x*, you can also create themes directly from CAD drawing and exchange files. Specifically, themes can be created from an AutoCAD *.dwg* or *.dxf* file, or from a MicroStation *.dgn* file. This offers you the ability to store and access data in its native format without the need for translation, enabling CAD users to continue to create and maintain data while extending the capability for ArcView GIS to display and query this data.

## Availability of ArcView GIS Extensions

With ArcView GIS 3.*x*, the functionality of ArcView is greatly extended through the availability of the optional Analyst extensions. These extensions—Network Analyst, Spatial Analyst, and many others—provide functionality previously only available within the ARC/INFO software. The Network Analyst extension provides a strong suite of tools with which to perform optimum routing, closest facility, and service area analysis. The Spatial Analyst extension adds powerful raster modeling functionality to the core ArcView GIS data model. This allows for sophisticated analysis with respect to density, proximity, and surface modeling.

## Spatial Database Engine

With ArcView GIS 3.*x*, the ArcView data model is extended to support connectivity to large databases storing spatial data. This connectivity is accomplished by using the Spatial Database Engine (SDE). SDE uses the power of an RDBMS, such as Oracle or Informix, to provide fast display and query of spatial databases comprised of hundreds of thousands of features. With SDE, ArcView GIS can be used as a front end for displaying and querying all data—spatial and tabular—and can allow the data to be stored in a single database.

## Book Structure

ArcView GIS is a program with immense capability and a wealth of features. This functionality is further extended with the release of ArcView GIS 3.*x*. ArcView GIS can be straightforward and easy to learn, if approached correctly. This book is organized by logical and functional sections to help you learn ArcView GIS quickly and efficiently.

Chapter 1 presents a description of ArcView GIS and the historical development of desktop mapping.

Chapter 2, "The Whirlwind Tour," contains a sample project in order to give you a sense of how an ArcView GIS project feels from beginning to end. Data for the sample project and for the exercises in subsequent chapters are recorded on the companion CD-ROM. In addition, we have saved the end results of the sample project and the exercises in "incremental projects" on the CD-ROM.

Chapters 3 through 10 focus on learning ArcView GIS's basic tools and functionality. Topics include projects and views; data display; data query; extending data; editing shapefiles; CAD and image themes; creating charts, graphs, and reports; and layouts (printing custom maps, charts, and reports). Exercises provide the opportunity to apply and experiment with what you have learned in respective chapters.

Chapters 11 through 14 focus on advanced topics. Chapter 11, "Beyond the Basics," covers overlay operations, advanced classification, hot links to data, and editing tables. An exercise is included that demonstrates these areas.

Chapter 12, "Advanced Functionality," introduces database themes, as well as the Network Analyst and Spatial Analyst. Chapter 13 covers optimization of project design, focusing on application-driven projects. Chapter 14, "ArcView Customization," discusses how to customize the interface and provides an introduction to Avenue, the programming language provided with ArcView GIS.

The final chapter, "ArcView in the Real World," presents case studies of ArcView GIS projects in government, business, and academia.

Four appendices contain information on installation and configuration, the transition from ArcView 2.*x* to ArcView GIS 3.0, a functionality quick reference, and a discussion of the nine MicroVision segments by Equifax National Decision Systems that are used in this book.

The glossary contains definitions of selected terms specific to GIS and ArcView GIS. Finally, the book ends with a general index.

---

# Typographical Conventions

> ✓ **TIP:** *Tips on functionality usage, shortcuts, and other information aimed at saving you time appear like this.*

> ➾ **NOTE:** *Information about features and tasks that is not immediately obvious or intuitive appears in notes.*

> ✗ **WARNING:** *A handful of warnings appear in this book. They are intended to help you avoid committing yourself to results that you may not have intended.*

The names of ArcView GIS user interface items, such as menus, windows, menu items, tool buttons, icons, and dialog window items, are capitalized.

> Access the Edit menu and select Paste. A copy of the theme will appear at the head of the view's Table of Contents.

User input as well as names for files, directories, variables, fields, themes, tables, coverages, and so on, are italicized.

> From the Project window, switch to the *$IAPATH\data* directory and add the *ihopad2.dbf* and *ihopcmp2.dbf* tables.

Emphasis is indicated by italics.

> In ArcView, the *destination table* is the table to which the fields of the *source table* will be appended.

General function keys appear enclosed in angle brackets. The Shift and Control keys are pressed at the same time as a mouse button or another key. Examples appear below.

<Shift>

<Tab>

<Esc>

<Ctrl>

<Enter>

Key sequences, or instructions to press a key immediately followed by another key, are linked with a plus sign. Examples follow:

<Ctrl>+s

<Shift>+<Tab>

## Companion CD-ROM

> **NOTE:** *The companion CD-ROM does **not** contain the ArcView GIS software. You must purchase and install ArcView GIS separately.*

The companion CD-ROM contains the following files:

- Data and project files (in ArcView GIS 3.*x* format) referenced throughout this book.
- A sample extension for use with ArcView GIS: the Navigation Tools Plug-In by Planet One GIS Software.
- Sample software and data files by Claritas, Inc.

The CD is organized into three main directories or levels, as described below.

- *avfiles*—ArcView GIS data and project files. The following subdirectories contain the files by platform type:
  - *unix*—UNIX
  - *mac*—Apple Macintosh
  - *windows*—Windows (3.1, NT, and 95) in uncompressed form
  - *zipfile*—Windows (3.1, NT, and 95) in compressed form
- *planet1*—Navigation Tools Plug-In, which contains subdirectories for the platforms below. (See "Installing and Using the Plug-in" below.)
  - *mac*
  - *unix*
  - *win16*
  - *win32*
- *claritas*—Sample software and data for use on Windows 3.1, Windows NT, and Windows 95 platforms. (See "Claritas Sample Software and Data Files" section below.)

The table below presents major directories and subdirectories of CD-ROM contents. Cumulative file size totals by subdirectory are included.

| **AVFILES** (contain the same files for all platforms) | |
|---|---|
| DATA | 11.2 Mb |
| DAYCARE | 1.79 Mb |
| PROJECTS | 2.1 Mb |
| WORK | (empty) |
| **PLANET1** | |
| MAC | 366.5 Kb |
| UNIX | 366.5 Kb |
| WIN 16 | 827.7 Kb |
| WIN 32 | 894.1 Kb |

| CLARITAS | |
|---|---|
| root directory files | 114.73 Kb |
| DEMO | 10.45 Mb |
| RESOURCE | 3.33 Mb |
| SAMPLES | |
| BOUNDARY | 763.5 Kb |
| HEALTH | 20.1 Mb |
| HIGHWAY | 136.1 Kb |
| LANDMARK | 11 Kb |
| REALEST | 17.21 Mb |
| RETAIL | 16.59 Mb |
| SOLUTION | 1.21 Mb |

# Installing the ArcView GIS Files from the CD-ROM

With regard to mounting and accessing the CD-ROM, some commands are very specific to the type of operating system and/or workstation you are using. Consequently, you should refer to the ArcView GIS *Installation Guide.* This guide shipped with your ArcView GIS CD-ROM.

Sections on transferring files from the CD-ROM to your hard disk according to operating system appear below.

## Windows 3.1, Windows NT, and Windows 95

ArcView GIS users on these platforms have the options of installing compressed files (a slightly less time-consuming procedure) or uncompressed files. Both are covered below.

### Compressed Files

1. Create a directory named *insideav* on your hard drive for the data and project files. You can issue the following command at the DOS prompt:

   ```
   md {drive:}\insideav
   ```

   Another option is to use File Manager to create the directory. (Select Create Directory from the File menu.)

3. Assuming the CD-ROM has been mounted as the d: drive, copy the *av.exe* file from the d: drive's *avfiles\zipfile* directory to the *insideav* directory on your system. You can enter the following command from the DOS prompt:

```
copy d:\avfiles\zipfile\av.exe {drive:}\insideav
```

   Or you can copy the file using File Manager. (Drag the *av.exe* file from the CD-ROM drive to the *insideav* directory on your hard disk.)

4. Run the following self-extracting command from the DOS prompt:

```
av.exe -d
```

   Or select the *insideav* directory in File Manager, choose Run from the File menu, and enter the above command.

5. To access the sample data with the accompanying ArcView GIS project files, you need to set the following environment variable on your system. For Windows 3.1 and Windows 95, add the following line to your *autoexec.bat* file:

```
SET IAPATH={drive:}\insideav
```

   where *{drive}* represents the letter of the drive on which you have installed the sample data and projects.

   For Windows NT, the environment variable is set from the Control Panel via the following steps: (a) From the Start menu, select Settings | Control Panel. (b) In the Control Panel window, open the System icon and click on the Environment tab. (c) Type *IAPATH* in the variable box and *{drive}\inside* in the value box, and click the Set button. If you are running ArcView when the environment variable is set, you must restart ArcView for the variable to take effect.

### Uncompressed Files

1. Create a directory named *insideav* on your hard drive for the data and project files. You can issue the following command at the DOS prompt:

```
md {drive:}\insideav
```

   Or use File Manager to create the directory.

2. Insert the CD-ROM in your CD drive. If you wish to copy files using the DOS prompt, go to step 3. If you prefer to copy files using File Manager, go to step 4.

3. To use the DOS prompt to copy the subdirectories and files from the *avfiles* directory on the CD-ROM to your *insideav* directory, use the command below. The command assumes the CD-ROM is mounted on the d: drive.

```
xcopy d:\avfiles\windows\*.* {drive:}\insideav /s
```

4. To use File Manager to copy the files from *avfiles\windows* on the CD-ROM to your *insideav* directory, select the *data* subdirectory on the CD-ROM and drag it to the *insideav* directory on your hard drive. Follow the same procedure for the remaining three subdirectories under *avfiles\windows* on the CD-ROM.

5. To access the sample data with the accompanying ArcView GIS project files, you need to set the following environment variable on your system. (See your system documentation for more information about how to do this.)

```
SET IAPATH={drive:}\insideav
```

## Apple Macintosh

1. Create a folder on your system and name it *insideav.*
2. Insert the CD-ROM in the CD drive.
3. Double-click on the CD-ROM's icon on your Desktop.
4. Open the *avfiles* folder on the CD-ROM. Inside you will find four subfolders titled *data, projects, daycare,* and *work.* Select the four folders and drag them to your *insideav* folder.

## UNIX

1. Mount the CD-ROM. Each brand of UNIX workstation has a different syntax. (Refer to the ArcView GIS *Installation Guide* for more information.)
2. Make a directory named *insideav.*
3. Change your working directory to *insideav.*
4. To copy the files and directories from the *avfiles/unix* directory on the CD-ROM, enter the following command line. This command

assumes that the CD-ROM is mounted as *cdrom*. Substitute your own CD-ROM path for */cdrom* below. The command line references a shell script on the CD-ROM.

```
/cdrom/install.sh
```

**5.** To access the sample data with the accompanying ArcView GIS project files, you need to set an environment variable on your system. If you are using the Bourne or the Korn shell, skip to step 7. If you are using the C shell, add the following command to your *.cshrc* file:

```
setenv IAPATH <path to insideav directory>/insideav
```

**6.** After you have edited the *.chsrc* file, type the following command line:

```
% source .cshrc
```

**7.** To set the environment variable using the Bourne or Korn shell, add the following commands to your *.profile* file:

```
$ IAPATH=<path to insideav directory>/insideav
$ export IAPATH
```

**8.** After you have edited the *.profile* file, type the following command line:

```
$ .profile
```

# Installing and Using the Plug-in

The plug-in tools for ArcView GIS from Planet One GIS Software are a group of time-saving software extensions for ArcView GIS written in Avenue. No programming experience is necessary to use the plug-in tools because they are packaged as extensions for ArcView GIS version 3.0. A sample plug-in, Navigation Tools, is provided on the CD-ROM in the *Planet1* directory. The Navigation Tools Plug-In implements an "overview" map and "spatial bookmarks."

Additional plug-in tools are available: the Data Dictionary, which catalogs and standardizes your spatial data holdings; the Image Registration Tool, which interactively creates an ESRI-compatible world file; and the Mail Merge Plug-In, which initiates a mail merge session with Microsoft Word based on the selected records of an ArcView theme. To try out the Navigation Tools Plug-In on the CD-ROM, follow the instructions below that are appropriate for your operating system.

## Windows 16-bit

Run the setup program in the *win16* directory called *setup16.exe*. Follow the instructions for installation. Typically, this process will copy the *p1navtls.avx* extension file to the *c:\esri\av_gis30\arcview\ext16* directory of your ArcView GIS installation. If, however, you have installed ArcView GIS in another location, you will need to copy the *p1navtls.avx* file to the *ext16* directory yourself.

After installation, note that an accompanying Windows help file has been copied to your system. It appears inside the new Planet One Program Group. To open this help file, double-click on it. This help file provides instructions for using the plug-in.

You can now load the Navigation Tools Plug-In from within ArcView GIS, as you would any other ArcView extension.

## Windows 32-bit

Run the setup program in the *win32* directory called *setup32.exe*. Follow the instructions for installation. Typically, this process will copy the *p1navtls.avx* extension file to the *c:\esri\av_gis30\arcview\ext32* directory of your ArcView GIS installation. If, however, you have installed ArcView GIS in another location, you will need to copy the *p1navtls.avx* file to the *ext32* directory yourself.

After installation, note that an accompanying Windows help file has been copied to your system. It appears inside the new Planet One Program Group. To open the help file, double-click on it. The help file provides instructions for using the plug-in.

You can now load the Navigation Tools Plug-In from within ArcView GIS, as you would any other ArcView extension.

## UNIX or Macintosh

For UNIX and Macintosh operating systems, there is no automatic setup program, and the documentation is provided as a series of HTML documents that you can view with any frames-compatible browser (such as Netscape).

To install the Navigation Tools Plug-In on either of these platforms, simply copy the *unix* or *mac* directory or folder from the CD-ROM to your hard drive. Then copy the *p1navtls.avx* extension file to the proper extension folder or directory in your ArcView GIS installation. To view the documentation, launch an HTML browser and open the *navtools.htm* file. All documentation is accessible through links from this page.

To purchase the Planet One Plug-In tools for ArcView GIS, contact SoftStore, Inc., by phone (toll-free) at 1-888-SOFTSTORE (1-888-763-8786) or 1-888-GEOBOOKS (1-888-436-2665); by fax at 1-505-474-5020; or by e-mail at *orders@hmp.com.* SoftStore, Inc., is a division of High Mountain Press, 2530 Camino Entrada, Santa Fe, NM, 87505-4835.

# Claritas Sample Software and Data Files

Claritas delivers solutions for precise, timely business decisions. Claritas data, software, support, and analytic expertise enable you to understand your customers and markets thoroughly to pinpoint your most profitable business opportunities. In addition, with specialized solutions and industry expertise in retail, healthcare, packaged goods, finance, telecommunications, real estate, and media, you will be working with seasoned experts who know your business.

The Claritas portfolio of GIS databases is among the most comprehensive in the world. It includes geographic, demographic, consumer, lifestyle, business and industry specific information. The company's experience in compiling and integrating data, and its expertise in applying information and analytic techniques, means that you will receive effective solutions to the challenges you face. Databases are available for whatever geographic level you need, including census, political, postal, media, and telecommunications. All databases are available in variety of media and formats to ensure plug-and-play usability.

Of course, the products you use are only as good as the company that supports them. Claritas's 25 years of experience, commitment to quality, premium value, and dedicated support for building your business are the reasons why over 97 percent of its clients say they would recommend Claritas to others.

Claritas can be reached at Claritas, Inc., 53 Brown Rd., Ithaca, NY 14850; (800) 234-5973; (607) 266-0425; *http://www.claritas.com*; or *info@claritas.com.*

## Claritas Solutions

Success in business requires fast, accurate decisions. Those decisions depend on information and software tools that can quickly reveal the best opportunities and the most critical risks for your business.

Now Claritas offers you the **solution**series—precision-crafted packages of data and software specifically designed for your industry. This innovative product line brings together the power of the industry's leading demographic and retail data with sophisticated software tools from the Claritas Precision Marketing Suite. With the **solution**series, you'll have everything you need to profile America's neighborhoods, businesses, retailers, and consumers.

The **solution**series delivers the information and tools you need to tackle many challenges:

- ❐ Trade area analysis
- ❐ Cannibilization analysis
- ❐ Competitive analysis
- ❐ Environmental impact
- ❐ List selection
- ❐ Market share and penetration
- ❐ Merchandising
- ❐ Product demand
- ❐ Sales forecasting
- ❐ Sales management
- ❐ Segmentation
- ❐ Site location
- ❐ Strategic planning
- ❐ Target promotions

### *Rely on some of the best names in the business*

In addition to Claritas's award-winning demographics, consumer expenditure estimates, and lifestyle segmentation data, the **solution**series integrates industry specific data and software tools from many other top providers:

- ❐ American Business Information, Inc.
- ❐ American Medical Information, Inc.
- ❐ Business Location Research
- ❐ Geographic Data Technology
- ❐ HCIA
- ❐ National Research Burea
- ❐ U.S. Bureau of the Census

Claritas has tailored a **solution**series for the healthcare, retail, and real estate industries. More industry versions are in the works. Within each **solution**series there are four levels of data and software. Each successive level delivers greater data diversity and detail, and more software functionality, as indicated in the matrix below.

| | 1 | 2 | 3 | 4 |
|---|---|---|---|---|
| *Attribute Data* | | | | |
| Demographics | x | x | x | x |
| Workplace profiles | | x | x | x |
| Consumer expenditures | x | x | x | x |
| Business information | | x | x | x |
| Data specific to your industry | | x | x | x |
| *Geographic Data* | | | | |
| Boundary files | | | x | x |
| Highways | | | x | x |
| City locations | | | x | x |
| Landmarks | | | x | x |
| *Software* | | | | |
| Claritas Connect | | x | x | x |
| Claritas Coder | | | | x |

For as little as $295, you can have accurate profiles of your retail markets. You can start with the complete **solution**series package immediately, or you can select just the components you need today and add on as your business grows. Choose the level that's right for you, and enjoy an unbeatable value.

**Level 1**: A generous demographic database customized for your industry.

**Level 2**: More comprehensive demographics and decision-critical data for your industry, such as business profiles, consumer spending estimates, workplace population, retail locations, traffic volumes, healthcare data, and more. Includes Claritas Connect for easy, 24-hour online access to all our Precision Marketing databases and reports.

**Level 3**: Even more detailed data, plus boundary and landmark files for mapping.

**Level 4**: The ultimate in data detail and software functionality! You get rich, robust data, plus Claritas Connect, mapping files, and Claritas Coder for online geocoding and point coding.

## Installing Claritas Demonstration Programs and Resource Files

The companion CD-ROM includes many resources to help you understand the **solution**series product line and to choose the level that best matches your needs and budget.

Install the **solution**series resources from the companion CD-ROM by following the steps below:

For Windows 3.1 and NT:

1. Insert the CD in your CD-ROM drive.
2. In the Windows Program Manager, select Run from the File menu.
3. Type *D:\CLARITAS\SETUP* (where *D* is your CD-ROM drive).
4. Click on OK.

For Windows 95:

1. Insert the CD in your CD-ROM drive.
2. Click Start on the Windows 95 Task Bar and select Run.
3. Type *D:\CLARITAS\SETUP* (where *D* is your CD-ROM drive).
4. Click on OK.

Follow the instructions in the installation program and choose the options that best suit your needs.

## Using Claritas Demonstration Programs and Resource Files

Once you have completed the installation, the following steps are recommended to fully appreciate the **solution**series files included on the companion CD-ROM.

1. Browse the **solution**series reference file. This simple help file is loaded with information on the benefits, content, and design of the **solution**series. The help file also includes layouts of all of the sample **solution**series data files on the companion CD-ROM. You'll find that Claritas has done its homework in compiling these databases and tailoring them to your needs. The finest data sources and software have been integrated to deliver analytic power to build your business.

2. Run the **solution**series slide show that was installed the companion CD. This Microsoft PowerPoint slide show illustrates how the **solution**series database and software components work together to tackle the tough business challenges you face.

3. Once you have inspected the contents of the **solution**series using the help file, take a look at the Claritas Connect demo which installs with the CD. You'll see how Connect can reveal new perspectives on your markets by allowing you to tap into its complete online library of databases and reports 24 hours a day. Samples of all Claritas Connect reports are included in the RapiData Reports help file, also available on the companion CD.

4. Now you're ready to take **solution**series data and boundaries for a test drive. The sample files on the CD cover all four **solution**series levels for each of three industries: healthcare, retail, and real estate. Files are formatted for use with ArcView. You need an installed and licensed copy of ArcView in order to map the sample data. Consult the "How to" section of the **solution**series reference file for more information on using the sample data.

5. The CD contains a Dr. Know-It-All Census help file. Dr. Know-It-All is a glossary of census geographic and demographic terms. It is an ideal reference for understanding the many nuances of demographic tabulations. Also included is the RapiData Report Guide—another help file which provides samples and pricing for all print reports

available through Claritas Connect and the Claritas call-in service. View these files at your leisure for an overview of the many Precision Marketing resources available to you.

## Claritas Connect

Claritas Connect, one of the software components in the **solution**series, can also be licensed separately. Claritas Connect is the perfect information resource to help you succeed in business today. It gives you access to the data you need to outpace your competition, take advantage of new opportunities, and perform critical business applications quickly and accurately. Claritas Connect can help you accomplish the following tasks:

- ❐ Conduct site and competitive analyses
- ❐ Identify market trends
- ❐ Define target markets
- ❐ Position products and services
- ❐ Develop merchandising and marketing strategies
- ❐ Assess acquisitions and consolidations, and more

Claritas Connect is High IQ software because it delivers "Information, *Quick.*" You get immediate online access to all the marketing intelligence you need to get a grip on today's marketing and consumer data explosion: Demographics... Business & Retail... Consumer Expenditure... Healthcare... Media... Financial Services... and more. In addition to Claritas proprietary data, such as PRIZM segmentation and award-winning local-area demographics, we also carry data from the country's leading suppliers.

### *Accessing data has never been so easy*

If you can use a mouse, you can use Claritas Connect. One click gives you access to the most current demographic, consumer, and business databases for every neighborhood and market in the U.S. All data, reports, and geographic lists are logically grouped in on-screen folders, making it easy to explore Claritas Connect's entire database collection.

Every step is easy: defining markets and selecting and integrating data (point, click). Choose from over 100 online preformatted reports or create your own folders in minutes, with data selections and study area definitions that you can re-use or modify later. You can even tailor the system's geography levels, default database formats, and more. When you're ready, Claritas Connect retrieves your data automatically online from the Claritas host. If you need it, help is always available through on-screen help and the Claritas technical assistance hot line.

### *From reports to maps, data in the form you want*

Whether you want printed reports, a database formatted for your favorite spreadsheet, or data and boundary files for your GIS mapping applications, Claritas Connect delivers. Data can be quickly exported to Lotus, Excel, Quattro, dBase, ASCII, or DBF formats. For mapping applications, both data and boundary files can be automatically integrated into a variety of GIS mapping packages, including ArcView. All levels of standard geographic levels are available including census, postal, media, and political geographic rosters. Claritas Connect also makes it simple to summarize information for study areas such as ring studies.

### *The power of online access*

Only Claritas Connect has the flexibility to deliver data online so that you can concentrate on how to use the data, not how to get it. Online, you harness the power of the Claritas host mainframe, which stores immense databases and processes your data requests and reports. Data are kept "fresh" through routine updates.

### *Data access for the budget-conscious*

With Claritas Connect, you can choose to pay only for the data you use or you can purchase an unlimited access license for data you use regularly. The license fee covers Claritas Connect software and basic upgrades, access to database updates, software and database documentation, and technical support. No other system is so flexible and so economical.

The companion CD-ROM includes a demonstration program for Claritas Connect. It will be automatically installed when you run *D:\CLARITAS\SETUP.EXE.* If you would rather run the demonstration from the CD-ROM, run *D:\CLARITAS\RESOURCE\CONNECT\CONNDEMO.EXE* (where *D* is your CD-ROM drive).

## Information Resource Files

As the leading provider of Precision Marketing information, Claritas has developed many information resources for users. Two such files have been included on the companion CD-ROM.

### *Dr. Know-It-All's Help file*

Dr. Know-It-All is a Windows help file that provides quick access to information about demographic and geographic terms. Did you know, for example, that census data for housing value are tabulated for "specified owner-occupied housing units" that only include one-family houses on fewer than 10 acres without a business or medical office on the property? That's why 1990 value data for Manhattan only totals 2,179 housing units, when there were actually over 785,000 housing units there in 1990. Drawing the right conclusion from demographic data requires that you understand the caveats inherent in the data. Dr. Know-It-All makes it much easier for you to do just that.

The companion CD-ROM includes a Dr. Know-It-All's help file. It will be automatically installed when you run *D:\CLARITAS\SETUP.EXE.* If you would rather run the demonstration from the CD-ROM, run *D:\CLARITAS\RESOURCE\DOCTOR\HELPFILE.HLP* (where *D* is your CD-ROM drive).

### *RapiData Report Guide*

Although maps are an effective tool for analyzing data, sometimes you only need printed reports showing the data in an easy-to-read layout. This is not always easy to produce with mapping software. That's why Claritas offers a full line of standard reports covering dozens of databases. RapiData reports can be ordered simply by calling 1-800-234-5973, or accessed online using Claritas Connect.

Included on the companion CD-ROM is the RapiData Report Guide. The RapiData Report Guide makes it easy to shop our portfolio of over 100 reports providing a wide variety of information about every American neighborhood. This Windows help file includes an introduction to each database as well as a listing of the geographic levels for which they are available. All reports are also available for summarized study areas such as ring studies. Most importantly, the RapiData Report Guide provides sample reports for each database so that you can choose the ones that fit your needs before ordering or accessing online.

The RapiData Report Guide will be automatically installed when you run *D:\CLARITAS\SETUP.EXE*. If you would rather run the demonstration from the CD-ROM, run *D:\CLARITAS\RESOURCE\RAPIDATA\RAPIDATA.HLP* (where *D* is your CD-ROM drive).

## On Directories and Operating Systems

After you have installed the CD-ROM files, the data and project files referenced throughout the exercises in this book will be organized into four subdirectories under a parent directory named *insideav* on your system. The four subdirectories are named *data, daycare, projects,* and *work*.

The *daycare* subdirectory contains files and further subdirectories of the data and ArcView GIS project files used in the exercise for Chapter 2.

The *data* directory contains data files used for exercises in Chapters 3 through 9.

The *projects* directory contains ArcView GIS project files used for exercises in Chapters 3 through 9. Throughout the exercises, you are notified about the existence of particular incremental versions of the work sessions. These files can serve as a convenient backup to your own work, or as a means to return to the defaults described in the exercises.

Finally, the *work* directory is where you should carry out the exercises in the book.

References to accessing files and directories throughout this book use the Microsoft Windows environment syntax. Regardless of your operating environment, however, you can perform the exercises in the book.

When you encounter a reference to a directory in the exercises, simply "convert" it to your operating system's syntax. For example, in several exercises you are directed to the *$IAPATH\data* directory. In the UNIX environment, you convert this directory to *$IAPATH/data*. If you are working in the Mac environment, this directory becomes *insideav:data*. (You would seek out the *data* subfolder in the *insideav* folder.)

# Chapter 1

# Introducing ArcView

ArcView is a sophisticated desktop mapping and GIS application that brings the power of geographic analysis to the average PC user.

Desktop mapping is built on the concept of spatial data, and spatial data is all around us. An estimated 80 percent of the data we use has a location component. For instance, it is highly likely that the data maintained by most businesses contain spatial elements, such as addresses, zip codes, or assessor's parcel numbers. With ArcView you can *visualize* site information, customer information, market demographics, and any other type of data with a spatial component.

Attributes of Tracts

| Tracts-id | Acres | [illegible] |
|---|---|---|
| 305 | 243.50 | 0.604 |
| 311 | 759.69 | 0.669 |
| 325 | 642.17 | 0.034 |
| 342 | 640.13 | 0.825 |
| 343 | 639.20 | 0.163 |
| 344 | 635.53 | 1.050 |
| 369 | 1293.64 | 0.607 |
| 374 | 636.98 | 0.647 |
| 370 | 634.64 | 0.807 |
| 398 | 624.51 | 1.159 |
| 399 | 665.31 | 0.696 |
| 418 | 642.03 | 0.483 |
| 414 | 660.18 | 0.539 |
| 430 | 635.30 | 0.382 |
| 431 | 640.71 | 1.050 |
| 286 | 304.87 | 0.981 |
| 306 | 2170.77 | 0.163 |
| 308 | 753.01 | 0.000 |



*Same data viewed in a spreadsheet and displayed on a map.*

The desktop mapping industry is enjoying phenomenal growth, and all signs indicate that this trend will continue well into the future. The following brief outline of the industry's background helps put the phenomenon into perspective.

# The Origins of Desktop Mapping

Desktop mapping is one component of a larger branch of information processing called geographic information systems (GIS). GIS arose from the need to incorporate the management of graphic and textual information into a single system. In GIS, the linking of graphic information in the form of a digital map, with textual information in the form of a tabular database, produces an "intelligent" or thematic map.

The concept of automated thematic maps is not new. It can be traced to the late 1960s when community planner Dr. Ian McHarg advanced new ideas in the integration of data used in planning decisions. McHarg envisioned a system whereby disparate pieces of data, such as zoning, slope, drainage, and planned communities, could be formed into a cohesive plan via the use of colored acetate overlays. Through visual interpretation of color combinations, optimal sites could be identified for development.

In *Design with Nature*, Dr. McHarg essentially anticipated the emergence of GIS. When addressing the problems inherent in the overlay process for thematic mapping, he wrote, "The mechanical problem of transforming tones of gray into color of equal value is a difficult one, as is their combination. It may be that the computer will resolve this problem, although the state of the art is not yet at this level of competence."

Computer analysts quickly addressed the problems of automating this manual overlay process. Early software systems required that graphics and textual elements be maintained and analyzed separately. The real breakthrough, however, came in the late 1970s with the emergence of software that coupled graphics and textual data into a single system. These integrated systems and the intelligent maps they provided made possible new analytical techniques in which the promise of data synthesis and analysis through GIS began to be realized.

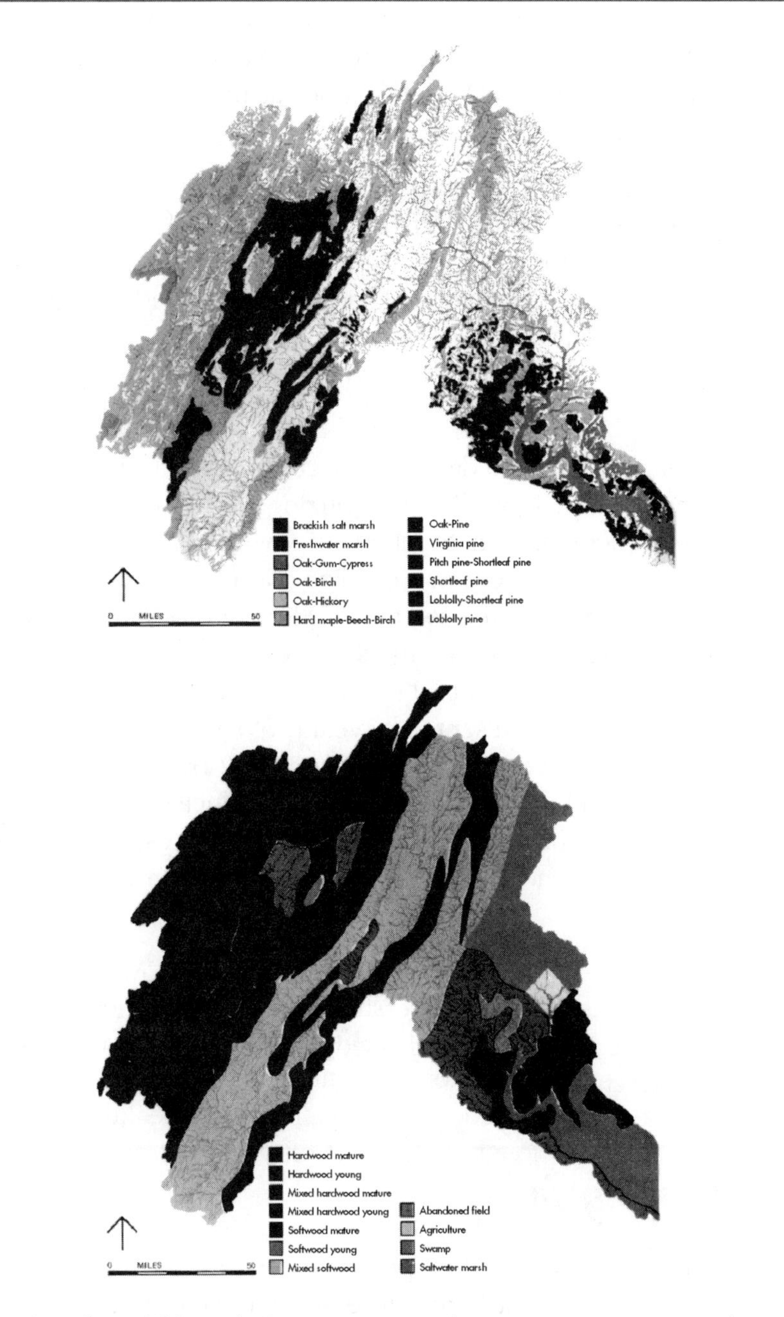


*Overlays for wildlife and plant associations from McHarg's Design with Nature (Garden City, NY: Doubleday & Company, Inc., 1971).*

Environmental Systems Research Institute (ESRI) was an early leader in this field. ESRI's ARC/INFO software, released in 1982, quickly became the dominant GIS software. Thousands of ARC/INFO users found, at their fingertips, a robust and powerful toolkit unequaled in the GIS field.

The advance of GIS in the early 1980s was still hampered by limited (and expensive) computing power, immature software algorithms, and the lack of a user-friendly front end. In addition, data input—a necessity for any GIS project—was difficult and time-consuming. Since the late 1980s, technological breakthroughs resulting in powerful yet affordable personal computers have led to a new class of Windows-based software, providing users with the power to perform analysis in a timely manner.

Advances in the availability of GIS data have occurred as well. The expense of custom capture of spatial data, typically by digitizing from paper maps, had previously limited GIS implementation to utilities, government agencies, and other large institutions with the resources to subsidize a long-term conversion program. With the 1990 census came the release of TIGER (Topologically Integrated Geographically Encoded Reference) files, arguably the single most significant event in the development of the commercial GIS world. TIGER files, and the subsequent release of related data files, represented a wealth of street and demographic data unparalleled in comprehensiveness and format. TIGER became the basis for market research and spawned hundreds of derivative products. Coupled with increasingly powerful hardware and improved graphical interfaces, TIGER has opened the door for smaller government and private users, who can now implement GIS in a cost-effective manner.

Today, GIS is a $3 billion-a-year industry, with an annual growth rate projected at 20 percent. The hurdles that block successful implementation are no longer technical. The only major constraint to increased usage is the capability to visualize how spatial issues affect our world.

## Desktop Mapping Today

As mentioned earlier, an estimated 80 percent of the data we use has a location component. By extrapolation, it is not unrealistic to expect at *least* one opportunity every day to produce a map linked to useful data. Below are a few examples of the types of questions that can be answered with the help of desktop mapping software.

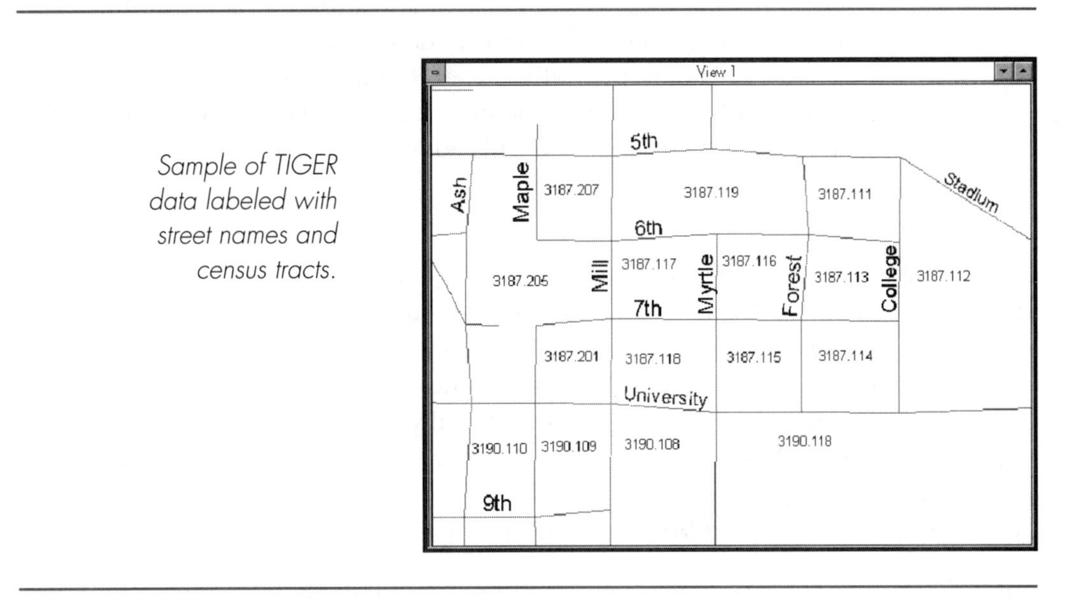


*Sample of TIGER data labeled with street names and census tracts.*

Developers David Z and Zena Y are interested in finding a site for a movie theater complex in the north valley of Seabrook. They notice three suitable properties while driving around in the area. The two decide that they need to examine more comprehensive information on possible sites, including acreage, population growth, and zoning.

David and Zena return to the office and start ArcView. They begin by overlaying a map of current land use in the north valley with a zoning map to identify all vacant commercially zoned properties. Eight of these land parcels are at least 40 acres in size, prime candidates for a shopping center that could anchor a theater complex. Next, Zena and David overlay a map showing population growth. Two of the 8 parcels are located in areas that have grown by over 20 percent in the last 5 years. At this point, the two return to their zoning and land use maps, and discover that one of the two parcels is adjacent to a large area of existing and planned multi-family residential units (i.e., sizable apartment complexes). Zena and David make a note to check the ownership of this parcel, and then add a title and legend to the completed map.

> A journalist, Harry A, learns from a source close to the Mudville city council that the federal government is planning to relocate the town. Mudville was recently flooded, and the plan specifies a move five miles to the east on federally owned land. Harry wants to know if the projected relocation site is the best possible alternative. He is familiar with desktop mapping, and calls Sally B, a consultant he has used on previous occasions.

When Harry arrives at Sally's office, a scanned aerial photo of the Mudville area has already been loaded into ArcView. Sally overlays the scanned image with a map of the 100-year floodplain. She confirms that the new site for the town is outside the floodplain. But Harry still wants to know if the proposed site would be superior to the current one in the long run. Sally says that something about the aerial photo makes her suspicious. She then overlays a soils map, and discovers that the soils at the proposed site are poorly drained and high in salts. Harry asks, "Is this really the best the federal government can do?"

Sally checks her land use map again, and notices a parcel about a mile farther to the east that is also federally owned. This parcel lacks the soil problems of the current proposed site. While Sally prepares map printouts of both the proposed site and the alternate site with appropriate titles and legends, Harry makes appointments with two federal officials and begins writing the first of a series of articles on the Mudville relocation issue.

> The Spicy Abode pizza chain decides to offer home delivery service in Bankersville, a city of 430,000. The company is faced with several daunting challenges, not the least of which is a well-positioned competitor, Pizza Dee-light. Key tactical objectives established by Spicy Abode are to mass merchandise the delivery service in support of all participating pizza franchise stores in the city, and to match the delivery time characteristics to which the market has become accustomed.

Spicy Abode management turns to ArcView and a commercial data provider to define service areas for each franchise store on a city street map. Next, all street addresses within each service area are identified on the map and coded for immediate retrieval.

The Spicy Abode management team decides to offer a single telephone order number for the entire market area. When a phone order comes in, the operator keys each delivery customer's address into a PC network. Each PC is loaded with ArcView and the city street map arranged by franchise service area. The operator then directs each order to the appropriate store location for preparation and delivery.

The purpose of presenting these examples is to demonstrate a wide variety of applications. Now think about the data you work with every day. If you can visualize the data—see the data displayed on a map—you are a prime candidate for desktop mapping.

# ArcView and Desktop Mapping

With its ArcView software, ESRI has targeted the desktop mapping market with a GIS product designed to be the layperson's entry into this technology. ArcView firmly lays to rest the impression, carried over from early PC-based GIS software, that easy-to-use equates with weak. ArcView's many strengths are outlined below.

## Power

Built on an object-oriented data structure, ArcView's data management and analysis are extremely flexible. ArcView can read data from ARC/INFO coverages and grids (ARC/INFO's spatial data structure), AutoCAD and Microstation drawing files, satellite imagery, scanned aerial photographs, dBase and INFO files, external databases such as Oracle and Sybase, and delimited text data. Additional data types can be supported as demand dictates.

The object-oriented data structure also provides great flexibility of spatial data modeling. For instance, a single ARC/INFO coverage can be modeled as several themes (map layers), each based on a different attribute associated with the coverage. No constraints are placed on how any of these themes are modeled.

## Flexibility and Customization

ArcView provides great flexibility of symbolization and map layout. A wide range of symbol and color selection is available, with the ability to import custom symbol sets as needed. The full range of symbols and palettes available for screen display can also be used for generating hard copy.

The Avenue programming language (which comes with ArcView GIS Version 3.0) gives you full control over every element of the ArcView environment. Through Avenue scripts, you can access the objects and classes from which ArcView is built. With Avenue, you can customize the interface, customize maps, and imbed macros in your application.

## Portability

The ArcView interface is consistent across PC and UNIX platforms. ArcView project files are stored in ASCII text file format, and they can be moved between platforms without translation or recompilation. Data are binary-compatible between PCs and workstations: an ARC/INFO workspace can be copied from a workstation or PC in native binary format and used without conversion. This feature positions the PC as a low-cost alternative to UNIX-based development.

## Network Savvy

ArcView can access data seamlessly across a heterogeneous network of PCs and workstations. Both ArcView data sets and project files can be distributed across the network. Additional ArcView nodes can be added to the network based on the requirements of the individual user, ensuring access to geographic data as well as other Windows- or UNIX-based applications.

## Integration

ArcView links seamlessly with the ARC/INFO GIS software, synergistically extending the power of each. ArcView can read ARC/INFO coverages and libraries without translation, and can link to related files in industry standard databases, including Oracle, Informix, and dBase.

# *The* INSIDE ArcView *Perspective*

While ArcView integrates admirably with existing ARC/INFO installations of any scale, its pricing and ability to run on a typical office PC clearly make it appeal to the large group of users who want to perform spatial analysis but have no prior GIS exposure. Included in this group are many businesses for whom previous GIS software packages were desirable but not cost-effective. With ArcView, the power of GIS and spatial modeling is now within reach.

ArcView is simply a very flexible and dynamic product. Rather than attempting to address all of ArcView's possibilities, this book projects a specific perspective or vision. Through discussion based on examples, it concentrates on ArcView as a stand-alone PC-based desktop application. Concepts are presented in a platform-independent manner. Exercises are drawn from a variety of business and government applications.

Note that this focus does not mean being confined to the basics. Many tasks of seemingly great complexity can be accomplished with ease—if you first know where you are heading. Chapter 2 follows a sample project from beginning to end, as a means to help you navigate through the ArcView toolkit.

Before you begin, a few housekeeping chores are in order. If you are already experienced in working with a Windows-type graphical user interface (GUI), skip the next section and proceed to "Windows, Icons, and Menus: The ArcView Interface."

# *Navigating Windows in ArcView*

ArcView runs in five different operating environments. In all these environments, the graphical user interface (GUI) is comprised of windows and diverse types of controls. This book assumes that you have a basic familiarity with your windowing environment. For those who are new to working with a GUI, the following description of basic windowing terminology may prove useful. For those requiring additional orientation to the use of the mouse and windows, go through the Windows tutorial available on your system, or through your system documentation.

➺ **NOTE:** *This book is oriented toward the Microsoft Windows version of ArcView. Users of ArcView on other platforms will find that while all text and exercises are applicable, functionality related to windowing and operating systems may differ. This includes the particular mouse button used for certain functions, and navigation through the underlying file system.*

## Clicking

Clicking refers to pressing and releasing a button on a mouse. Typically, clicking is used to select a program feature, access a menu item, or activate a tool or button icon. In Microsoft Windows, all clicking is done with the left-hand mouse button (or right-hand mouse button if you are using the mouse with your left hand).

## Double-clicking

Double-clicking refers to quickly pressing and releasing a mouse button twice in succession. Double-clicking is commonly used to select a program feature or activate a function associated with that feature.

## Mouse Button Functions

Actions that are performed with a mouse button—clicking, double-clicking, selecting, and dragging—are collectively referred to as mouse button functions. These functions are a combination of the operating environment and the application software.

## Mouse Pointer

The mouse pointer is a symbol on the screen that corresponds to your movement of the mouse. The symbol can change with its location on the screen, but the most common pointer symbols are an arrow and an I-beam, which looks like a large capital I.

## Dragging

Dragging refers to moving the mouse pointer to an object, pressing and holding down the primary mouse button, moving the pointer to a new location, and releasing the mouse button. Dragging is commonly used to resize or reposition objects on the screen.

## Menu

A menu is a list of application functions. Typically, menu titles appear in a horizontal bar at the top of your screen. The bar containing the menu titles is called a *menu bar*. To pull down a list of menu options, move the mouse pointer to the menu title, and then click the primary mouse button (usually the right mouse button). Another way to look at a list of menu options is to press a function key (e.g., <Alt>, <Ctrl>, <Shift>) followed by a character key. The character key to use for a particular menu is usually indicated by an underlined character in the menu's title.

## Icon

Icons are graphical image shorthand for program functions. Icons (little pictures) appear either on buttons or by themselves. Clicking on an icon, or a button containing an icon, activates a particular program function or operation.

## Pulldown Menu

When you click on a menu title in the menu bar, a box containing a list of menu functions and options appears under the menu title. The box is called a pulldown menu. Pulling down a menu is somewhat similar to pulling down a window shade.

## Status Bar

The status bar is a narrow bar at the bottom of your screen. Its purpose is to provide cues about program functions, as well as information during program processing. Within many applications moving the mouse pointer to a button containing an icon will trigger the display of a brief description of the button's function in the status bar. The status bar may also inform you of the progress of an operation currently executing (e.g., *30% complete, 60% complete*).

## Dialog Window

Dialog windows appear when you choose a menu option or button for which the program requires additional information from you. Assume you want to print a page. First, you activate the print menu. At this point you see a dialog window requesting additional information, such as the number of copies and the number of the page you want to print. You can choose the defaults, or input your choices via the keyboard.

## Popup Window

In some applications, holding down the non-primary mouse button (usually the left mouse button) in a view window brings up a popup window just below the mouse pointer. This window contains choices appropriate to the current state of the view.

## Resizing

You can use the mouse to drag the side or corner of a window to a new location. When you release the mouse button, the window size is adjusted to the new dimensions. This is referred to as resizing the window.

## Repositioning

When you drag a window by its title bar (the bar across the very top of a window), the window is moved to the new location. This is referred to as repositioning the window.

## Make Active

When multiple windows are present on your screen, any action you take will be performed on the window that is active. To make a window the active one, move the mouse pointer anywhere within the window and click the right mouse button.

### Maximize

To maximize a window, click on the Maximize button in the upper right corner of the window. The window will be resized to fill the entire screen.

### Minimize (Iconify)

To minimize or iconify a window, click on the Minimize button in the upper right corner of the window. The window is collapsed into an icon that is displayed at the bottom of the screen.

### Close

To close a window, click on the Close button at the upper right corner of the window (For Windows 3.1 and Windows NT, double-click on the Close button in the upper left corner.).

### Restore

Double-clicking on an iconified (minimized) window will restore the window to its previous size.

### Bring to Front or Forefront

If multiple windows are present on the screen, clicking in the title bar of a window that is partially covered will make that window active. In the process, the window will be brought to the forefront of the screen (in front of all the other windows). If a window is completely covered, simultaneously pressing the <Alt> and <Esc> keys will bring the hidden window to the front of the screen.

## *Windows, Icons, and Menus: The ArcView Interface*

ArcView organizes your mapping project and the tools available to you within a system of windows, menu bars, button bars, tool bars, and icons. The entire Arc-

View environment and GUI are contained in the main application window. All user interactions take place in this area, as well as the display of ArcView output.

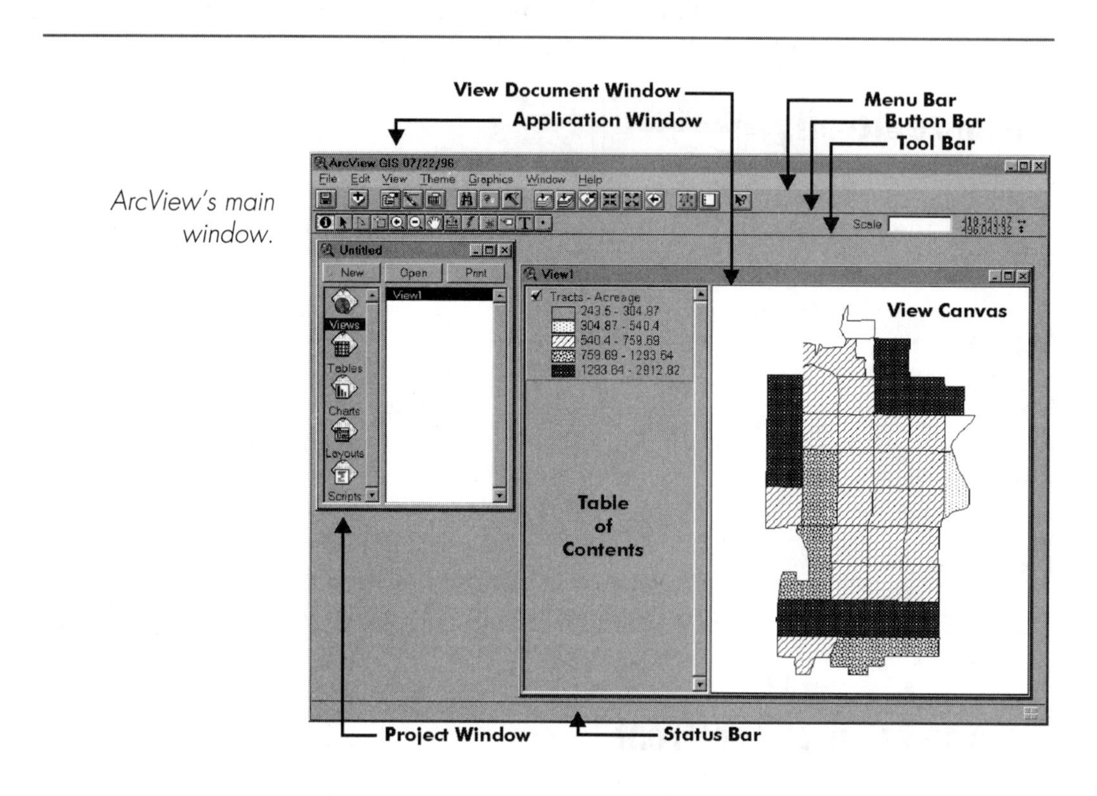


ArcView's main window.

Additional ArcView window elements.

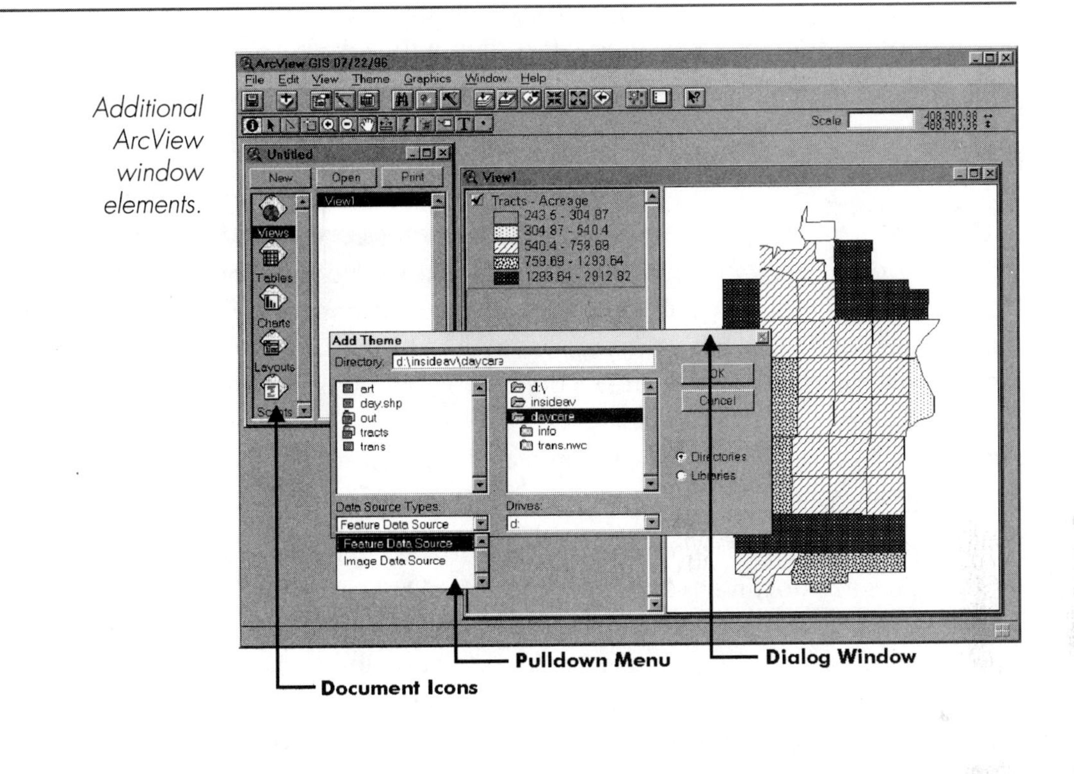


There are several ways to interact with ArcView. Pulldown menus are the most conventional method. Choosing menu items communicates commands to ArcView by executing embedded scripts (macros or mini-programs). Menus can be pulled down and selected in one mouse operation, or pulled down with one mouse click and selected with a second.

Actions are also conveyed with buttons and icons. A single mouse click activates the button bar and tool bar icons. Clicking on an icon may produce an immediate action or bring up an additional input window, depending on the tool. These additional input windows are known as *dialog windows.*

A helpful feature in navigating the icons is the *status bar* at the bottom of the application window. When the mouse pointer is placed over an icon, a brief description of the icon's function is displayed in the status bar. For example, if the pointer is placed over the Add Theme icon (the plus sign), "Inserts New Themes Into The

View" is displayed in the status bar. The status bar can be invaluable to new users and to those who find icon symbols less than intuitive.

ArcView also displays "tool tips." These text descriptions of a tool or button's function appear inside small yellow boxes, which pop up when you move your mouse cursor over a button or tool and pause.

ArcView presents all of its features in a series of nested windows. The top level or main window is the *project window.* The project window organizes all files and documents that you generate, access, and input when using ArcView, including spatial data, associated tabular data, scanned images, charts, graphics, and scripts. An ArcView project organizes these documents in multiple related windows, and these windows are all accessible from the project window. When an ArcView project is saved and reopened, the project is returned to the exact state of the last save, with all windows, associated graphics, and tables displayed as before.

There are five document categories used in ArcView: *views, tables, charts, layouts,* and *scripts.* Each document category contains a list of available documents created as part of the current project. Each document (e.g., a specific view or table) can be opened by double-clicking on the name of the document from the document list, or by selecting the document name and then clicking the Open button. Other documents and files associated with each document will be displayed in separate windows. For example, when a theme is displayed in a view window, a table and a chart associated with the theme may also be displayed in other windows.

> **NOTE:** *It is important to note that the tools available from the menu bar, button bar, and tool bar will change depending on which document type is active. The tools available for working with views are different from tools available for tables, as are the tools for charts, layouts, and scripts.*

Do not feel intimidated if this organization sounds foreign or unfamiliar. Even experienced GIS professionals are bound to spend some time learning the vernacular. Entire chapters in this book are devoted to ArcView projects, views, tables, charts, and layouts. Scripts are addressed in Chapter 14. For the moment, just be sure you have this section marked because there will be frequent references to the above terms.

# Chapter 2

# The Whirlwind Tour

Viewing your data spatially is a central concept of GIS. For many people trained in the "flat" world of spreadsheets and databases, the spatial mindset required by GIS is not at all intuitive.

GIS requires users to think more holistically than other software applications. In spreadsheets and databases, you can enter data and postpone decisions about how to conduct analysis without worrying about whether you will ultimately be able to obtain the answers you seek. But GIS is different. There are countless possibilities for how data might be organized and modeled. Unless you know where you are going from the outset, you may never complete your projects effectively.

This chapter uses a sample project to demonstrate how ArcView is applied to this process. Within ArcView, a project is the primary structure used to organize the files of your work session, such as digitized maps, tabular data, and presentation graphics. You will be introduced to the "look" of ArcView, and you will see how an ArcView project feels from beginning to end. The program topics and functions covered below are discussed in detail in subsequent chapters.

As you thumb through this book, you will notice that exercises have been interspersed throughout the chapters. Most of the exercises are short because they have been tailored to illustrate specific concepts. The sample project exercise in this chapter is lengthy because it will be used to illustrate an entire ArcView project. Follow along, and if you get lost, do not worry. You can find incremental versions of the sample project on the companion CD-ROM, so you can always check to see where you should be at any point along the way. (If you installed the *projects* directory from the CD onto your hard disk, you can access the incremental versions from your system.)

# The Sample Project

In this example, Sharon Y seeks to open a new daycare facility. Ordinarily, where would she start? If she were like most small business owners, she would sell her general concept to investors first. Then she would contract with a commercial real estate broker to locate available properties that match her needs. The basic flaw in this arrangement is that the entrepreneur and the investors are postponing a decision about the most important aspect of their venture: "location, location, location."

An alternate arrangement would be for Sharon to carry out locational analysis at the outset; that is, identify the ideal location, inquire about likely cost, and then ask her investors for the necessary funding. This is a win-win situation for everyone: Sharon is better prepared to acquire a winning location, and her investors can be more confident about supporting a successful venture. One could argue that Sharon's "optimal" location may not be for sale. Many professionals' experience, however, indicates that real estate can be acquired from property owners who are not currently in the market if they are approached by an interested and motivated buyer. Thus, in our hypothetical world, Sharon kicks off her venture with locational analysis.

The first step in Sharon's new venture is to identify the most promising locations in her market area. For a daycare business, an entrepreneur might wish to focus on the following variables:

- Population distribution
- Number of children age six years or less, as a proportion of the total population
- Average household income

By classifying various neighborhoods according to these criteria, ArcView can be used to visually portray how promising various locales may be for a new daycare center. Sharon can also begin to identify competitors' market areas by adding existing daycare facilities to a map created in ArcView.

Subsequent steps could involve examining the most promising locations more closely. For example, once the primary criteria have been explored, Sharon can study secondary criteria, such as the number of working mothers with children under age six as a proportion of the total population, local occupation distribution, educational attainment, population projections, and the locations of major employers.

With GIS mapping as support, Sharon will be much more effectively positioned to initiate her property search. Instead of analyzing the suitability of available properties, she can direct her energy toward analyzing the availability of suitable properties.

Throughout the remainder of this chapter, you will act as Sharon's GIS support by using ArcView to help her identify suitable sites for her business. We could tell you how to use ArcView, but GIS makes more sense—and is a lot more fun—when you get in there and do it yourself. We encourage you to perform the analysis along with us.

## Exercise 1: A Sample Application

This sample application is straightforward. You need to identify the most promising locations in which to open a new daycare center in Tempe, Arizona. You will prepare a map locating Sharon's potential customers and the existing competition. Your data sets include demographic data from the 1990 census, with total population by census tract as well as population age six or less; Yellow Pages listings for daycare centers with street addresses; and spatial data sets for Tempe.

The spatial data sets are derived from the U.S. Census Bureau's TIGER files, and consist of (1) a file representing census tracts (called a *coverage)*, and (2) a street map (called a *street network* or *net*) of Tempe, which has been digitized. The street net file contains codes that represent address ranges (e.g., 1400 N. 15th Ave. to 1500 N. 15th Ave.). Both spatial data sets include tabular data consisting of codes that represent geographic locations. (See the insert, "Cleaning TIGER Street Nets.")

> **NOTE:** *In the following exercise, the spatial data sets (the Tempe census tracts and street net) are stored separately from the non-spatial or tabular data (the demographic data and competing daycare locations). The two types of data files will be subsequently linked together*

*using a locational field, such as a census tract number or a street address. This separation of spatial data and attribute data is typical of most GIS projects.*

Begin by starting ArcView. Drag the bottom right corner of the application window down and to the right to fill the entire screen.

Your first task is to prepare ArcView for importing data files into the project.

1. Click on the Views icon, and select New. A view window is opened with the default title of *View1*.
2. Drag the bottom right corner of the view window to fill the available space in the project window. Because the *View1* window will display graphics and analysis results, it should be as large as possible.

Before proceeding with the exercise, you need to set the working directory for the project. Click on the project window (labeled *Untitled*) to make it the active window. If necessary, move or close the view window to make the project window visible. From the Project menu, select Properties. The default setting for *Work Directory* is set as *$HOME*. Change the name to *$IAPATH\work*, and click on OK to accept the change. Alternatively, the working directory can be set from the view window. Select Set Working Directory from the File menu and enter the new working directory in the dialog window.

If you previously closed the view window, click again on the view window (*View1*) to make it active.

> **NOTE:** *The* $IAPATH *variable references the directory in which you installed the* INSIDE ArcView *sample data. If you have not set the* $IAPATH *variable yet, refer to the instructions in the Introduction for installing the sample data from the CD-ROM.*

You are now ready to add the spatial data sets or, in this case, Tempe geography. In ARC/INFO, these data sets are referred to as *coverages*. In ArcView, spatial data sets—coverages and shapefiles—are called *themes*. When spatial data is imported into ArcView, it is transformed into a theme. Think of themes as maps and map overlays. You will now proceed to import four themes into ArcView.

1. On the button bar, click on the Add Theme icon.

*Add Theme button icon.*

A dialog window is opened for additional input.

ArcView GIS 07/22/96

File Edit View Theme Graphics Window Help

Scale

Untitled

New

View1

Views

Tables

Charts

Layouts

Scripts

Add Theme

Directory: d:\insideav\daycare

art

day.shp

out

tracts

trans

d:\

insideav

daycare

info

OK

Cancel

Directories

Libraries

Data Source Types:

Feature Data Source

Drives:

d:

*Adding themes to the view.*

2. In the Add Theme dialog window, click on the Directory box until you see the daycare directory installed from the CD-ROM (*$IAPATH\daycare*). Four coverages are listed: *art* (arterial street net for Tempe), *out* (outer boundary of Tempe), *tracts* (census tracts for Tempe), and *trans* (street net for Tempe), as well as one shapefile, *day.shp*. Select the four coverages.

✓ **TIP:** *After clicking on the first selection, hold down the <Shift> key while clicking on additional selections. If you do not hold the <Shift> key down, each theme selected replaces the previously selected theme.*

3. After all four coverages are selected, click on OK. The dialog window is cleared and the themes are added to the view. A view can be conceptualized as a collection of map overlays (themes).

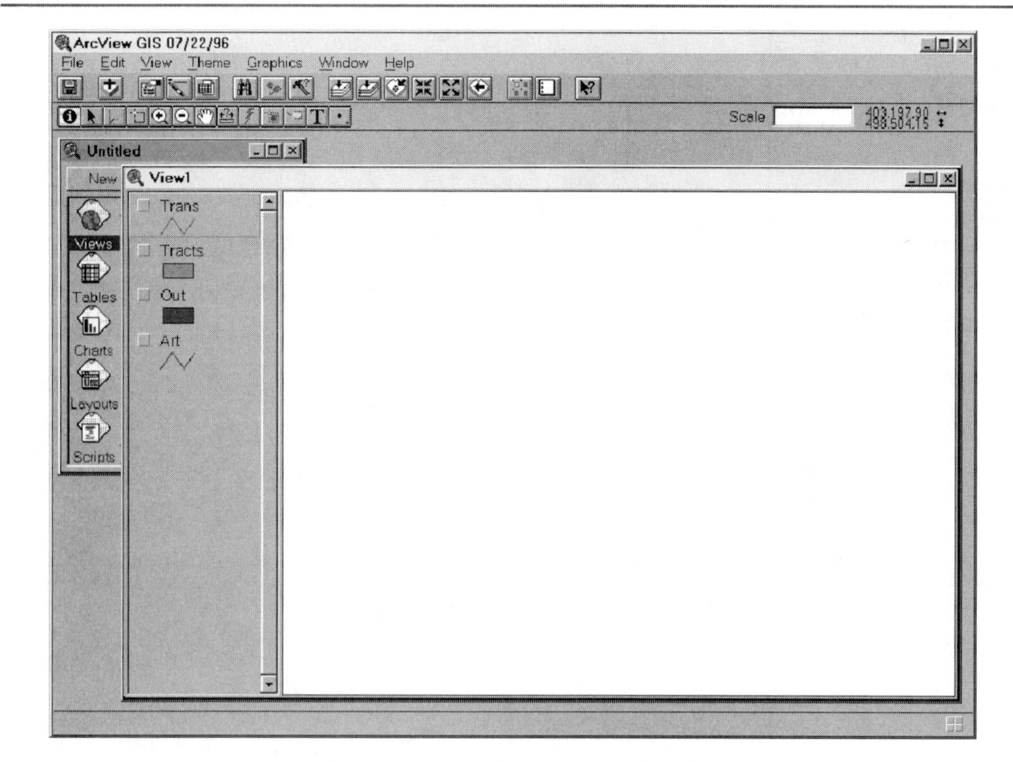


View1 with the four themes added.

Take a peek at the raw data by clicking on the check box adjacent to the *Tracts* theme. The census tracts for Tempe are drawn in the graphics display area of the view window. By default, the theme is drawn in uniformly shaded areas. The demographic data you will add later will be displayed over the census tract geography.

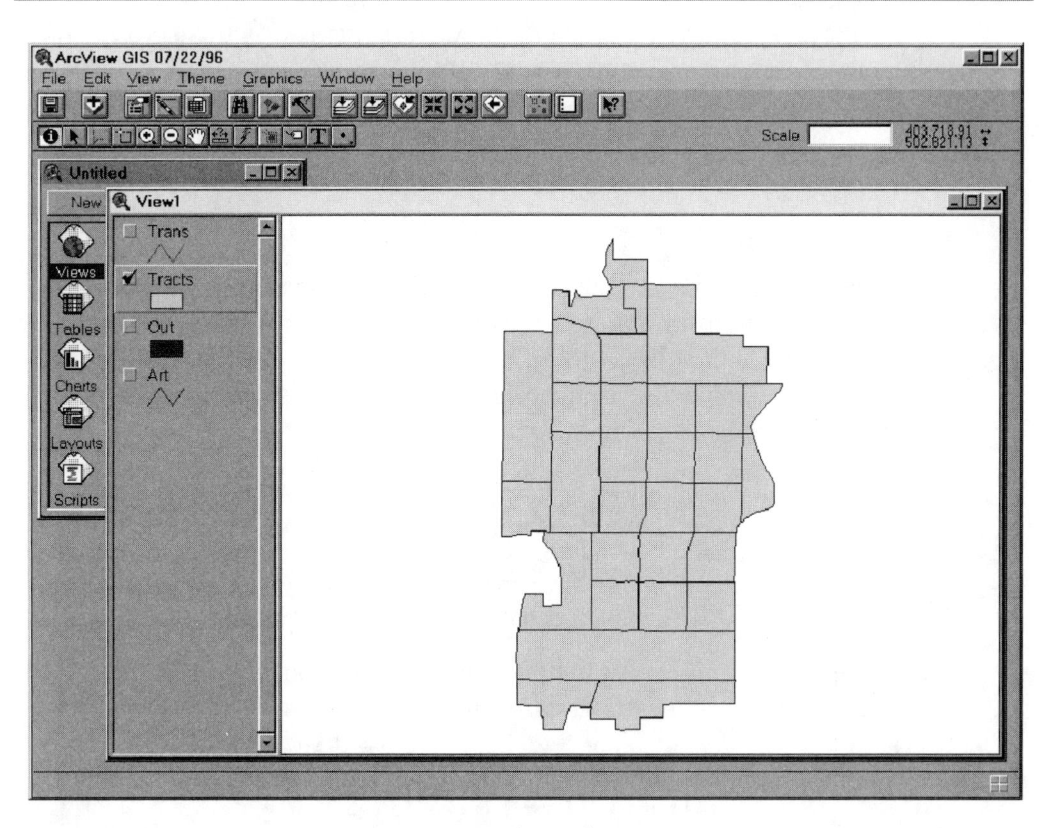


*Census tracts theme for Tempe.*

## View Navigation Basics

The first thing you may wish to do upon displaying a new theme in a view is to take a closer look. Nestled in the View graphical user interface are the tools to do just that—zoom in, zoom out, and pan from one area to another.

*Zoom In and Zoom Out icons from the View button bar.*

The Zoom In and Zoom Out buttons on the View button bar provide the basic functionality of zooming in and out on a view. Clicking on these buttons will cause the display to be zoomed in or out on by a factor of two, centered on the extent of the original display. These buttons provide a fast means of changing the extent of a view.

*Zoom to Previous Extent button icon on the View button bar.*

The Zoom to Previous Extent button on the View button bar allows the view to be redrawn at the previous extent.

*Zoom In, Zoom Out, and Pan tool icons on the View tool bar.*

The Zoom In and Zoom Out tools from the View tool bar provide more expanded functionality than the Zoom In and Zoom Out buttons described above. Instead of immediately zooming in or out, the Zoom In and Zoom Out tools become active when clicked on: The result is dependent on the action you subsequently take in the view. Clicking on the Zoom In tool to make it active, followed by clicking on a point in the display, causes the view to be zoomed in on by a factor of two, centered at the point of clicking. You can also use the mouse to drag the shape of a box. The display is redrawn containing the extent described by the box. Zooming out works by clicking or dragging in much the same manner.

The Pan tool is used to change the extent of the display without zooming in or out. When you click on the Pan tool, the mouse cursor changes to resemble a hand. Click on a point in the view and drag while holding the mouse button down to move the map to a new location. Releasing the mouse button results in the display being redrawn with this new extent.

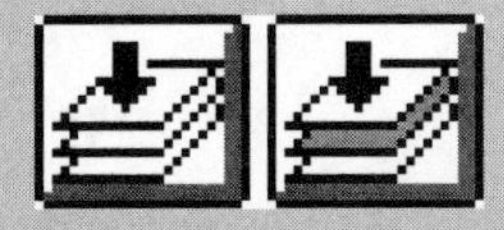

*Zoom to Full Extent and Zoom to Active Themes button icons from the View button bar.*

You may want to return easily to the full extent of the map after zooming in to a selected area. The Zoom to Full Extent button will redraw the display zoomed to the extent of all themes in the view, or the full map extent of the view.

You can also use Zoom to Active Themes to zoom to the full extent of only those themes currently *active* in the view. Clicking on an entry for a theme in the Table of Contents causes that theme to become active. The theme is highlighted in the Table of Contents and appears as a raised box. If the extent of this active theme is less than the full extent of all themes in the view, clicking on the Zoom to Active Themes button will cause the display to be zoomed to the extent of the active theme. If more than one theme is active, clicking on the Zoom to Active Themes button will cause the display to be zoomed to the combined extent of the active themes. (To make more than one theme active, hold down the <Shift> key while clicking on additional themes in the Table of Contents.)

Next, you need to import the tabular data sets. These tables contain demographic data and the locations of competitive daycare centers. In preparation for this exercise, the tabular data have been entered in ASCII text files with the fields separated by commas. You should be able to import data from almost any source using this format. ArcView can also import data directly from dBase format (*.dbf*) and from INFO's files, the default database supplied with the workstation version of ARC/INFO.

> ✓ **TIP:** *When ArcView imports a comma-delimited ASCII text file into its tabular format, the program searches the first line of the text file for field names. Thus, the first line of the text file should contain the names for each field separated by commas. Field names can include spaces, and need not be preceded or followed by double quotation marks.*
>
> *The data in the fields are recorded from all lines except the first line of the text file. The fields in the data lines (records) are separated by commas. A field containing numeric values that must be treated as a character string should be enclosed in double quotation marks. Finally, the ASCII file must be given a* .txt *extension in order for it to be listed in ArcView's Import menu.*

To import the tabular data, take the following steps:

1. Make the project window active by clicking on the title bar—at this point it is labeled *Untitled.* Next, click on the Tables icon in the project window.

> **NOTE:** *The GUI in the ArcView application window is context-sensitive. Only menus, buttons, and tools appropriate to the active document are presented. When the project window is active, there are fewer applicable actions, and fewer applicable choices, than when a view or table window is active.*

2. Pull down the Project menu from the menu bar, and select Add Table. As seen in the following illustration, the Add Table dialog window is opened.

3. Navigate to the *$IAPATH\daycare* directory.

4. Pull down the List Files of Type menu and select Delimited Text (**.txt*). Two files are listed: *day.txt* and *maricopa.txt.* Select both (with <Shift>+click) and click on OK.

   The text files are imported and displayed in tabular format. To examine the data, you can stretch and scroll these windows as you wish.

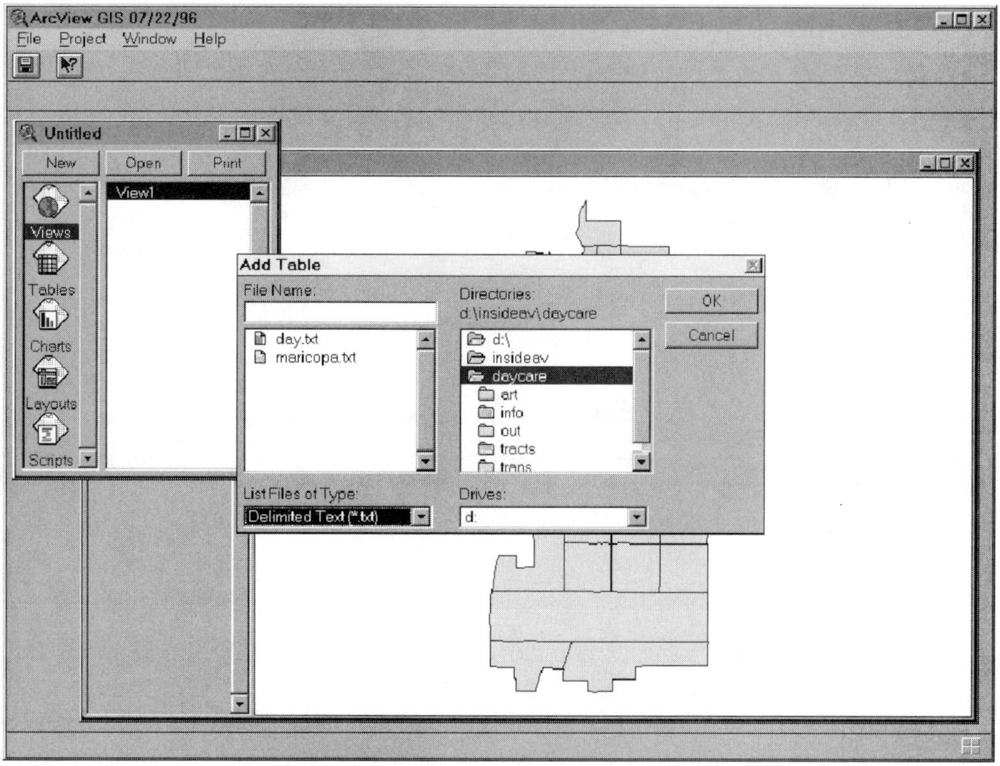

*Adding tables to View1.*

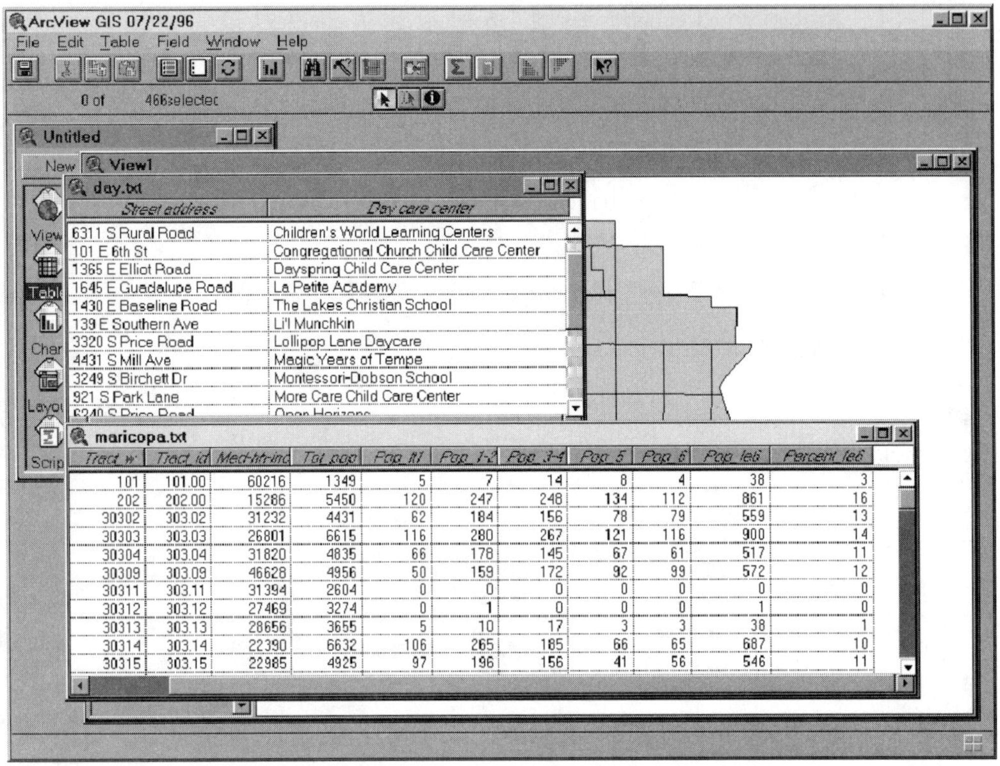

*A closer look at the tables.*

**5.** The *day.txt* table can be closed because you will not need it until later. Return to the spatial themes by clicking on the Views icon in the project window, and opening *View1*. (For your convenience, we have saved incremental versions of the exercise project files on the CD. These incremental projects serve as a reference point in case you need to return to our defaults. For the daycare exercise, the incremental projects have been stored in the *\insideav\daycare* directory. The incremental project to this point has been stored as *ch2a.apr*.)

At this point you have imported all of the raw data into ArcView. Now you need to link data together. The first step is to *join* the two tabular data files to the two spatial data themes so that the tabular data can be spatially displayed. You will begin with the demographic data in order to display it against the census tract theme.

Tabular data can be linked to a theme via a field that identifies the geographic location. Take the following steps to join the demographic tabular data to the census tract theme.

1. Click on the *Tracts* theme in the legend of the *View1* window in order to make it the active theme. (When active, the legend for the theme appears raised in the legend area of the view window.)
2. Select the Open Theme Table icon from the button bar. (Refer to the text displayed in the status bar for confirmation that you are making the proper selection.)

---

*Open Theme Table button icon.*

---

The following illustration shows the attribute table for the *Tracts* theme, displayed as *Attributes of Tracts* in a new window.

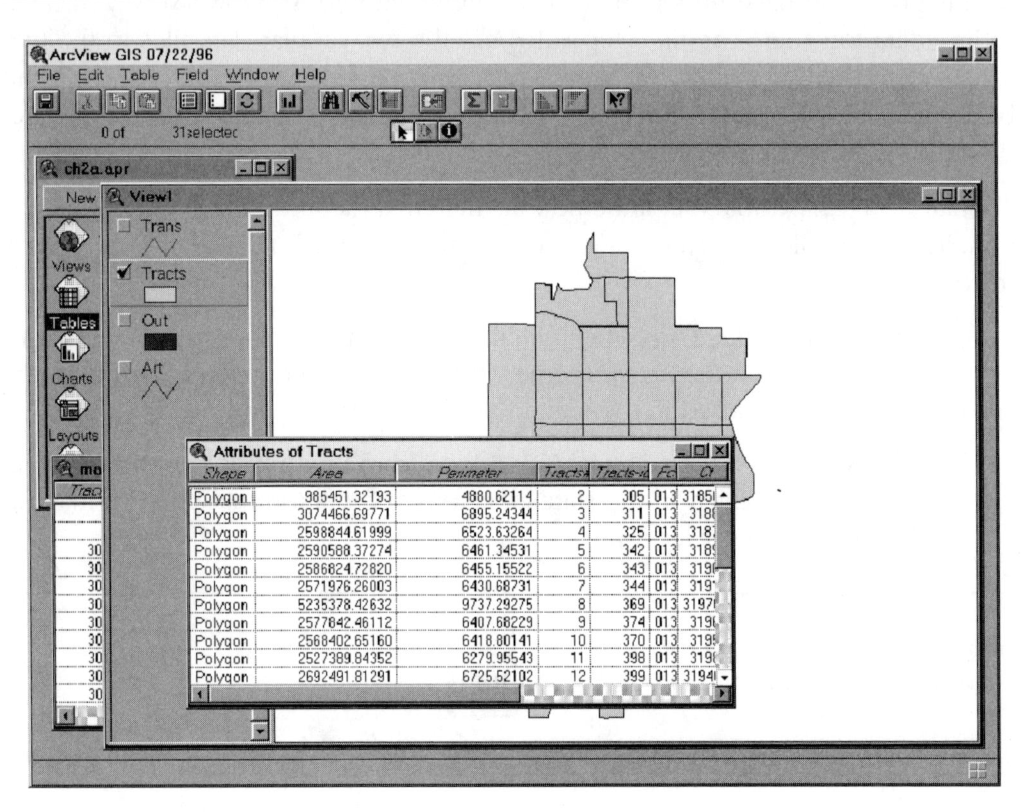


Opened Attributes of Tracts table.

3. Bring the window displaying the *maricopa.txt* table to the foreground, and stretch and position it so that you can view its window and the Attributes of Tracts window simultaneously.

4. To be sure that none of the records in either table have become accidentally selected, check the tool bar. It will indicate the number of records selected (it should show 0). If there are any selected, you need to deselect them before joining. Do so by clicking on the Select None button.

**5.** To join two tables, a field that contains the same values must exist in both tables. For a one-to-one association to be made, these values must be unique to both tables. In this project, the unique field common to the two tables is the census tract number. In the *Attributes of Tracts* table, this field is named *Cti*, and in the *maricopa.txt* table, it is named *Tract_w*. For each field, click on the column heading in the appropriate window. Each name will be highlighted when selected.

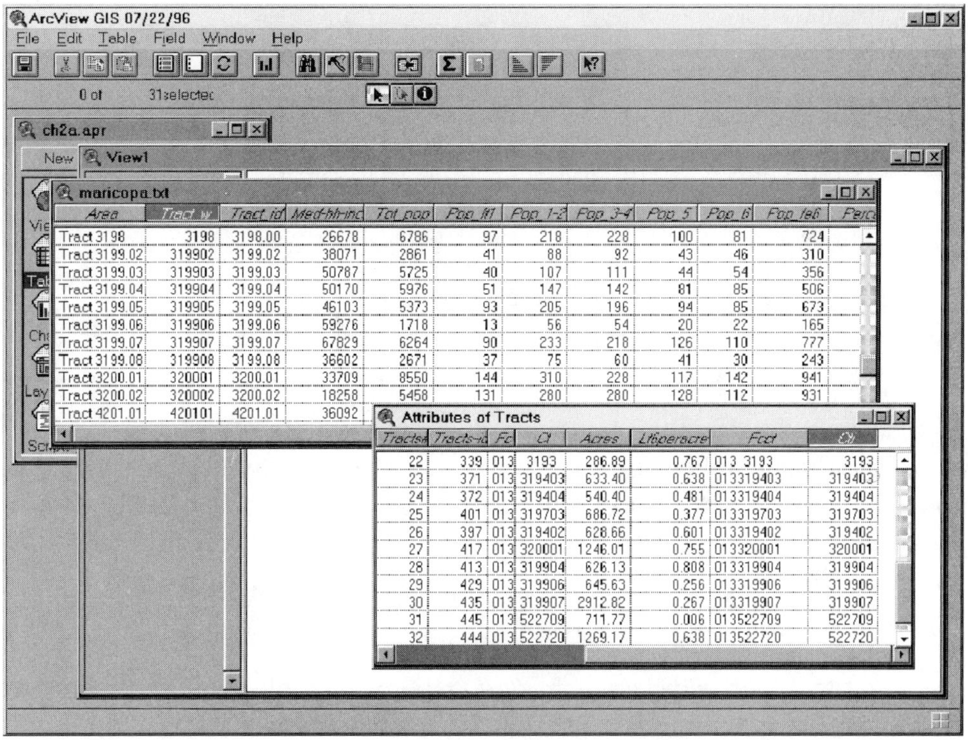

ArcView GIS 07/22/96

File Edit Table Field Window Help

0 of 31 selected

ch2a.apr

View1

maricopa.txt

| Area | Tract_w | Tract_id | Med-hh-inc | Tot_pop | Pop_lt1 | Pop_1-2 | Pop_3-4 | Pop_5 | Pop_6 | Pop_le6 | Perc |
|---|---|---|---|---|---|---|---|---|---|---|---|
| Tract 3198 | 3198 | 3198.00 | 26678 | 6786 | 97 | 218 | 228 | 100 | 81 | 724 | |
| Tract 3199.02 | 319902 | 3199.02 | 38071 | 2861 | 41 | 88 | 92 | 43 | 46 | 310 | |
| Tract 3199.03 | 319903 | 3199.03 | 50787 | 5725 | 40 | 107 | 111 | 44 | 54 | 356 | |
| Tract 3199.04 | 319904 | 3199.04 | 50170 | 5976 | 51 | 147 | 142 | 81 | 85 | 506 | |
| Tract 3199.05 | 319905 | 3199.05 | 46103 | 5373 | 93 | 205 | 196 | 94 | 85 | 673 | |
| Tract 3199.06 | 319906 | 3199.06 | 59276 | 1718 | 13 | 56 | 54 | 20 | 22 | 165 | |
| Tract 3199.07 | 319907 | 3199.07 | 67829 | 6264 | 90 | 233 | 218 | 126 | 110 | 777 | |
| Tract 3199.08 | 319908 | 3199.08 | 36602 | 2671 | 37 | 75 | 60 | 41 | 30 | 243 | |
| Tract 3200.01 | 320001 | 3200.01 | 33709 | 8550 | 144 | 310 | 228 | 117 | 142 | 941 | |
| Tract 3200.02 | 320002 | 3200.02 | 18258 | 5458 | 131 | 280 | 280 | 128 | 112 | 931 | |
| Tract 4201.01 | 420101 | 4201.01 | 36092 | | | | | | | | |

Attributes of Tracts

| Tracts# | Tracts-id | Fc | Ct | Acres | Lifeperacre | Fcct | Cti |
|---|---|---|---|---|---|---|---|
| 22 | 339 | 013 | 3193 | 286.89 | 0.767 | 013 3193 | 3193 |
| 23 | 371 | 013 | 319403 | 633.40 | 0.638 | 013319403 | 319403 |
| 24 | 372 | 013 | 319404 | 540.40 | 0.481 | 013319404 | 319404 |
| 25 | 401 | 013 | 319703 | 686.72 | 0.377 | 013319703 | 319703 |
| 26 | 397 | 013 | 319402 | 628.66 | 0.601 | 013319402 | 319402 |
| 27 | 417 | 013 | 320001 | 1246.01 | 0.755 | 013320001 | 320001 |
| 28 | 413 | 013 | 319904 | 626.13 | 0.808 | 013319904 | 319904 |
| 29 | 429 | 013 | 319906 | 645.63 | 0.256 | 013319906 | 319906 |
| 30 | 435 | 013 | 319907 | 2912.82 | 0.267 | 013319907 | 319907 |
| 31 | 445 | 013 | 522709 | 711.77 | 0.006 | 013522709 | 522709 |
| 32 | 444 | 013 | 522720 | 1269.17 | 0.638 | 013522720 | 522720 |

*Attributes of Tracts and maricopa.txt tables prepared for joining.*

**6.** Click on the title bar for the *Attributes of Tracts* window to make this table the primary table for the join. (When joining tables, the attribute table for the spatial theme should always be the primary table.) Pull down the Table menu from the menu bar and select Join; the status bar of the application window will show the progress of the operation until the join is complete.

When the join is complete, the *maricopa.txt* file will be closed and the *Attributes of Tracts* table will contain the fields from the *maricopa.txt* table columns, to the right of the spatial theme columns. You can scroll through the table to examine the results of the join.

*Resultant joined Attributes of Tracts table.*

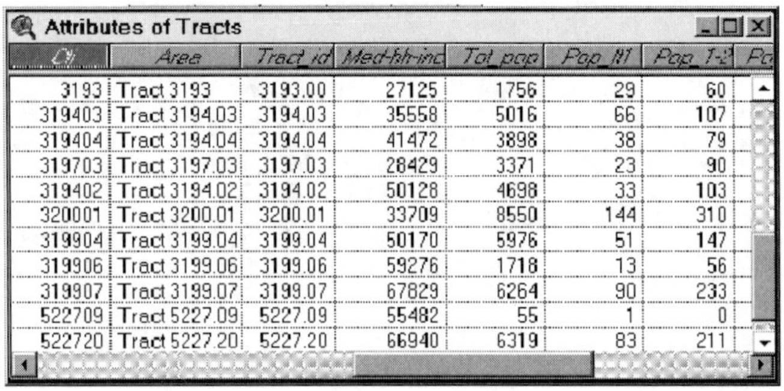

Attributes of Tracts

| Ct | Area | Tract_id | Med-hh-inc | Tot_pop | Pop_lt1 | Pop_1-2 | Po |
|---|---|---|---|---|---|---|---|
| 3193 | Tract 3193 | 3193.00 | 27125 | 1756 | 29 | 60 | |
| 319403 | Tract 3194.03 | 3194.03 | 35558 | 5016 | 66 | 107 | |
| 319404 | Tract 3194.04 | 3194.04 | 41472 | 3898 | 38 | 79 | |
| 319703 | Tract 3197.03 | 3197.03 | 28429 | 3371 | 23 | 90 | |
| 319402 | Tract 3194.02 | 3194.02 | 50128 | 4698 | 33 | 103 | |
| 320001 | Tract 3200.01 | 3200.01 | 33709 | 8550 | 144 | 310 | |
| 319904 | Tract 3199.04 | 3199.04 | 50170 | 5976 | 51 | 147 | |
| 319906 | Tract 3199.06 | 3199.06 | 59276 | 1718 | 13 | 56 | |
| 319907 | Tract 3199.07 | 3199.07 | 67829 | 6264 | 90 | 233 | |
| 522709 | Tract 5227.09 | 5227.09 | 55482 | 55 | 1 | 0 | |
| 522720 | Tract 5227.20 | 5227.20 | 66940 | 6319 | 83 | 211 | |

At this juncture, the demographic data is associated with the census tract polygons. You can now make a thematic map. A thematic map displays a set of related geographic features. Typically, a classification scheme has been applied so that the map displays non-spatial data associated with the geographic features.

1. Close the *Attributes of Tracts* window and click on the title bar of the *View1* window (the bar across the top) to make it active.

2. Now you can classify the *Tracts* theme. Double-click anywhere on the *Tracts* entry in the *View1* legend (except the check box) to bring up the Legend Editor.

3. For Legend Type, select *Graduated Color*, and for Classification Field, select *Pop_le6* (the number of persons age six or under by census tract). By default, five quantile classes are generated. Click on Apply to apply this classification to the theme.

*Default classification on Pop_le6.*

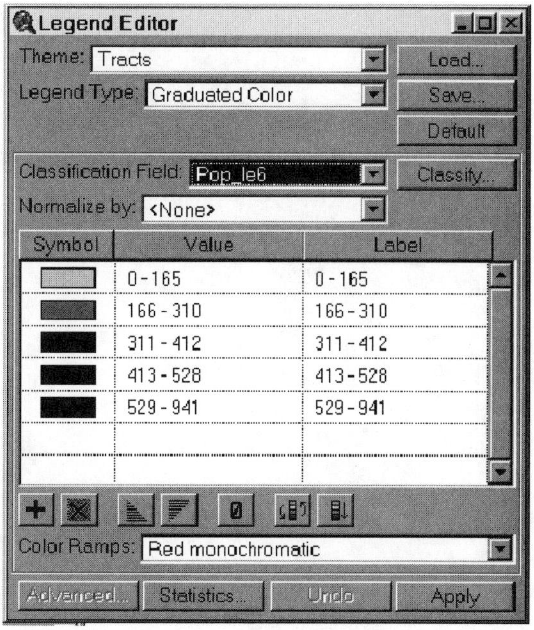

| Symbol | Value | Label |
|---|---|---|
| | 0 - 165 | 0 - 165 |
| | 166 - 310 | 166 - 310 |
| | 311 - 412 | 311 - 412 |
| | 413 - 528 | 413 - 528 |
| | 529 - 941 | 529 - 941 |

4. By default, the Red Monochromatic color ramp is applied. To more clearly display the class breaks, select Orange Monochromatic from the Color Ramps pulldown list and click on the Apply button. Additionally, you can double-click on the first symbol to bring up the Symbol Editor. From the Symbol Editor, you can change properties such as outline width, fill pattern, and color. Experiment with color choices until you have a satisfactory combination. Click on the Apply button in the Legend Editor to apply these changes to the map.

5. The screen is redrawn with your new choices. If you do not like the final result, try again. If you like your results, go ahead and save them now. We have saved our choices for you to reference, if necessary. (The incremental project has been saved as *ch2b.apr.*)

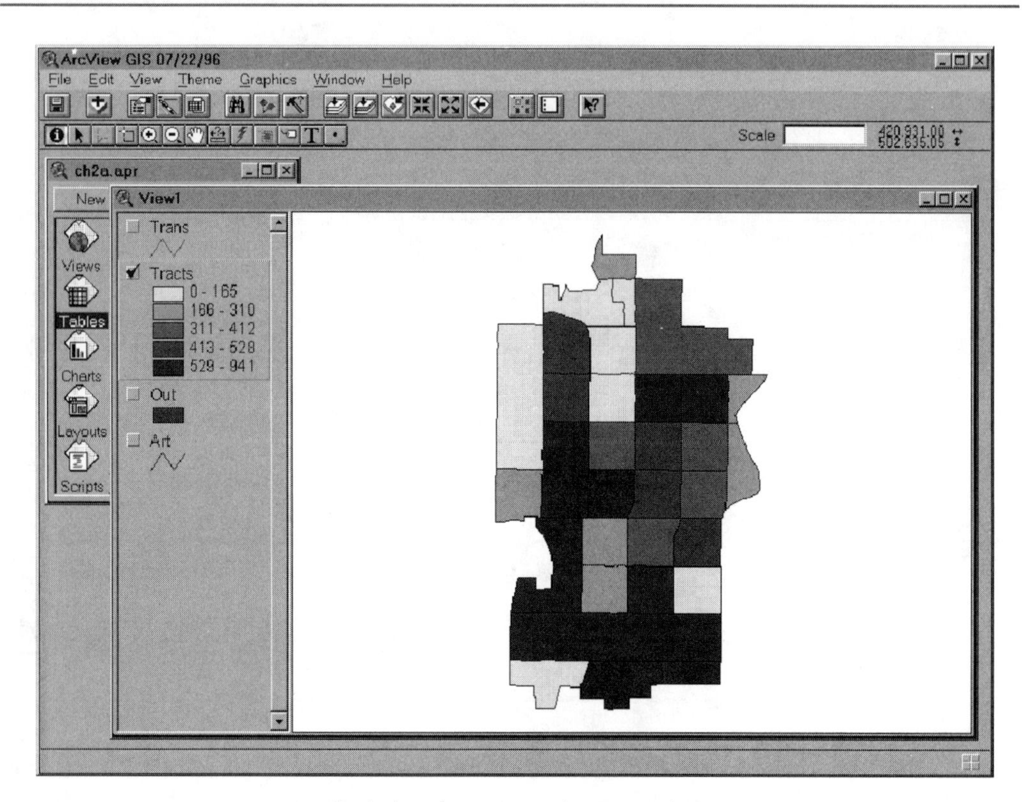


*Applied classification on the Tracts theme.*

The demographic data has been imported, joined to the *Tracts* theme, and classified. Your initial task of locating concentrations of children under age six is complete. Now you can proceed with the data on competitors.

You can locate competitors by matching their street addresses to a street network (map). This process of address matching is referred to as *geocoding*. One of the themes you initially imported was the street net theme for Tempe, illustrated in the figure below. Although this street net theme is coded with the attributes necessary for address matching in ArcView (i.e., street names and address ranges), the theme must first be made *matchable* before you can geocode a data file containing street addresses against it.

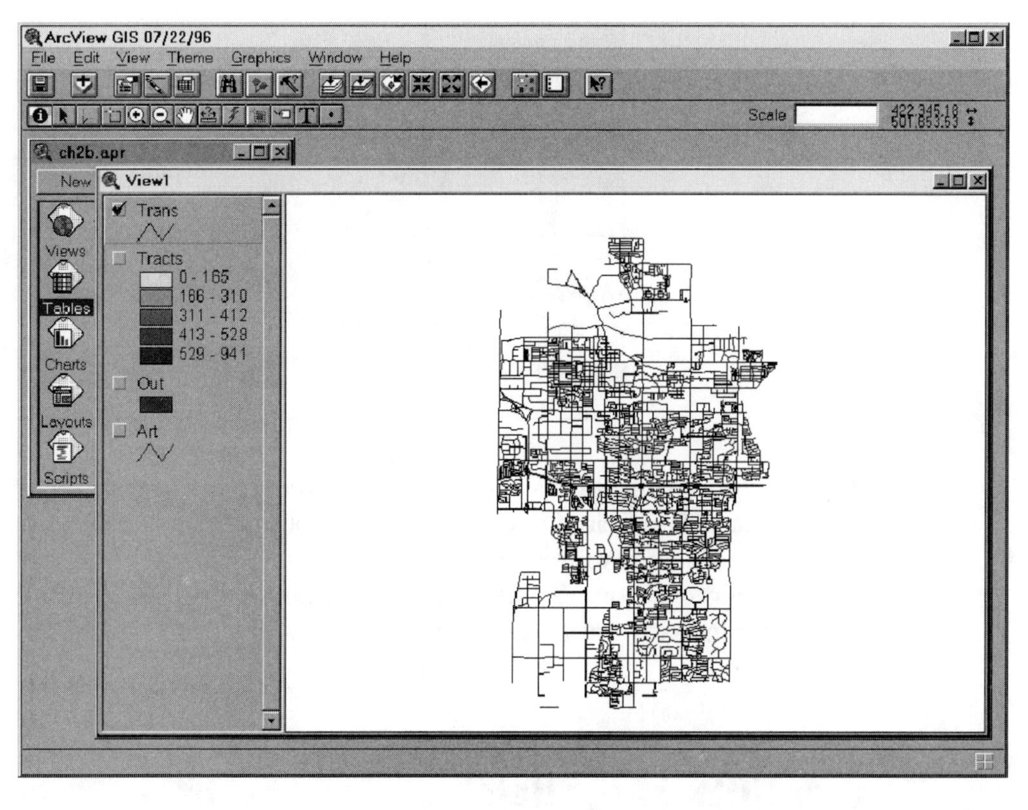


*Street net theme (Trans) for Tempe.*

To make the theme matchable, take the following steps:

1. Make the *Trans* theme active by clicking on its entry in the legend area of the *View1* window.
2. From the Theme pulldown menu on the menu bar, select Theme Properties.
3. The Theme Properties window contains a number of theme operations. Scroll down the icons on the left side of the window until you find a mailbox icon labeled Geocoding. Click on this icon and then identify the fields to be used in address matching.

*Theme Properties dialog window.*

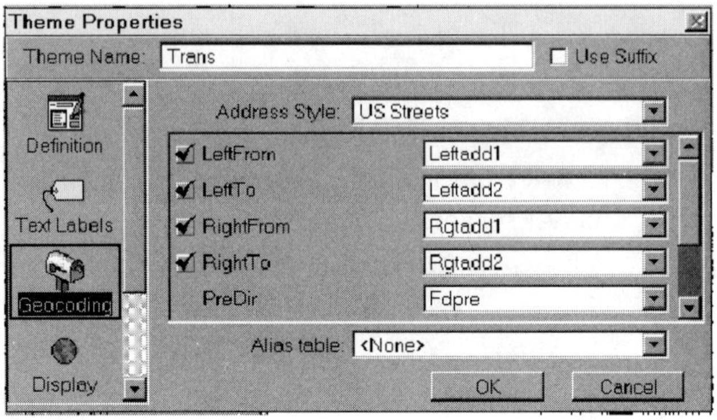

4. Because you are using a standard TIGER street net, ArcView recognizes the standard field names present. Accept the default choices, as well as the default Address Style, *US Streets*, by clicking on the OK button. Then click on Yes when you are asked if you want to build geocoding indexes using the US Streets address style. The geocoding indexes are then built, and the theme will be matchable.

## Cleaning TIGER Street Nets

In Exercise 1, when you geocoded the address data to the street net, you obtained a 100% match the first time. Although there is no reason to expect a nightmare when you attempt this on your own, the variability in address formatting and the dynamic nature of an urban street net dictate that a few hiccups on the way to a successful match are to be expected. Errors in the street net and in your data file are, effectively, inherent to the geocoding process.

The most readily available street net for the United States is the TIGER street net, available from the U.S. Census Bureau. Priced at approximately the cost of distribution, TIGER is a bargain, but it is not without problems.

The primary problem with TIGER is basically one of completeness. TIGER was created for use during the 1990 census data collection effort. The cutoff point for entering new street segments occurred prior to 1990. If you are working in a rapid growth region, you can be sure that many street segments will be missing. In addition, the existing street segments may be coded with incorrect address ranges or may lack address ranges altogether.

In the course of preparing Exercise 1, five of the 23 street addresses did not match (were not located) on our first attempt. Upon examining the TIGER street net for Tempe, we found one missing street, which we digitized and coded manually, and four streets with missing or incorrect address ranges, which we fixed manually. We neither searched for nor fixed any additional coding errors in the street net. You can verify this by examining the attribute table for the entire street net.

A host of vendors have come forth to provide better and cleaner TIGERs. The vendors have fixed the coding errors, added the missing address ranges, and in many cases digitized additional street segments not present when TIGER was created. You will pay a little more for these TIGERs, but it may be well worth it.

The second source of error is in the address field of the database you are attempting to geocode. A full discussion of geocoding and address formats takes place in subsequent chapters, so suffice it to say that spelling errors in street names, incorrect street types, and inclusion of additional data in the primary address field, such as apartment or suite numbers, can reduce the accuracy of matching. You can use ArcView's Edit window to correct these errors on the fly, but that is no substitute for having clean and properly formatted data from the outset.

You are now prepared to match the addresses from *day.txt*, the table of competing daycare centers. The first step in this task is to create an *event theme* from this table. (See the insert titled "Why an Event Theme?" for more information about event themes.)

To create an event theme, take the following steps:

1. Make the *View1* window active, or open it from the project window if it has been closed.

2. From the View pulldown menu, select Geocode Addresses. The Geocode Addresses dialog window is initialized.

*Geocode Addresses dialog window.*

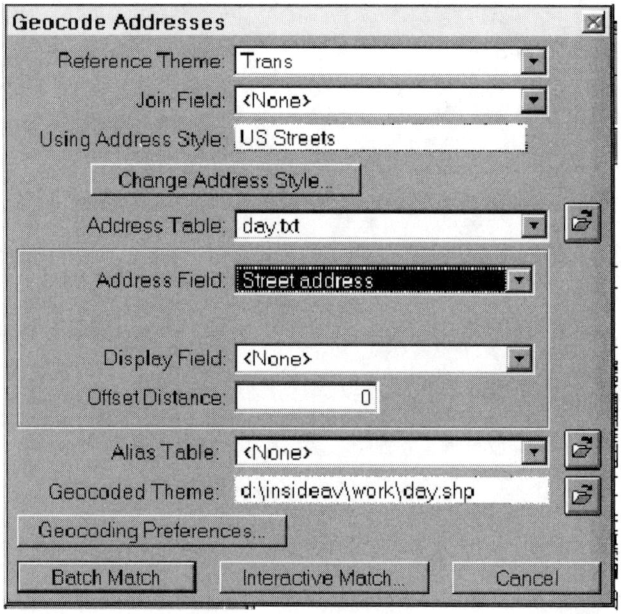

3. You are then asked to identify the reference theme (the theme containing the address-coded street net). To do so, select *Trans* from the pulldown list.

4. Leave the Join Field set to *None*, because you are electing not to physically join the two tables. From the Address Table pulldown list, select *day.txt,* the table containing the address events. For Display Field, accept *None*, the default, and accept the default Offset Distance of *0.*

**5.** From the Address Field pulldown list, select *Street address.*

**NOTE:** *For address matching, the entire address should be contained in a single field.*

**6.** For Alias Table (an alias table lets you associate place names with a street address), accept *None*, the default. For Geocoded Theme—the name for the shapefile that will be created—save the theme in your working directory, with the name *day.shp*. This shapefile will contain the point locations of all the locatable addresses in your data table.

**7.** Now that you have completed all inputs, click on Interactive Match to begin geocoding.

## Why an Event Theme?

In the sample project from Exercise 1, you worked with both spatial data sets (census tracts and a street net) and tabular data sets (demographic data and competitors' addresses). When working with the demographic data, the link between the spatial and tabular data sets seemed clear. You joined the two tables on a common attribute and, as a result, were able to display the demographic data spatially; that is, to map it by census tract.

However, when preparing to work with address data and join it to the street net, it becomes evident that the relationship is not so clear. Before this link can be performed, an "event theme" must be created from the address data. (An event theme is a theme based on X,Y events; address events; or route events. See the Glossary for more information.) What is so different about addresses? For the answer to this question, we need to look under the hood briefly and examine how data is stored and modeled in ArcView.

The original data structure beneath ARC/INFO, and one that is carried forward into ArcView, is a georelational data model. In this model, the spatial data—the lines, points, and polygons that are used to represent the real world—are stored in one set of files, and the attribute data associated with these features is stored in another set of files. The attribute data are linked to the spatial data in a one-to-one relationship. The *relate* concept is paramount: while the data model can be extended through multiple relates, one-to-many relates, external relational databases, and the like, it is still built upon, and limited by, the foundation of relational database technology.

ArcView is new software, built from the ground up using an object-oriented data model. As a result, ArcView is much more flexible about data types and linkages than ARC/INFO, but the underlying framework remains. ArcView still "likes" to see a one-to-one relationship between spatial features and the attributes associated with these features.

Address geocoding is too unwieldy to maintain a discrete node in a street net for every address. Instead, each street segment is coded with address ranges, and the software interpolates along the segment to locate a specific address. These interpolated points are determined on-the-fly while the data file containing street addresses is geocoded against the street net.

In the ArcView world, however, you want to preserve the results of this linkage for future analysis. To make the outcome of on-the-fly interpolation more lasting, ArcView creates a *shapefile*, a new spatial data set comprised of the point locations for each located address as well as the link back to the address record in the data file. The shapefile is not an ARC/INFO coverage but, in ArcView, it works virtually the same way. The outcome is yet another instance of our basic data model (a spatial data set, or series of geographically distributed points) linked to an attribute table (address records and other associated data). Like any other ArcView theme, you can display, classify, and model it however you wish.

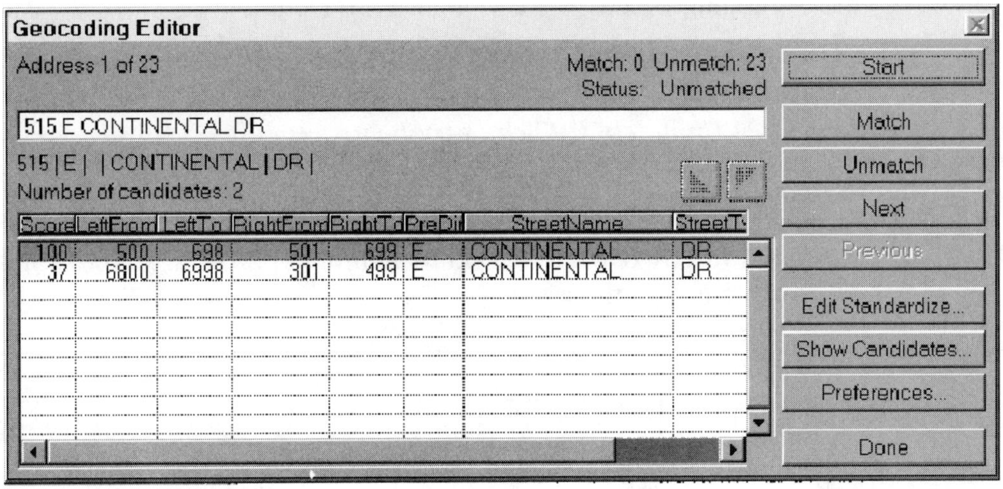

Geocoding Editor

Address 1 of 23

Match: 0 Unmatch: 23

Status: Unmatched

515 E CONTINENTAL DR

515 | E | | CONTINENTAL | DR |

Number of candidates: 2

| Score | LeftFrom | LeftTo | RightFrom | RightTo | PreDir | StreetName | StreetT |
|---|---|---|---|---|---|---|---|
| 100 | 500 | 698 | 501 | 699 | E | CONTINENTAL | DR |
| 37 | 6800 | 6998 | 301 | 499 | E | CONTINENTAL | DR |

Start

Match

Unmatch

Next

Previous

Edit Standardize...

Show Candidates...

Preferences...

Done

*Geocoding Editor dialog window.*

At the start of geocoding, as shown in the previous illustration, the first record from the data file is presented along with all possible street segment candidates in which to locate this address. ArcView indicates the likelihood of a match by assigning a score from 0 to 100. (By default, the threshold for determining a successful match is a score of 50.) If there are multiple candidates, they are listed by descending score value, with the best match listed first. If there is no match, the field is blank. If you are lucky, one candidate will be shown, for which the matching score is high. In our example, the first matching street segment is *501–699 E. Continental Dr.* with a score of 100. This is most definitely a match.

**8.** At this point you can proceed in one of two ways: step through the data file record by record to evaluate all matches, or let the software match as many as it can without user interaction. The success and high score of the first match indicates that the address field is properly formatted. Rather than step through record by record (using the Match button), match the entire file by clicking on the Start button.

9. In our sample application, all 23 records match. Accept the results of the geocoding process by clicking on the Done button.

10. At the completion of geocoding, the Geocoding Editor dialog window displays the last address matched. The status bar should report *Match: 23 Unmatch: 0.* Click on Done to accept the results.

11. The geocoded points for each address are now created. At completion, the Re-match Addresses dialog window is presented. Because all 23 addresses matched with a Good Match (a score of 75 to 100), click on Done to accept the results.

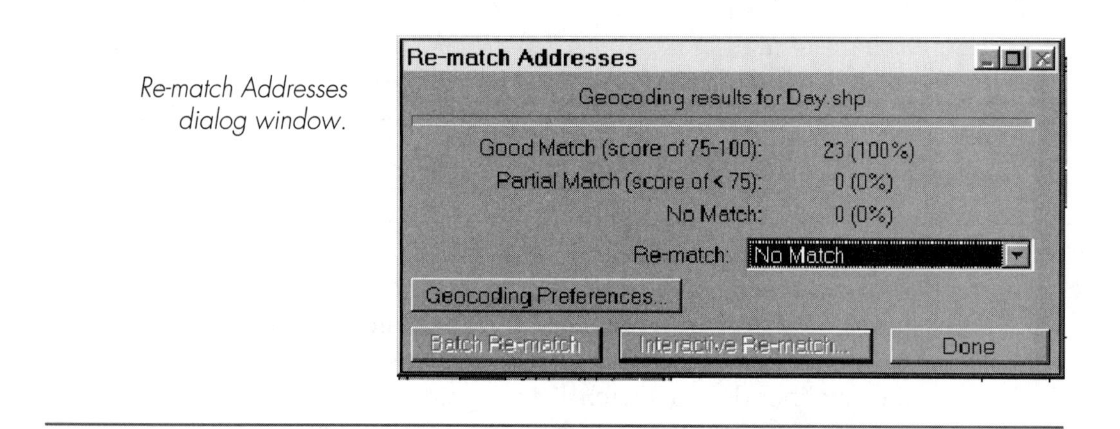


*Re-match Addresses dialog window.*

12. At this point the address matched event theme is added to the view. This theme is like any other in that it can be viewed, queried, and classified. Use the Symbol Editor to change the color of the dots if you want more contrast. (The incremental project has been saved as *ch2c.apr.*)

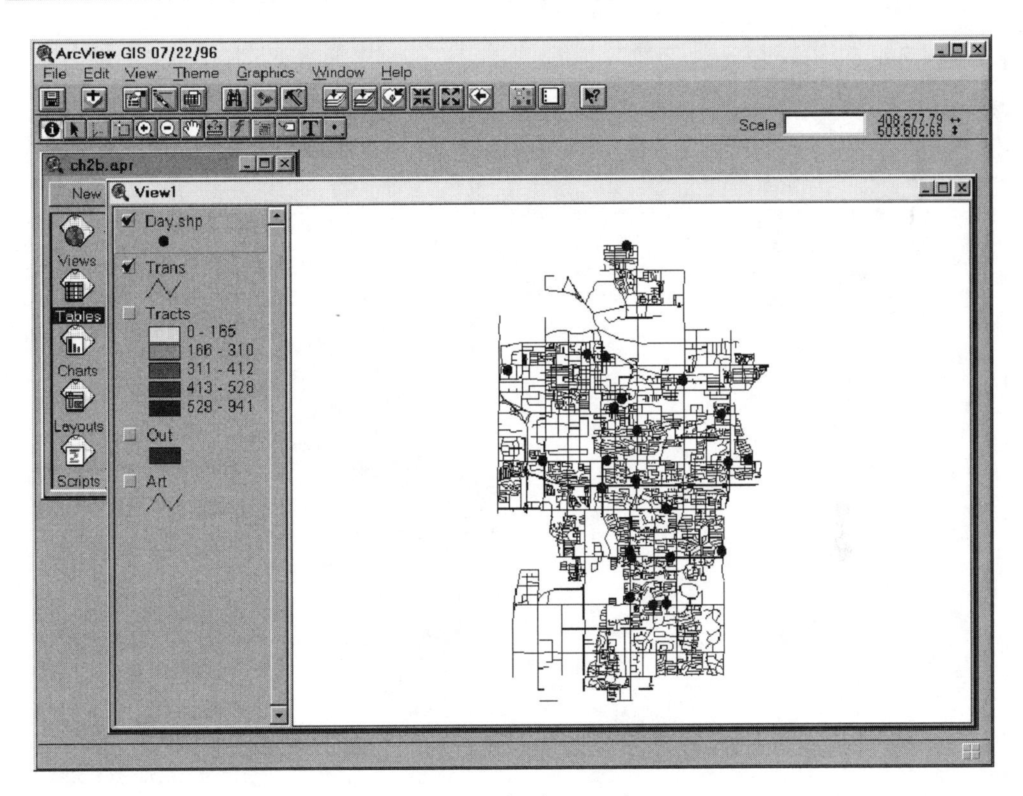


Address matched event theme.

You are now ready to display the locations of existing daycare sites against the five classes of population age six and under. Click on the check box to display the population theme alongside the daycare sites, as shown in the next illustration.

> **NOTE:** *The drawing order of themes in ArcView is controlled by their positions in the Table of Contents. ArcView themes draw from bottom to top. To change the drawing order, click on a theme in the contents. Then, while holding the mouse button down, drag the theme to a new position in the contents. Release the mouse button when you have the theme where you want it. The display will redraw automatically.*

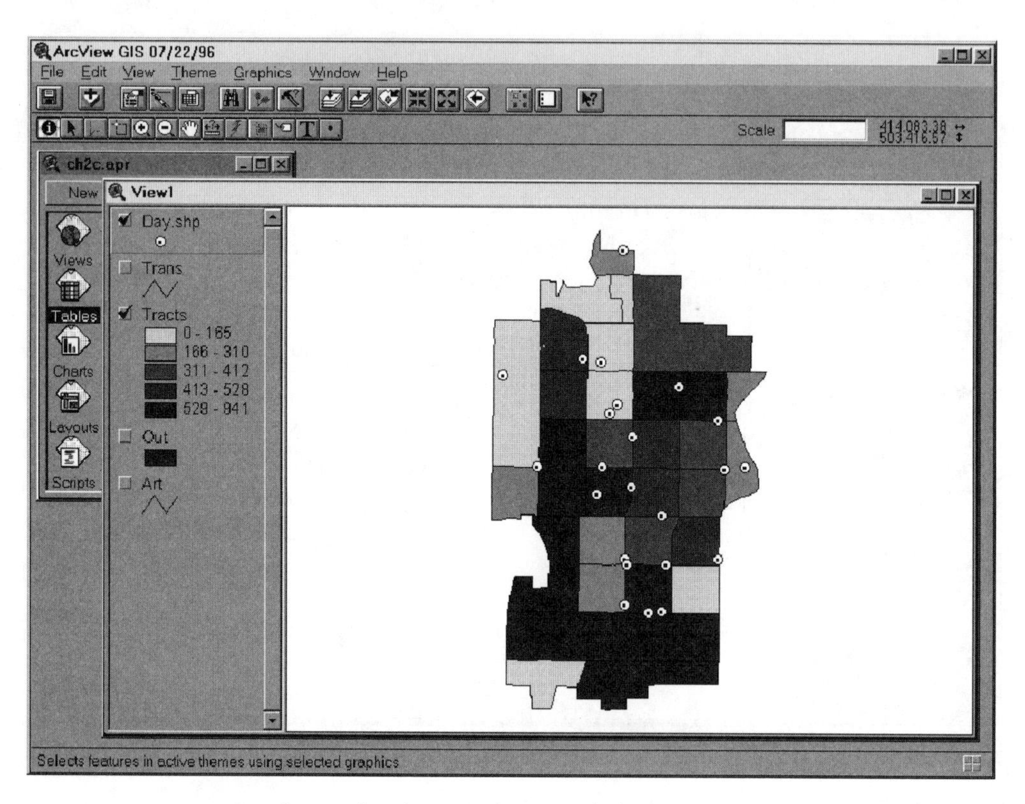


*Displaying the daycare theme with the Tracts theme.*

The darkest areas are those with the largest number of kids age six and under. The location of existing daycare sites suggests several promising areas for the location of a new center. However, you are not yet ready to make recommendations to Sharon Y. The demographic numbers are raw counts, and the census tracts are not all the same size. While the raw counts are useful because Sharon needs a minimum service base regardless of population density, it is still necessary to know more.

The next step is to display the population age six or younger on a per-acre basis; that is, to *normalize* (see Glossary) the values for population age six or younger by area. Fortunately, this functionality is built into the Legend Editor.

1. Display the *Tracts* theme if it is not currently displayed in the view.
2. Double-click on the *Tracts* theme in the Table of Contents to bring up the Legend Editor. The previously applied classification is displayed.
3. Note the *Normalize by* pulldown list directly below the *Classification Field* pulldown list. It is currently set to <*None*>. From the list, select *Acres*. The class breaks now fall at 0.163, 0.483, 0.696, 0.825, and 1.218. Click on Apply to apply the classification to the theme. Note how the distribution changes as the theme is redrawn.

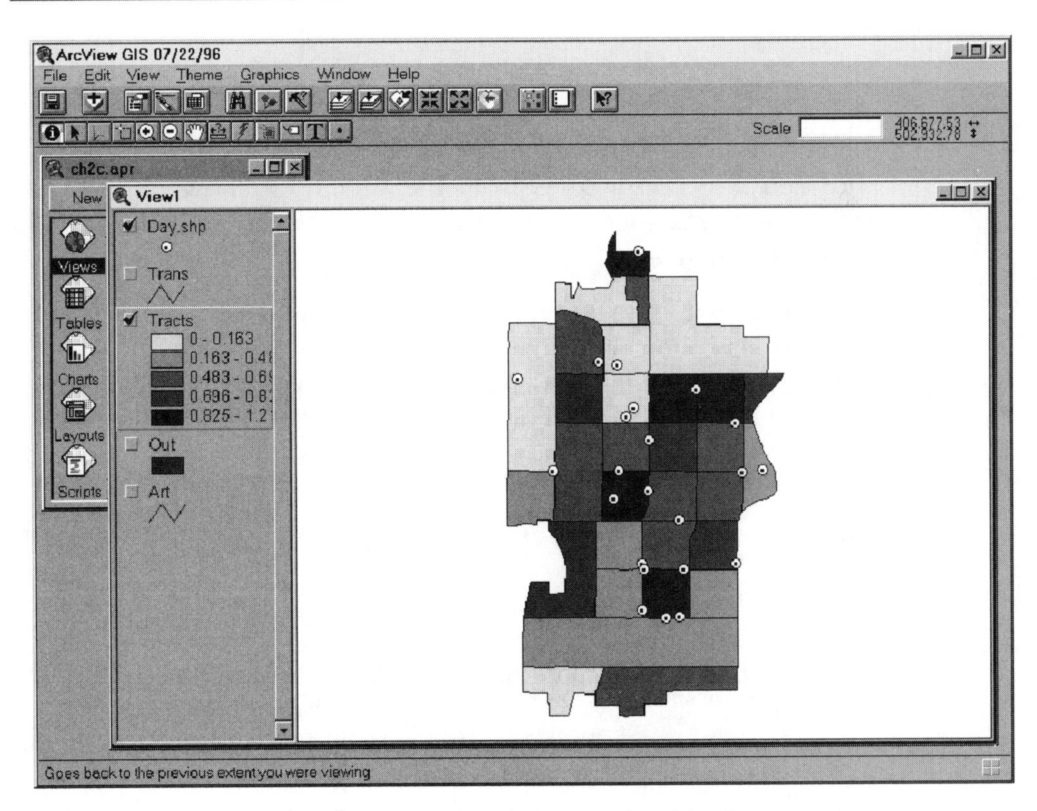


*Classification on Pop_le6, normalized by Acres.*

While this goes a long way toward making the data interpretable, another field in the demographic data set, percentage of total population age six and under, might be useful. This field will allow you to identify tracts with the largest proportion of youngsters. Follow the steps below to display proportional concentrations of children age six and under.

1. Click on the *Tracts* theme in the legend area to make it active.
2. Pull down the Edit menu from the menu bar, and select the Copy Theme option. This copies the theme onto the clipboard: select the Paste option from the Edit menu to create a second instance of this theme in the view.
3. An exact copy is added to the top of the Table of Contents. To differentiate the copy from the original, rename the copy. Click on the theme entry in the Table of Contents to make it the active theme and then select Properties from the Theme pulldown menu. For Theme Name, change *Tracts* to *Tracts - by Percent.* Click on OK to accept the change.
4. At this moment, the second theme is indistinguishable from the first with the exception of the names. To remedy the situation, double-click anywhere on the new theme in the Table of Contents (except on the check box) to bring up the Legend Editor. Classify the theme as you did the first time around, this time selecting *Percent_le6* from the *Classification Field* pulldown list, and *None* from the *Normalize by* pulldown list. An initial classification of five classes appears, the highest of which is *11 - 13*. For the current analysis, you want an either/or classification. Select *Classify* (see Glossary) from the Legend Editor. In the Classification dialog window, select *Quantile* from the *Type* pulldown list, and *2* for the number of classes. Click on OK to accept.

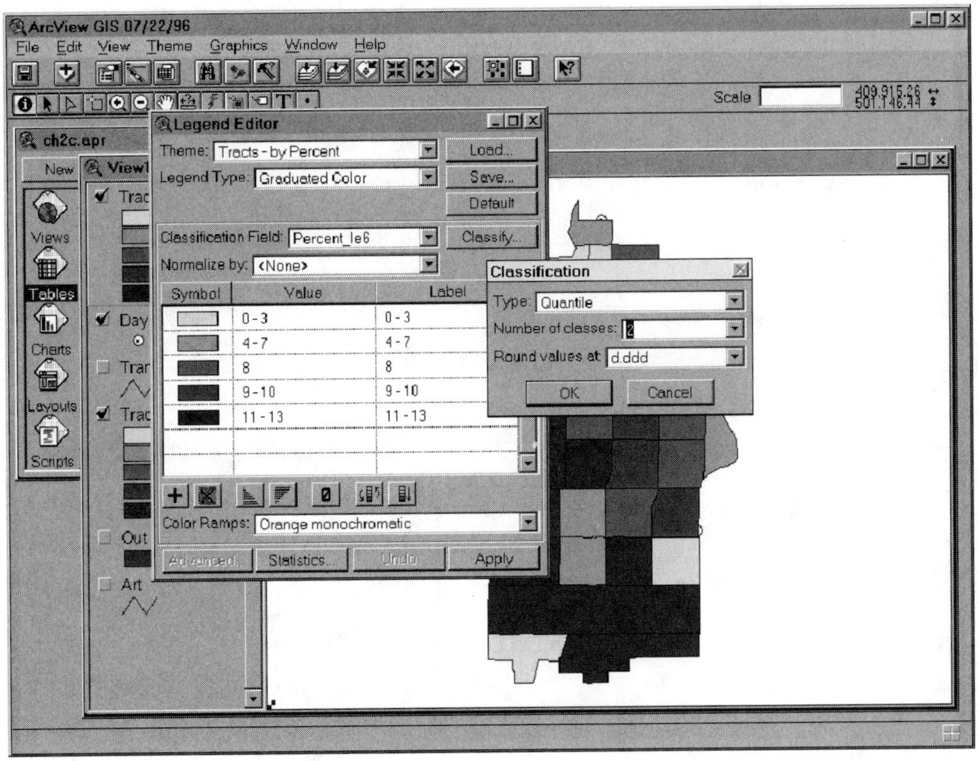

*Preparing to generate two classes on Percent_le6.*

5. There are now two classes in the Legend Editor: *0 - 9* and *10 - 13*. For a threshold of ten percent, click on the *Value* cell and enter the new class range of *0 - 10* to replace *0 -9*. Click on Apply. Note that the labels for the Table of Contents were updated as the new class ranges were entered.

6. Change the symbology for the two classes. Double-click on the first symbol to bring up the Palette choices. For the first, choose the upper left symbol pattern, the open square. This is a transparent symbol and will allow other data to be seen through it. For the second symbol, the objective is to select a bold, diagonal hatch pattern with a transparent background that will be easily distinguishable from the solid shades of

the population theme. Select a bold hatch pattern from the fill palette. Switch to the color palette by clicking on the the paintbrush in the palette window. Select Foreground in the Color pull-down, and click on black. Select Background in the Color pull-down and click on the transparent selection (the box containing an X at the upper left corner of the palette). Click on Apply to apply the new symbol selections to the theme.

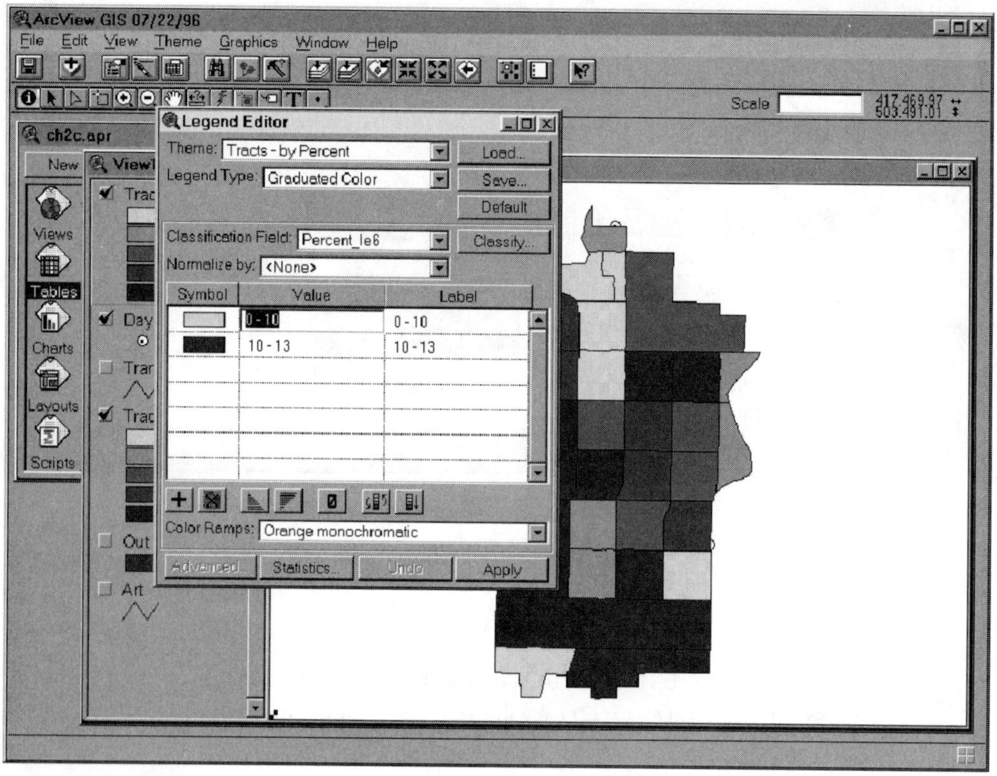

Revised classes for Percent_le6.

7. Click on, or activate, the *Tracts - by Percent* theme in the Table of Contents to display the new theme along with the previous themes. You should see the map that appears in the following illustration. With

the exception of two blocks, all blocks with the highest raw counts for population age six years and under also have a high proportion (greater than ten percent) of the population in this age group. South Tempe may well be a favorable area for Sharon's new daycare center. (The incremental project has been saved as *ch2fin.apr.*)

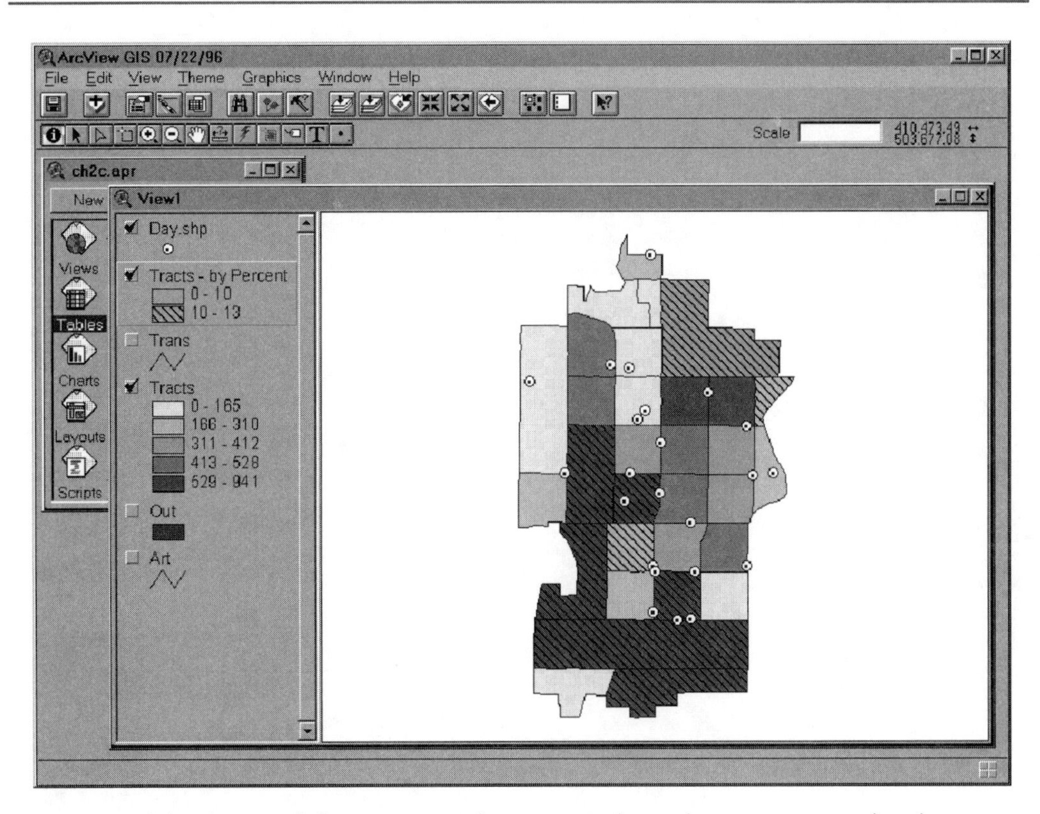


*Final distribution of daycare center locations and population age six and under.*

Before finalizing the proposal for Sharon's location options, you might want to explore additional variables, such as household income, or you may want to convert the population of age six and under to a per acre figure. You have the data you need to extend the analysis. Before finalizing the proposal for Sharon's location options, you may wish to explore additional variables, such as household income.

## Chapter 3

# Getting Started: Projects and Views

In Chapter 1, you were introduced to ArcView and the fundamentals of GIS. In Chapter 2, you took a whirlwind tour. At this point, we suspect that either the light bulb has clicked on or you are totally lost. We hope it is the former, but if the latter applies, take heart—the rest of the book proceeds at a slower pace and covers the material in greater detail.

Where to begin? When you are lost or just want more information, ask for help. With a pulldown Help menu and the Help tool icon, assistance with ArcView is always just around the corner.

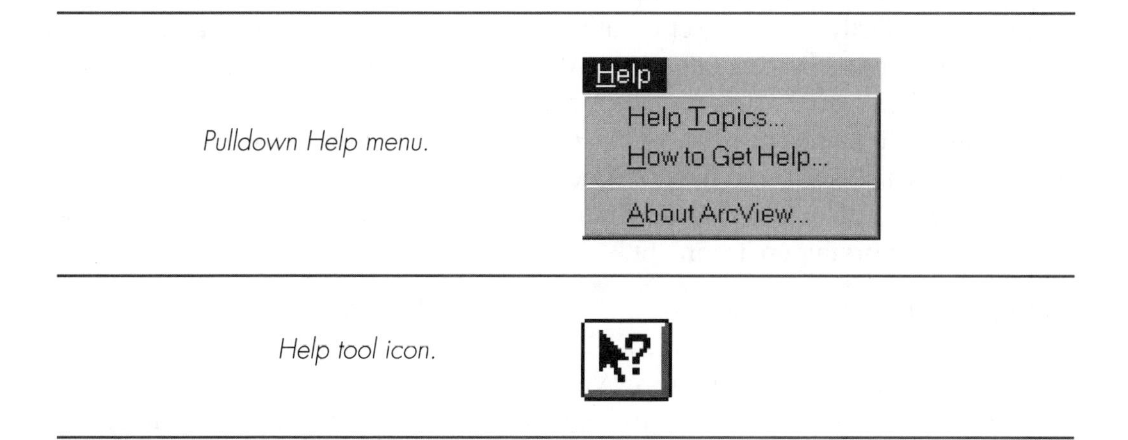


*Pulldown Help menu.*

*Help tool icon.*

Like many Windows-based products, ArcView's help system provides guidance and support at both novice and expert levels. The system presents ArcView's features and provides examples of usage. The Help tool icon is used primarily to obtain information about a specific tool or menu option. Clicking on the Help tool, followed by clicking on the tool in question, brings up a help topic for that tool. Alternatively, you can click on a window or element with the Help tool to bring up a help topic related to that window or element.

The pulldown Help menu is particularly useful when you are not certain which function to perform, how a function operates, or which options you have. The Help menu allows you to browse or search the help system via a keyword or phrase. The help system, like the ArcView application, conforms to the Windows 95 look and feel across all platforms.

Also remember that whenever you are inside ArcView's help system, you can print out help topics for later reference.

## The First Step: Opening a Project

Efficient navigation in ArcView requires understanding the program's organizational structure. The highest organizational unit in ArcView is the *project.* An ArcView project is similar to a working file in that it allows you to group all program components—*views, themes, tables, charts, layouts,* and *scripts*—into a single unit.

An ArcView project is stored as a project file. Basically, this file stores the ArcView environment exactly as it was set up at the time you executed a save, and enables you to recall the session at a later time. All ArcView components, including joined tables, logical or spatial reselects, geocoded event themes, as well as references to all the graphic files and maps on your screen, are saved in the project file. All these components will be covered in subsequent chapters. The key point is that an ArcView project, which is stored as an ASCII file with an *.apr* extension, is the repository for your working environment.

Project files are *dynamic,* meaning the steps that lead to the views of your data are stored rather than the data itself. As a result, the project stays current with your data. The next time you open an ArcView project, any changes in your data will be reflected in your maps. Database operations, such as tabular joins or logical queries, will be performed anew. In this way, ArcView ensures that what you see is always updated. The ArcView project is an excellent tool for tracking and modeling a dynamic database.

To create a new project from scratch, begin by selecting New Project from the File menu. The Open Project selection from the same menu (the File menu) is used to open an existing project.

> ✓ **TIP:** *Because the UNIX version of ArcView is started from the command line, you can specify the name of the project to open when starting ArcView (e.g.,* arcview myproj.apr*).*

As demonstrated in Exercise 1, the first steps to take with a new ArcView project are opening a *view* (essentially a map) and adding data. To open a view, click on the Views icon from the project window and select the New option. A view is opened with the default name of *View1*.

Once a view is open, you can add data to it. In ArcView, data fall into two main categories: spatial and tabular.

Spatial data are also referred to as *digital cartographic data*. Think of these data as specialized graphics in which digitized geographic elements are coded with the absolute coordinates that locate the elements on Earth. Spatial data are depicted in vector and raster forms. ArcView supports both.

Within *vector* data, map elements are stored as a series of X,Y coordinates that define lines, points, and polygons. ArcView supports vector data in five formats:

1. ARC/INFO coverages
2. ArcView shapefiles
3. AutoCAD .dwg files
4. AutoCAD .dxf files
5. MicroStation .dgn files

ARC/INFO data support is especially important because ARC/INFO can convert data from myriad other sources, which are then translatable to ArcView.

> **NOTE:** *The shapefile is ArcView's native data format, and has been optimized for performance within the program. See "Working with Shapefiles" in Chapter 9 for more information.*

Within *raster* data, a geographic area is divided into rows and columns, effectively creating a data grid. Each grid cell, or *pixel*, is coded with a value representing information about the location being depicted. The most familiar raster data type is satellite imagery. Raster data are also referred to as *image* data. Raster formats supported in ArcView include the following:

- ❒ ARC/INFO grid data
- ❒ TIFF
- ❒ ERDAS .lan
- ❒ ERDAS Imagine
- ❒ BSQ, BIL, and BIP
- ❒ Sun raster files
- ❒ Run-length compressed files
- ❒ JPEG

Once you have spatial data, you will want to link descriptive or tabular data to your graphics. Typically, tabular data are imported in order to provide additional information about an existing spatial data set or *theme* (geography). Tabular data may be geographic in nature, having a locational component such as an address or zip code, or it may be informational, containing additional attributes associated with features in a theme, such as soil properties or land use descriptions. (See insert titled "Where to Get Data.") Standard tabular formats supported in ArcView include the following:

- ❒ dBase files
- ❒ SQL servers such as Oracle, Sybase, Ingres, and Informix
- ❒ ASCII tab- or comma-delimited text files
- ❒ INFO tables

## Where to Get Data

To a large extent, your data sources will depend on where you work and how many of your colleagues are already using ARC/INFO or other ESRI products. ArcView users will generally obtain data from the following five sources:

1. The ArcView software. ArcView ships with a large amount of sample data, including the digital five-digit zip code boundaries for the entire United States.
2. The ARC/INFO users within your own organization. Not only will they be a likely source of existing data, they will also be able to provide powerful data conversion capabilities as well as the ability to capture data from scratch by digitizing or scanning.
3. Public agencies, such as local utilities and federal, state, county, or municipal governments. These agencies are traditional GIS strongholds. Many have been creating data for years. In addition, because of government policy regarding public access, this data is often available at nominal expense.
4. Colleges and universities. While universities do not typically provide a GIS production shop with an extensive data catalog, they can serve as a valuable source of expertise for GIS data capture and conversion for specific projects. Student interns are often available to assist, not only in initial data capture, but in the subsequent analysis as well.
5. Private data vendors. These firms may be regional or national in scope and may range from basic providers of geographic data to full service consulting shops capable of handling all aspects of GIS project management. See the ArcData catalog or visit ESRI's web site at www.esri.com for a listing of vendors.

While public agencies are a likely first place to look—particularly if you need government information such as census, planning and zoning, or natural resource data—you may find yourself on your own with regard to locating data sources and ensuring that the data are provided in a usable format. Private consultants and data resellers, by providing government as well as proprietary data on a value-added basis, can help ensure that data formatting, such as map coordinates and map projections, is consistent across all data sets, and that the associated tabular data is formatted appropriately for joining to your own tabular data.

ESRI provides data and consulting services, as does Equifax/National Decision Systems (NDS), the suppliers of the market and demographic data used in the examples in this book. Many other private suppliers are also available, including Wessex, Claritas/National Planning Data Corporation, Etak, Geographic Data Technology (GDT), and Business Location Research (BLR). In addition, many national and regional consulting firms are now providing GIS data and services, thereby increasing the likelihood that you can find a suitable firm close to your place of business. The *ArcData* catalog—available free from ESRI at 1-800-GIS-XPRT (447-9778) or from the ESRI web site at www.esri.com—provides a comprehensive listing of spatial data available to support a wide range of ArcView applications.

Finally, do not overlook your local ARC/INFO users group. Many state and regional user groups are very active and can either help you locate data or provide short-term expertise. Contact your regional ESRI office for information about user groups active in your area.

# Adding Data

Whether you are adding spatial data (themes) or tabular data (tables), the dialog window looks very similar. The functions, however, are accessed from ArcView quite differently.

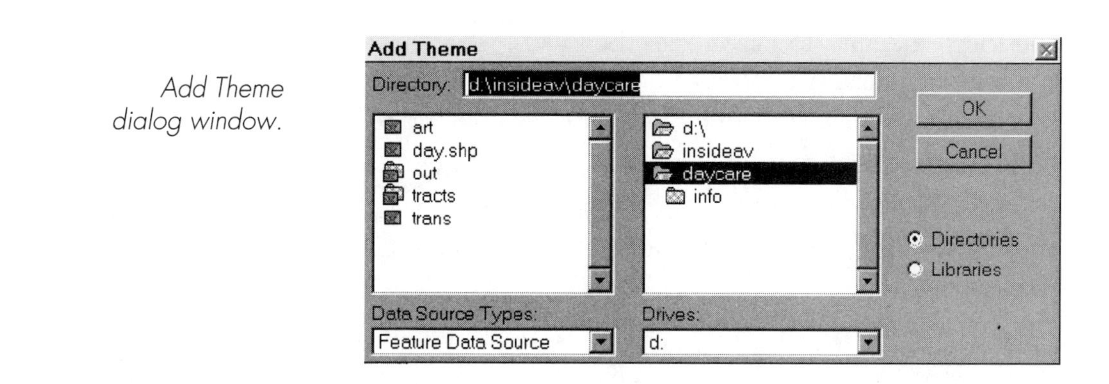


Add Theme dialog window.

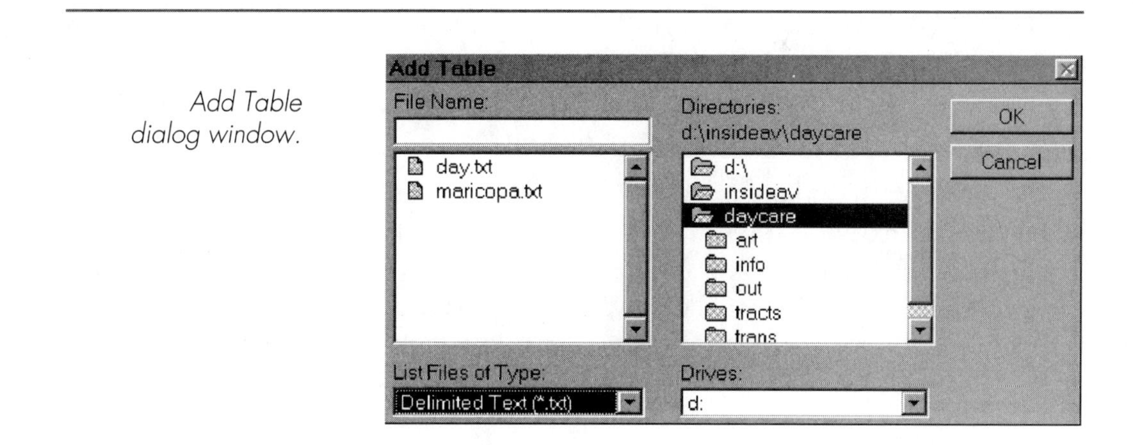


Add Table dialog window.

The Add Theme option is available only when you are working with views (maps). Both new and existing maps are opened via the Views icon in the project window. Once a view is open, new geographies (themes) can be added to the map via the Add Theme icon on the button bar or via the View menu on the menu bar. The corresponding dialog window prompts you to identify spatial files using the standard Windows conventions for requesting drives and directories, with two additional choices: a toggle between Directories and Libraries, and a drop-down list for assigning the data source type.

The toggle defaults to Directories, but also gives you the option of working with Libraries, a special ARC/INFO storage format. (For more information about Libraries, see the "ARC/INFO Libraries" section in Chapter 4.)

Selecting a data source type allows you to view the available vector data files and grid and image (raster) files. Vector data include ARC/INFO coverages and ArcView shapefiles. (See "Importing and Exporting ARC/INFO Data" later in this chapter.)

Unlike graphic data, tabular data are accessed from the project window rather than from the active view. To understand this, it is worth reviewing the difference between ArcView views and tables.

Recall that views are equivalent to maps. Maps contain several layers of data, such as streets, political boundaries, lakes, and points of interest. In ArcView, each layer becomes a theme. All combined layers or themes are represented in a single view. Views are not exclusive. The same theme can be present in several different views. However, only one view can be active at a time.

There is no corresponding structure for *tables.* Tables are stored in a common area. There are no links that represent how tables appear or how tables are combined or juxtaposed with other tables. Hence, adding tables is accomplished from the project window, rather than the view window. To add a new table, select Add Table from the Project menu. You can also add a table by clicking on the Tables icon in the Project Window's scrolling icon list and then clicking on the Add button.

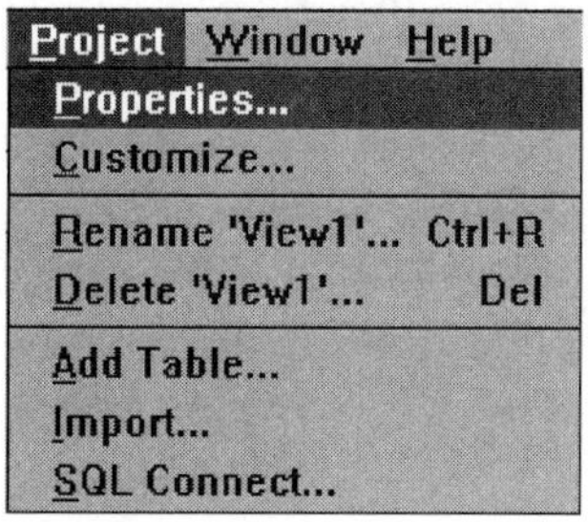

*Adding tabular data with the Project pulldown menu.*

Project Window Help
Properties...
Customize...
Rename 'View1'... Ctrl+R
Delete 'View1'... Del
Add Table...
Import...
SQL Connect...

The Add Table dialog window presents a pulldown list that allows you to specify the file type you wish to open. The following types of files are available:

- dBase (.dbf)
- Delimited text (.txt)
- INFO

If you do not select a tabular data file type, ArcView assigns dBase as the default. To open a specific tabular data source, highlight the file name and click on OK.

## More About Themes

In ArcView, themes are the basic building blocks of the system. Themes can represent essentially any spatial data set; that is, features with locational attributes. Data sources that can be represented as themes include ARC/INFO coverages and images, as well as tabular data with locational components, such as latitude/longitude coordinates or street addresses. An ArcView theme is comprised of one feature type. The primary feature types in ArcView, as supported by ArcView shapefiles, are points, lines, and polygons. Through ARC/INFO coverages, additional feature types are available, such as annotation and labelpoints.

Note that ArcView themes need not represent all features from the original data source. For example, an ARC/INFO coverage may be comprised of both line features and polygon features, such as a census coverage containing both street attributes and census tract attributes. An ArcView theme could represent either lines or polygonal elements from the ARC/INFO coverage. When working with ARC/INFO data that contains more than one class of feature, the cover name is preceded by a folder icon as well as the default envelope icon. By clicking directly on the folder icon, the cover folder is opened, and the available feature types for the cover are displayed.

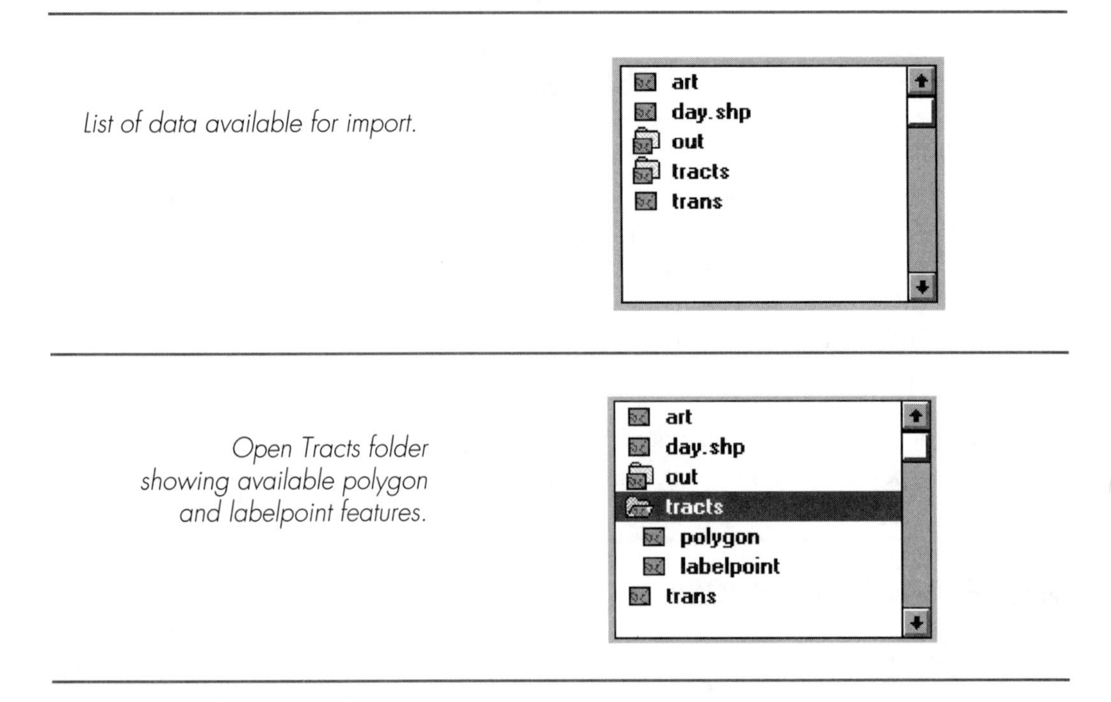


*List of data available for import.*

*Open Tracts folder showing available polygon and labelpoint features.*

Annotation features from ARC/INFO coverages are also fully supported. This includes annotation stored in annotation subclasses with text attribute tables (TAT), as well as earlier versions of annotation without attribute tables, stored in text files. If text attributes, such as size, symbol, or color, have been defined within ARC/INFO, ArcView retains these properties.

While annotation themes can be added from an ARC/INFO coverage, they cannot be created directly within ArcView. Alternatively, text graphics can be created within a view and linked to a specific theme. In addition, converting an annotation theme from an ARC/INFO coverage to a shapefile will result in a polygon theme in which each polygon defines the extent of the source annotation.

A theme might also be created from a subset of feature elements in the original data source. By using query techniques, you can create a theme that represents only a subset of the total features in a data source. For example, parks or recreational areas could be extracted from an ARC/INFO land use coverage. You will explore this topic further in Chapter 6.

When you work with themes, you can set properties that describe how the themes are named, how they appear, and other special processing rules. To access the dialog window for setting theme properties, click on the name of the theme in the Table of Contents to make it active. Next, click on the Theme Properties icon from the button bar, or select Properties from the Theme menu. Theme properties are listed below. (Setting theme properties is part of the exercise at the end of this chapter.)

- ❒ Theme name
- ❒ Selected theme features
- ❒ Minimum and maximum scales at which the theme will be displayed
- ❒ Field and position for labeling theme elements
- ❒ Theme hot links to other data sources or applications
- ❒ Theme geocoding properties

## More About Views

Views organize themes. If themes are the individual players on your ArcView team, views are the stadia in which they perform.

Each view window contains a graphical display area (on the right) and a scrolling Table of Contents (on the left) that lists all themes present in the view. Each theme's entry in the Table of Contents contains the name of the theme, the symbol legend for the theme, and a check box. If the check box is turned on, that theme is currently set to draw in the view. A scroll bar is available if the list of themes is longer than the view window.

The drawing order for the themes in a view is determined by the position of the themes in the Table of Contents. Themes at the bottom are drawn first; themes at the top are drawn last. Thus, if you want a particular theme to be "on top" of another, position it accordingly in the Table of Contents. You can do this by clicking on a theme's entry and dragging it to a new position among the themes listed in the Table of Contents.

Many of ArcView's operations, such as selecting and identifying features, or zooming in and out, are configured to work on an active theme. To make a theme active, click on its name in the Table of Contents. The theme will appear to be highlighted, or raised, in the legend.

Similarly to themes, the properties of views can also be set. To set view properties, select Properties from the View menu. The resulting dialog window contains the following properties:

- Name of the view
- Map units
- Display units
- Map projection

# Saving Your Work

There are three ways to save an existing project as you work: select Save from the File menu, click on the Save Project icon in the button bar, or press <Ctrl>+s.

ArcView projects can be saved at any time. Upon issuing a save, environment components (themes, tables, charts, layouts, and scripts) and all dynamic aspects of the components (joined tables, logical queries, and thematic display) are written to the project file. If you close the file and reopen it, these elements are restored to their exact saved state.

Note that you can also create a new project from an existing one by saving the project under a different name. Click the project window title bar, choose Save Project As from the File menu and enter the new project name.

The concepts in this chapter are critical to establishing a strong understanding of ArcView. Some of these concepts will be reinforced in Exercise 2.

# Exercise 2: Opening a Project

Because this is a new project, you will start from scratch. Open ArcView, and from the opening File menu, select New Project. When the new project is open, click on the Views icon to make it active. Then select New to open a new view.

A view window is created with the default title of *View1*. Resize the application and view windows as desired to accommodate the graphics that will be displayed during this exercise.

Set the working directory to the current workspace:

1. Click on the project window to make it active.
2. Select Properties from the Project menu. For Work Directory, enter *$IAPATH\work*.
3. Click on OK to accept the change.
4. Click on the *View1* window to make it active.

The current project involves locating potential markets for a client. Three data sets are available to accomplish this: demographic, cable viewership, and restaurant locations.

The first task is to load the spatial data (geography) you will be using to locate and reference the tabular data. For this project, you will use two spatial themes: census block groups and the census street net. The census block groups will be used to display demographic and cable viewer data, which has been aggregated to the block group level. The street net will be used as a reference for displaying restaurant data, which has been coded with X,Y coordinate locations.

## Importing and Exporting ARC/INFO Data

The initial step to setting up an ArcView project is loading data, such as ARC/INFO coverages. Typically, these coverages have been created in ARC/ INFO on a UNIX-based workstation. How to transfer a coverage composed on a UNIX platform to a PC depends on the data source and, to a lesser degree, the maintenance needs for your data. To explain the options properly, we need to get technical.

An ARC/INFO coverage consists of data files that contain spatial data and the associated attribute data. For the workstation version of ARC/INFO, the attribute data is stored in the INFO database. For each ARC/INFO coverage, a directory holds the data files. In addition, an INFO directory holds the data file templates for the attribute data. The ARC/INFO *workspace* is composed of both directories.

ArcView can directly access an ARC/INFO workspace. Access can occur through a network of workstations and PCs, or the workspace can be copied as a unit to the PC. File names, however, must adhere to the DOS 8.3 standard. (This standard dictates that a file name contain eight characters or less, and that the extension be three characters or less, for example, *filename.dbf.*)

As of ARC/INFO 7.0, all new workspaces are created to conform to the DOS 8.3 standard. In addition, ARC/INFO's CONVERTWORKSPACE command converts a pre-7.0 workspace to the 7.0 format. Note, however, that the standard imposed in version 7.0 only ensures that the internal file names adhere to the DOS 8.3 standard. It is the user's responsibility to ensure that all ARC/INFO coverage names are eight characters or less in length.

An ARC/INFO workspace can be copied without prior conversion from a UNIX workstation to a DOS-based PC. (Note that associated ASCII text files may need to be passed through a UNIX-to-DOS utility.) However, once you have installed the files on a PC, you will have no utilities within ArcView with which to maintain the ARC/INFO workspace. Due to the existence of a common INFO directory containing attribute table templates for all covers in the workspace, if an ARC/INFO cover needs to be updated, the entire workspace will need to be updated elsewhere and then reinstalled as a unit.

An alternative means of data exchange is via the IMPORT utility supplied with ArcView. The IMPORT utility reads an ARC/INFO EXPORT file—the ARC/INFO exchange format—and converts it to a PC ARC/INFO format workspace. In PC ARC/INFO, data files for each ARC/INFO cover are stored in a separate directory. Coverage attribute tables are stored in dBase format. Along with table templates, the coverage attribute tables are stored in the coverage directory. Because there is no common INFO directory, each ARC/INFO coverage is stored in a single self-contained directory. Consequently, it is possible to update individual coverages in a workspace without having to re-create the entire workspace.

Certain limitations are inherent in ArcView's IMPORT utility under Windows. IMPORT does not directly support double-precision ARC/INFO coverages (if the interchange file was created from a double-precision coverage, the coordinates in the output cover will be truncated to single precision). Next, extended data types from ARC/INFO 7.0 (routes and regions) are not supported; neither are ARC/INFO node attribute tables (NAT) and annotation (TAT) tables. ARC/INFO covers with polygons containing more than 5,000 vertices, while acceptable in ARC/ INFO 7.0, will cause the ArcView IMPORT utility to abort. In addition, certain INFO item types are not supported, and any INFO REDEFINED items will be dropped.

Another option is the additional importing utility provided with the Windows NT and Windows 95 versions of ArcView. This utility, IMPORT71, imports ARC/INFO interchange files into the 7.0 ARC/INFO format. This utility fully supports all ARC/INFO feature types, including routes, regions, and annotation subclasses. Additionally, double-precision coverages are supported, and the size limits present in the original IMPORT utility have been removed.

➺ ***NOTE:*** *The IMPORT utility supplied with the UNIX version of ArcView is the same IMPORT utility supported within UNIX ARC/INFO. Accordingly, all ARC/INFO coverage feature classes and INFO items are supported.*

The ArcView shapefile format can serve as an alternate exchange format when maintaining ARC/INFO workspaces on both platforms is not necessary. Shapefiles offer the advantage of uniform bi-directional translation between platforms, without the limitations mentioned above. However, certain limitations still remain. Shapefiles can be created only from a single feature class of an ARC/INFO coverage; coverages containing multiple feature classes, such as arc and polygon attributes, must be translated into separate shapefiles. In addition, associated cover attribute tables beyond the feature attribute table, such as look-up tables, must be exported and transferred separately. Next, if you wish to maintain coincident workspaces across platforms so that a single ArcView project file can be used on both platforms, the data must be maintained on the ARC/INFO platform in the shapefile format. This procedure necessitates redundant data storage.

Which method is best for you? The choice depends on your data requirements, the need to maintain ARC/INFO coverages, and the ability to maintain a mirror of your PC ARC/INFO workspace on a UNIX workstation. While maintaining your ARC/INFO workspace in UNIX format makes maintenance and updating more difficult, it does allow you the full range of ARC/INFO data types. In addition, workspaces can be copied back to the workstation as needed without conversion.

**NOTE:** *As mentioned previously, the ArcView shapefile format offers ease of data exchange at the expense of maintaining redundant data storage. In addition, dealing with ARC/INFO coverages containing multiple feature classes may complicate data exchange.*

If you obtain your data from a third-party provider—and the data has been formatted for use in ArcView, and the vendor has already provided for maintenance of your data sets—none of the above necessarily applies.

Steps for loading spatial and tabular data appear below.

1. Click on the Add Themes icon from the button bar.
2. Navigate to the *$IAPATH\data* directory and select the *blkgrp* and *trans* shapefiles. Display the resulting themes by clicking on the check box next to the theme name in *View1*'s Table of Contents. Through the Theme Properties dialog window, rename the *Trans.shp* theme *Trans*, and the *Blkgrp.shp* theme, *Blkgrp*.

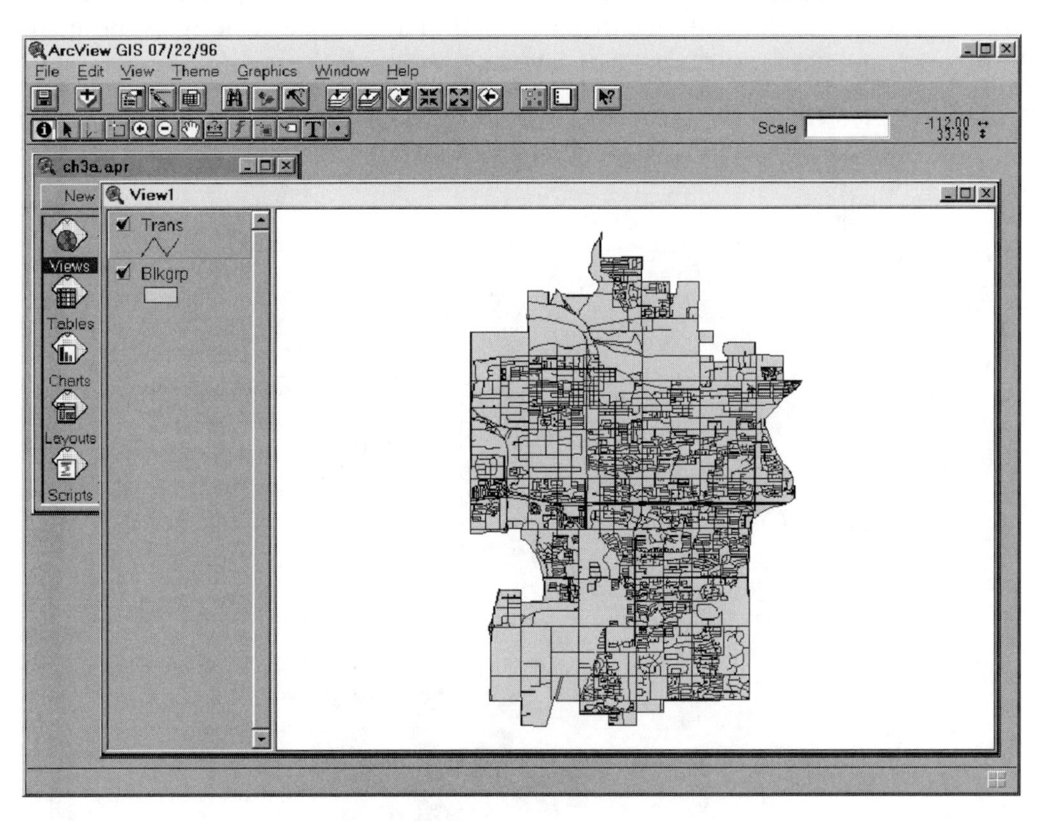


*Trans and Blkgrp themes displayed.*

Feel free to call up the Legend Editor by double-clicking on a theme name in the Table of Contents, if you wish to change the theme's color or symbology.

3. Once you are satisfied that the spatial data has been imported successfully, add the tabular data. Click on the project window (*Untitled*) and select Add Table from the Project pulldown menu. Navigate to the *$IAPATH\data* directory and set the file type to display dBase (**.dbf*). From this list, select *demog.dbf*, *cable.dbf*, and *restrnt.dbf*. (Remember, after selecting the first file with the mouse, you can add to your selected set by holding down the <Shift> key while clicking on your selection.)

4. Click on OK to add these tables to the project. As the tables are added, they are opened for display, each in a separate window. (The incremental project has been saved as *ch3a.apr*.)

**NOTE:** *In the current and subsequent exercises, the incremental project files have been stored in the* \insideav\projects *directory.*

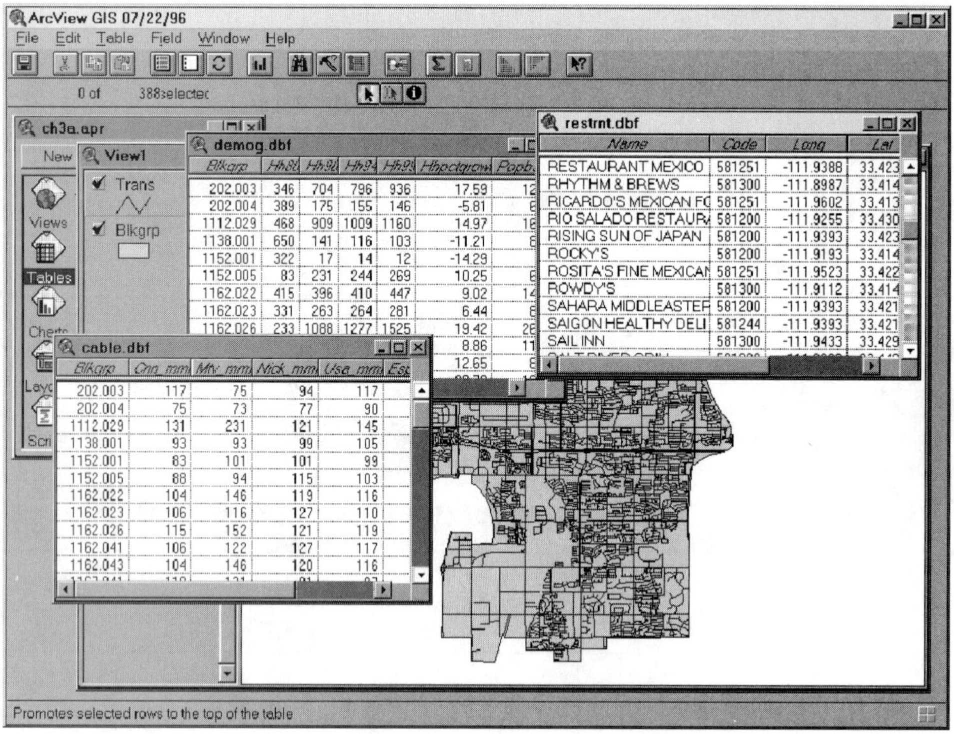

Cable.dbf, demog.dbf, and restrnt.dbf tables.

The links between the tabular data and the spatial themes—the starting point for our geographic analysis—will be established in Exercise 3 (Chapter 4). Before closing the project, take a moment to examine certain properties of theme and view displays.

## Theme Display

As mentioned earlier, the drawing order for themes is bottom to top, on the theme listing in the Table of Contents. To change the drawing order, click on the theme entry in the Table of Contents and, while holding down the mouse button, drag the theme to a new position.

Change the drawing order of the themes in this exercise by first clicking on the check boxes to be sure that both the *Trans* and *Blkgrp* themes are displayed. Click and drag the *Blkgrp* theme so that it is at the top of the list. Note that the solid shading for the *Blkgrp* polygons is drawn over the street net from the *Trans* theme.

There are two ways to display the *Blkgrp* theme on top of the *Trans* theme while allowing the *Trans* theme to be seen. One method involves changing the shade symbol from a solid to a hatched pattern, and the other is to change the theme symbology so that only the polygon outlines (block group boundaries) are drawn. The symbology for a theme is changed by accessing the Legend Editor.

To open the Legend Editor, double-click on the *Blkgrp* theme entry in the Table of Contents. This action brings up the Legend Editor window. To change the symbol pattern and color, double-click on the symbol's cell in the legend. This brings up the Palette Editor. Note that the current shade pattern selected is the solid black pattern, which produces a solid fill of the selected color. Select one of the hatched patterns. Then apply this change by clicking on the Apply button in the Legend Editor window. The theme is immediately redrawn with the new symbol, and the *Trans* theme can be seen.

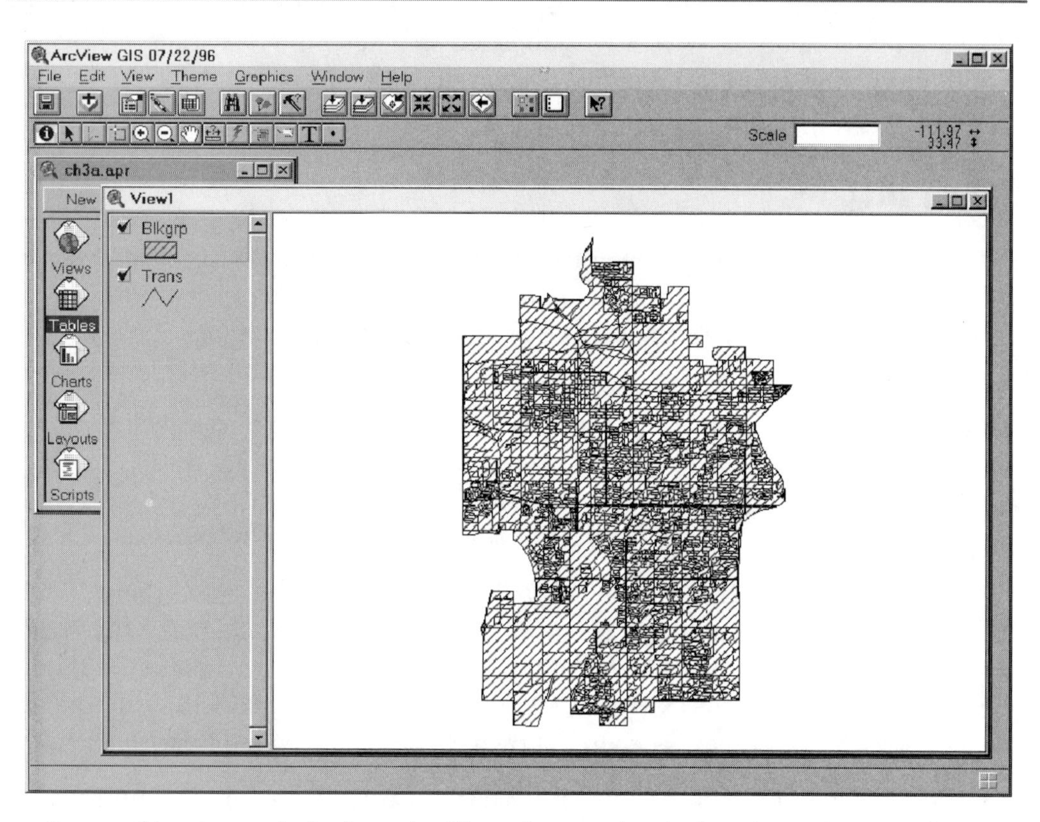


*Diagonal hatching, which allows the Blkgrp theme to be displayed over the Trans theme.*

To change the symbol color, click on the brush icon in the Palette Editor window to switch to the Color Palette. Click on the desired color, and apply as before.

If the *Trans* theme cannot be seen through the *Blkgrp* theme, access the Color Palette and verify that the background color for the hatch pattern is set to transparent. To do this, select Background from the Color pulldown list. If it is not already selected, click on the box at the upper left corner of the palette, the one containing an X. This box represents transparency in the palette of color choices.

To draw theme outlines only, click on the far left icon in the Palette window to return to the Palette Editor. Select the clear box symbol at the upper left corner.

*Clear box symbol displayed in the upper left corner of the Color Palette window.*

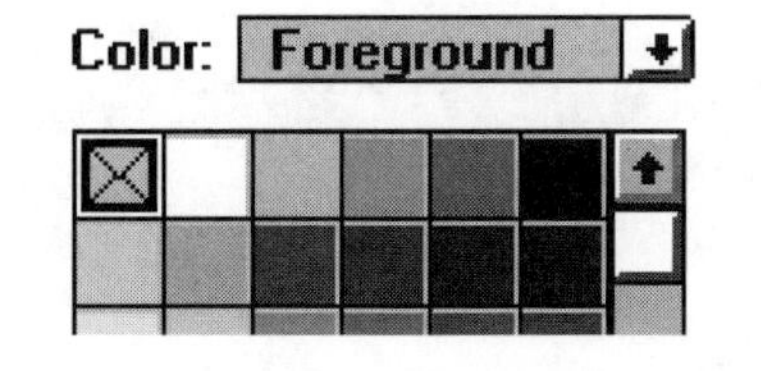


This symbol pattern is transparent, which allows the polygons from the theme to be drawn without shading. At the bottom of the Palette Editor from the pulldown list for Outline, select 2. The line weight is changed to a double width.

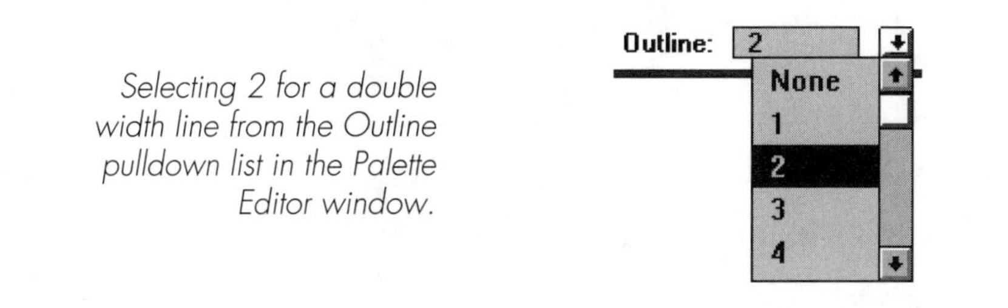


*Selecting 2 for a double width line from the Outline pulldown list in the Palette Editor window.*

The polygons are now drawn with a bold outline. To change the color, return to the Color Palette and select Outline from the Color pulldown list (which defaults to Foreground). Selected color choices will now apply only to the polygon borders.

*Selecting Outline from the Color pulldown list in the Color Palette window.*

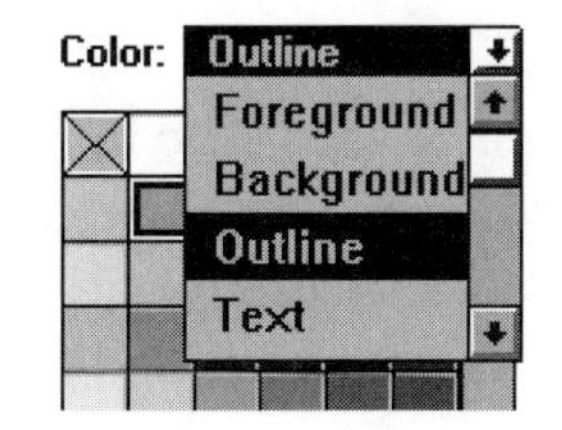

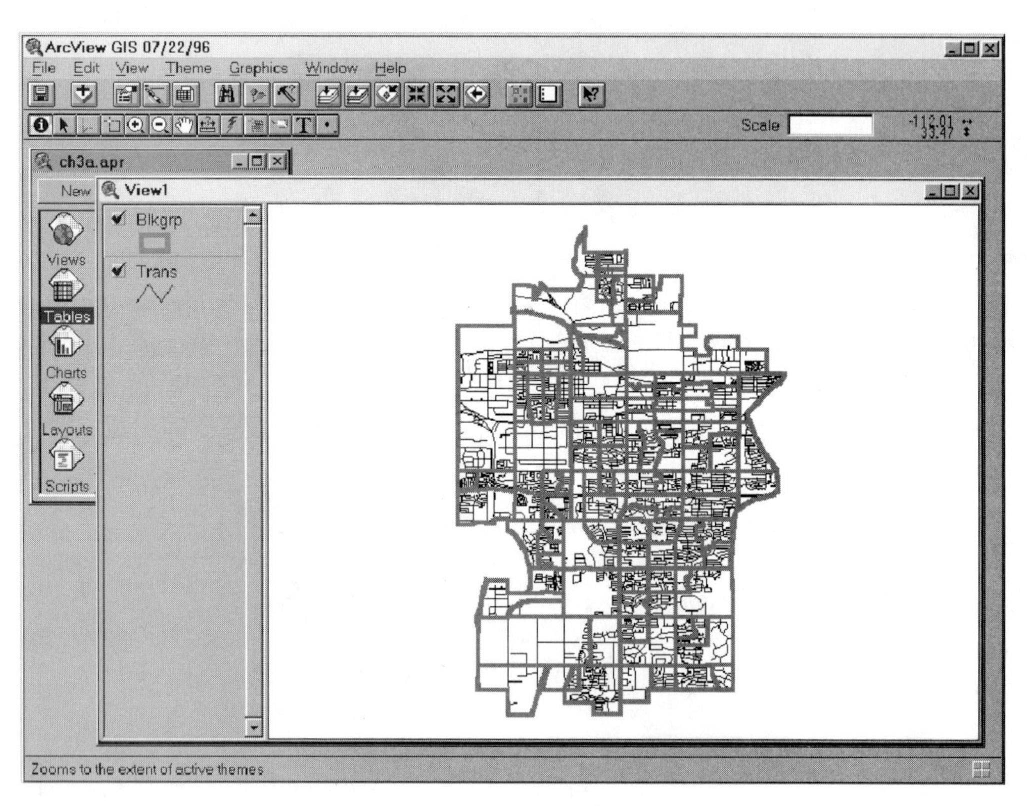


*View1, with the Blkgrp theme displayed as outlines only.*

# View Display

When you initially open a view, the map units are not set. For all themes to be displayed together, the themes you add must share the same coordinate system as the view. After you have added themes to the view, you should then set the view's map units.

Select Properties from the View pulldown menu on the menu bar. A scrolling list appears for Map Units. (Map units are the units in which the coordinates of your spatial data are stored.) The list defaults to Unknown. Click on the down arrow to activate the list.

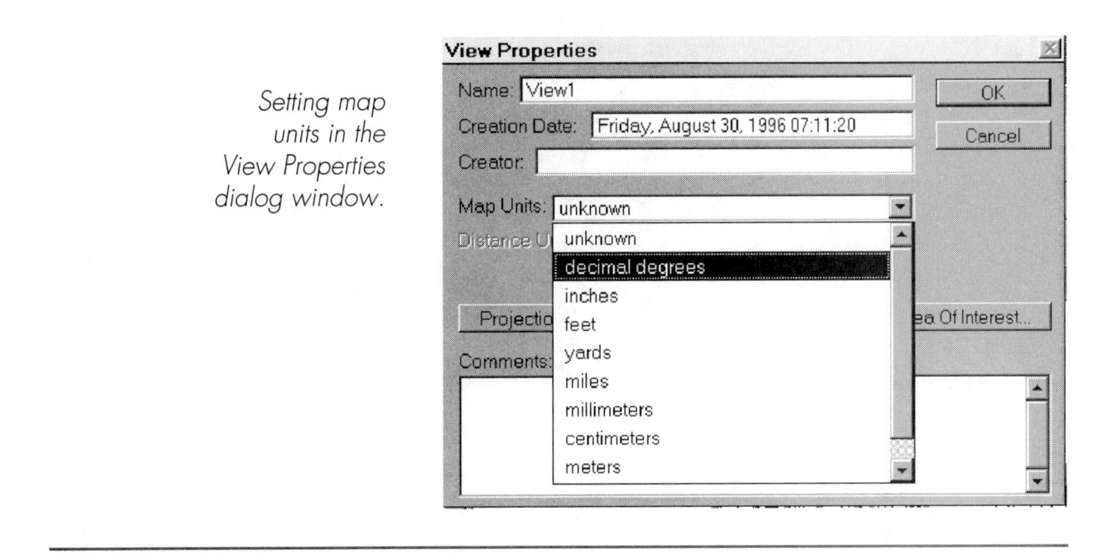


*Setting map units in the View Properties dialog window.*

The themes you have added are stored using latitude and longitude coordinates. Latitude and longitude are expressed in degrees, minutes, and seconds, and are commonly calculated as decimal degrees. Select *decimal degrees* from the list, and click on OK.

Note that a numerical value is now displayed in the Scale box on the tool bar. Click on the Zoom Out tool from the tool bar and click in the display area to zoom out from that selected point. As the map is redrawn, note how the scale value increases. Now click on the Zoom to All Themes icon from the button bar. The display is redrawn to show the full extent of the theme, at a scale of approximately 1:175,000, depending on the size of the view window.

Knowledge of the map scale can be important in two areas. First, the Scale box allows you to control the display by direct input. Click on the Scale box and change the scale number to read *50000*. After you press <Enter>, the map is redrawn at the new scale, either zooming in or out from the center of your previous display. Second, the scale can be used to control the display of a specific theme. For example, click on the *Trans* theme to make it active, and select Properties from the Theme pulldown menu to access the Theme Properties dialog window. Click on the Display icon at the left side of the window. You are presented with two fields in which to enter minimum and maximum scale values.

*Setting the maximum display scale in the Theme Properties dialog window.*

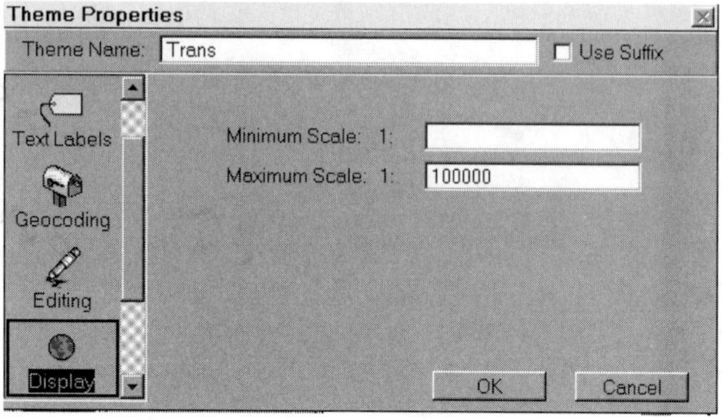

Key in *100000* for the maximum scale and click on OK. The display is redrawn. Zoom out to the full extent of the theme by clicking on the Zoom to All Themes icon from the button bar. Note that even though the theme is clicked on in the Table of Contents, the theme is not drawn. The value displayed in the scale bar shows approximately 175,000, which is greater than the 100,000 value you entered for the maximum display scale. Therefore, this theme is not displayed. In this manner you can create scale-dependent themes. Accordingly, thematic display can be limited when the display would not serve a useful purpose, such as when a map is zoomed to the extent that the amount of detail renders the display illegible or the redraw time extensive.

You have now explored the basics of adding themes and tables to a view, as well as the basics of controlling the properties with which themes are displayed. Remember to save your work before exiting, because in the next exercise you will continue where you left off in this exercise. (The incremental project has been saved as *ch3fin.apr*.)

## Chapter 4

# Extending Data

In Chapter 3, we focused on the basics of themes and views as well as importing data into ArcView. Before you can go on to maps, queries, and analysis, there is still more setup work to do. As demonstrated in Exercise 1 (Chapter 2), linking tabular data to spatial themes and creating a geocoded event theme were necessary to make the data usable. In this chapter you will learn additional basic information about extending data.

We must first take a peek under the hood to deal with data formatting of map projections. ArcView requires all spatial data sources to share a common coordinate system. If you have obtained all your data from a single source, such as a third-party data provider, or if you are using ArcView with an existing ARC/INFO installation, this issue has most likely been taken care of for you. However, if you are gathering data on your own from several different sources, we encourage you to study the insert titled "Defining a Common Ground: Dealing with Map Projections."

## Defining a Common Ground: Dealing with Map Projections

Within digital cartographic data, map elements are stored as a series of X,Y coordinates that represent a location on Earth's surface. In Exercise 2, you brought in spatial data and set map units to latitude/longitude expressed as decimal degrees. This activity, however, merits some discussion. While you can locate a point on Earth with great accuracy, representing the same point on a map is still an approximation. When attempting to use a flat surface (a map) to represent a curved surface (Earth), distortion is inevitable.

Geographers have developed a number of *map projections* to reduce distortion. While a full discussion of map projections is beyond the scope of this text, suffice it to say that each projection strives to resolve the distortion problem in a manner that satisfies the needs of a particular group of users. Some projections are very accurate over short distances; others are better for long distances or for representing large areas of the planet. Some projections are optimal for representing distances on a map, and others are optimized to accurately represent area.

The coordinate system used in this book's exercise sample data is latitude/longitude, often referred to as *geographic coordinates*. Strictly speaking, this system is not a map projection but rather a reference grid. Distances along the grid are uniform on the Y axis, traveling toward the poles. In contrast, distances along the grid on the X axis become increasingly shorter as the poles are approached. Because the grid is uniform, it is relatively easy to mathematically transform geographic coordinates to another map projection. As such, latitude/longitude coordinates are useful as a common reference system for storing digital spatial data when the ultimate use of the data is not known in advance. All GIS and desktop mapping software, including ArcView and ARC/INFO, can use data that is stored in latitude/longitude form.

Two points to remember about making map projections:

- Not all available digital data are stored in the same map coordinates and projections.
- ArcView requires that all digital data share the same map coordinates and projections.

In the United States, in addition to latitude/longitude, digital data is likely to be found in the Universal Transverse Mercator (UTM) or State Plane Coordinates map projections. Both projections divide the United States into a series of zones, each with its own relative X and Y coordinates. For example, the state of Arizona is covered by UTM zones 11 and 12, and the State Plane zones of Arizona West, Arizona Central, and Arizona East. Both projections are very accurate when mapping regional extent.

The UTM projection is widely used by public agencies, particularly those involved in natural resource management such as the U.S. Geological Survey and the Bureau of Land Management. The State Plane Coordinate system is commonly used by land surveyors, public works departments, and state and local transportation departments.

ESRI's ARC/INFO software can transform data to and from all major map projections. ArcView cannot. If possible, request that your data be projected onto a common coordinate system before delivery so that you can avoid seeking a data conversion source at a later date.

In addition to needing map projections that fit the Earth's surface to a flat map, you also need to compensate for the fact that the Earth is not a true sphere. The establishment of a network of precisely located points with respect to both location and elevation (i.e., horizontal and vertical) is referred to as *geodetic control.* With geodetic control, the accurate location of points on the Earth's surface is accomplished with the aid of a *datum* that accurately describes the shape of the Earth. In the United States, the two datums in general use are the North American Datum of 1927 and the North American Datum of 1983.

The vast majority of federal and regional mapping projects—and, consequently, the majority of digital cartographic data derived from these projects—were carried out using the 1927 datum. At present, however, the 1927 datum is being adjusted to the 1983 datum. (Between the 1927 and 1983 datums, the shape of the Earth was redefined by adjusting all coordinates that dictate the shape.) In turn, corresponding digital databases will need to be adjusted as well. This is particularly significant for mapping projects that use global positioning system (satellite) survey units. All points located using GPS are tied to the 1983 datum.

Lest you think that the difference between the 1927 and 1983 geodetic control devices is insignificant to all but surveyors, the distance between the same points in the 1927 datum as opposed to the 1983 datum can be in excess of 90 feet. This difference can certainly be significant when locating features such as water mains or property lines. If your mapping project requires high levels of accuracy, you will need to ensure that the datum is specified, as well as the time when the digital data was obtained.

## Joining Tables

The ability to join tables based on a common item is one of the most important functions in database management. Simply stated, it allows for non-redundant data storage and simplifies database maintenance.

For example, a data file that contains the inventory of a large parts warehouse can include a price code field for each item. A look-up table can then be prepared associating each price code with the current price. Editing the much smaller look-up table ensures that price changes will be subsequently applied to each item referencing particular price codes. The two major benefits that derive from the use of joined tables are (1) when changes occur, only one file has to be updated; and (2) because the link between joined tables is dynamic, subsequent views of the joined table will reflect changes after a file has been changed.

The strengths of the above model also apply to GIS database design. For instance, a theme of digitized land parcel boundaries can be prepared and coded with the assessor's parcel number. The theme can be joined via the same field to tabular data from the assessor's office. In this way, frequent changes to parcel attributes, such as assessed value or last sale, can be kept separate from the digitized parcel boundaries, which change much less frequently. The join ensures that changes to the joined table will automatically be associated with the spatial theme, keeping mapped attributes current and also allowing for temporal change mapping.

ArcView's Join function is particularly robust, allowing tables from dissimilar sources to be joined and stored as a *virtual table.* The source file that is to be joined to the spatial theme attribute table can be a dBase or INFO table, a table from an RDBMS (relational database management system) such as Oracle or Sybase, or a delimited ASCII text file. Once these tables are imported into ArcView, they are stored in the same internal format and are available for further manipulation, including joining to spatial theme attribute tables.

The mechanics of how to perform a join are straightforward, and they will be covered in Exercise 3 at the end of this chapter. A related concept, namely the relationship between the source table and the destination table, remains to be covered.

In ArcView, the *destination table* is the table to which the fields from the *source table* will be appended. The destination table is typically the attribute table for the spatial theme. The results of a join are accurate only if there is a one-to-one correspondence between records being matched from the source to the destination table; that is, only one unique record in the source exists for each record in the destination. The relationship may be one-to-one, as in the link between a land parcel theme and the associated parcel data. The relationship may also be a many-to-one relationship, as in the link between a land use theme and the look-up table explaining each land use code.

If there are many records in the source table that link to the destination table, only the first record from the source table will be joined to the destination table. In this situation, you should be *linking* rather than joining the tables.

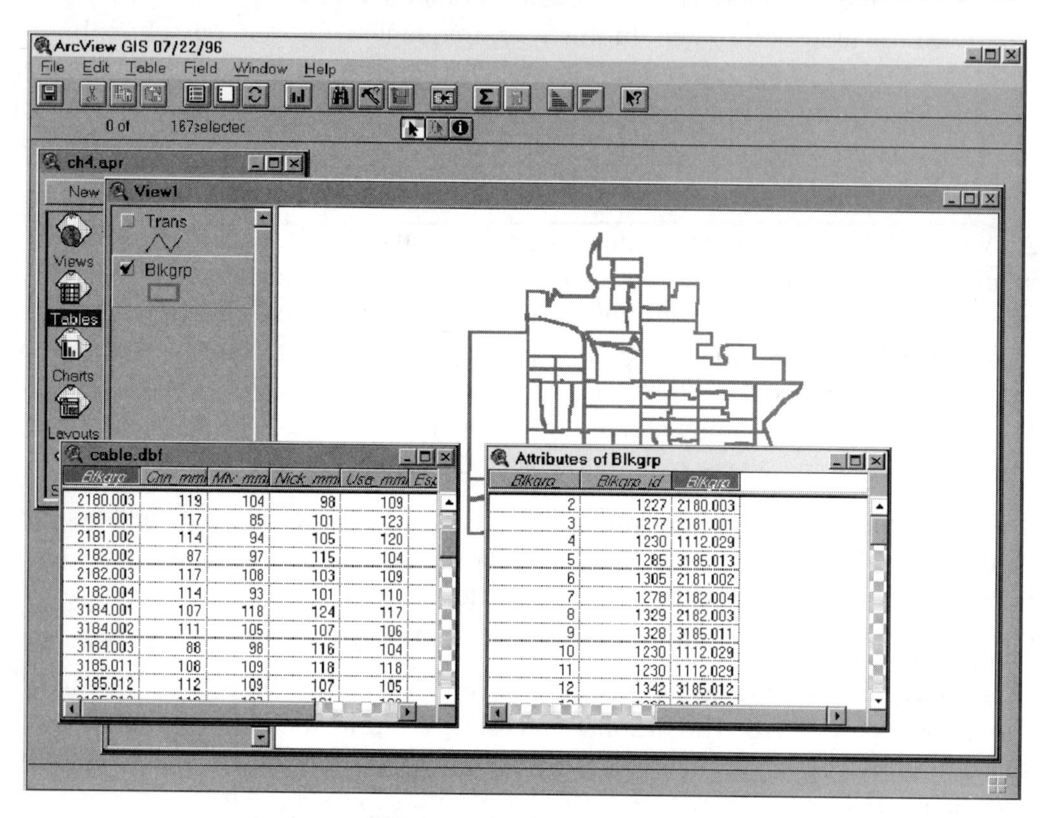


*Attributes of Blkgrp and cable.dbf tables displayed, showing join field Blkgrp with common values.*

# Join Versus Link

The Link function is used when a many-to-one relationship exists between a source table and a destination table. Linking the two tables ensures that all records from the source table are associated with, but not actually joined to, the destination table. The many source table records will subsequently be available when querying the spatial theme. For example, clicking on a land parcel could select all records from a linked table comprised of past owners of that parcel.

**NOTE:** *Unlike the Join function, Link merely establishes a link between tables rather than joining the tables into a new virtual table. As such, fields in the linked table are not available for thematic query or analysis. Unless a many-to-one relationship exists between the source table and the destination table, you should always use Join.*

# Event Themes

In Exercise 1 (Chapter 2), you may recall how you located daycare centers against a street net theme geocoded by street address. The resulting point theme was one example of an ArcView *event theme.*

In ArcView, an event theme is constructed from an *event table* that contains geographic locations. These locations can be absolute, such as X,Y coordinates, or relative, such as street addresses.

When an event theme is created, the geographic locations from the event table are converted into an ArcView–supported spatial data format. For a table that contains absolute locations, ArcView associates a point on the theme with every X,Y coordinate pair. For a table that contains relative locations, ArcView creates a *shapefile*—ArcView's native spatial data format—that contains point or linear features corresponding to the location of each entry in the event table. In brief, ArcView translates each feature from a relative location to an absolute location, and makes these features available for subsequent query and analysis.

ArcView supports the following event categories:

- XY events
- Route events
- Address events

## XY Events

XY event tables contain the exact location of point features using X,Y coordinates. The map coordinate system should correspond to the spatial themes against which these events are to be displayed. Commonly used systems include latitude/longitude, UTM, and State Plane Coordinates.

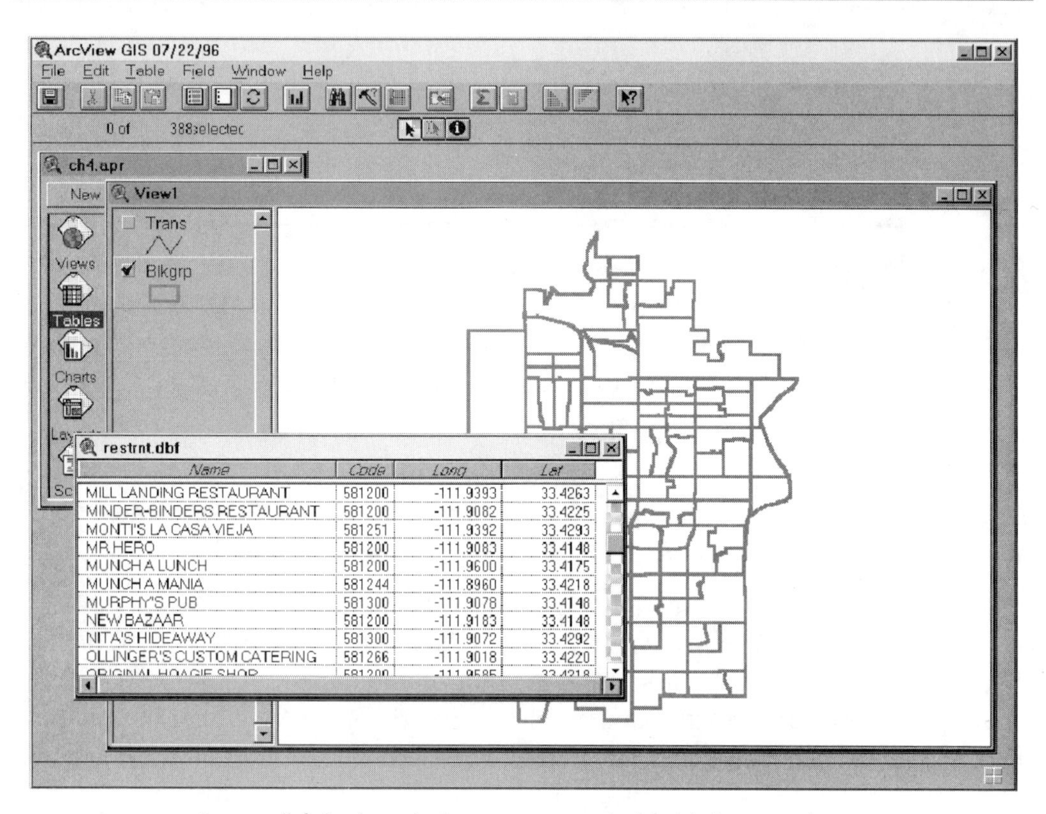


*Restrnt.dbf displayed, showing geocoded fields Long and Lat.*

**NOTE:** *If you are entering latitude/longitude coordinates, verify that the coordinates are properly coded as negative or positive to match the geographic quadrant in which they are located. For example, the longitude values for all locations in the northwest quadrant, which includes the United States, should be entered as negative values. Typically, these values are left unsigned when data is gathered; thus a longitude-latitude (X,Y) coordinate pair entered as 112.30 33.12 should properly have been coded as –112.30 33.12. Importing these coordinates unsigned would result in a rather strange distortion—the resulting theme would be the inverse or mirror image of what was intended.*

# Route Events

Route event tables contain the relative location of features along a *route system.* Route systems are most commonly associated with road networks, and are referenced as a distance from a known starting point, such as 12.3 miles from the beginning of Route 5.

> **NOTE:** *A route system must be built in ARC/INFO before route event themes can be added to ArcView.*

Route events can take the following forms:

- *Point events* are features that are located at specific points along a route, such as accident locations.
- *Linear events* are features that are located along a specific segment of a route. A pavement test section occurring from milepost 12.5 to milepost 13.8 along Route 5 is an example of a linear event.
- *Continuous events* are features that are located continuously along a route. An inventory of pavement condition that has been done by condition class of good, fair, or poor, and then coded as to the location where the condition class changes is an example of continuous event data.

# Address Events

Address event tables contain a locational identifier. The most common locational identifier is a street address along a linear street network. Address events can also be created against a polygon or point theme, such as a table containing zip+4 data.

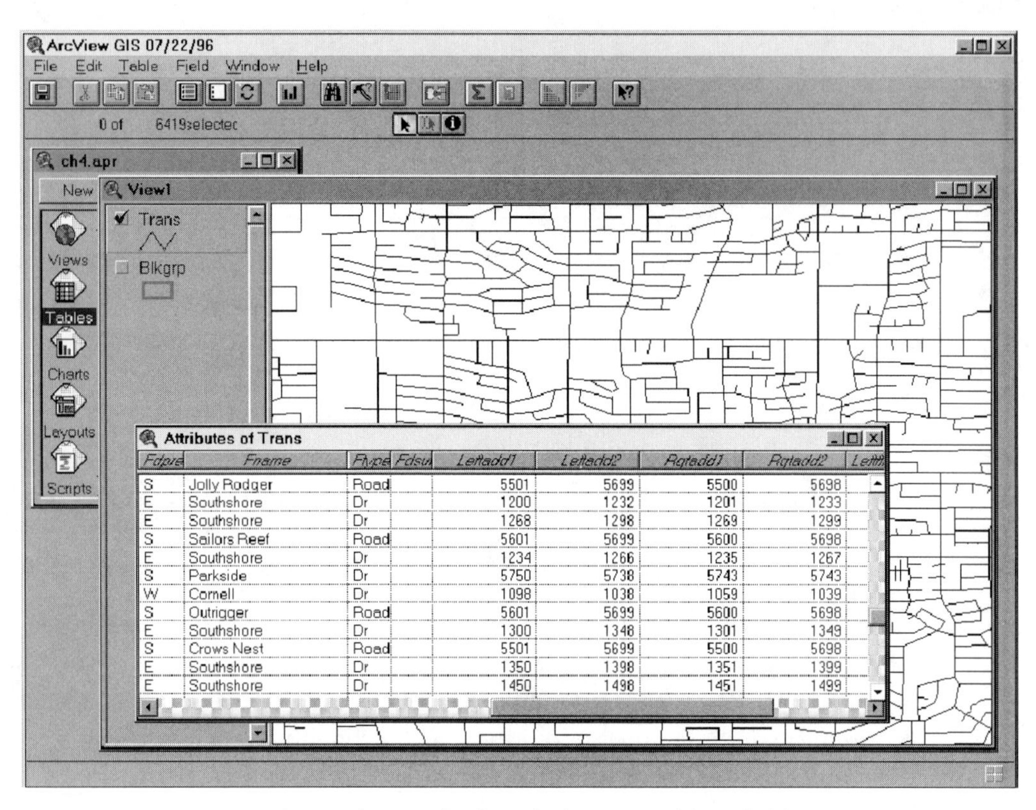


*Attributes of Trans displayed, showing address fields.*

Exercise 1 included the basics of creating an event theme by matching street addresses against a geocoded street network. Regardless of the format of your particular address field, the steps followed in geocoding are the same:

1. Add the required theme and table to the view. The spatial theme can be an ARC/INFO coverage or an ArcView shapefile containing the requisite address fields for geocoding. The event table is comprised of fields containing address information. (Note that the table containing the addresses for geocoding does not need to be added to the project before creating the event theme.)

2. Make the theme *matchable.* First, you need to make the theme active and set the theme's geocoding properties. In the Geocoding Properties dialog window, which you can access by choosing the Geocoding option from the Theme Properties window, select the address style and identify those fields that contain the theme's address components.

3. Add the event theme. In the Add Event Theme dialog window, accessed from the View menu, specify the name of the event table and the field in the table that contains the address of the event. This action will call up the Geocoding Editor dialog window, which will allow you to process the records in the event table singly or in batch, and edit records as necessary in order to ensure a match.

4. After completing the first address matching, you can re-match all addresses not matched in the first pass. At this time, you can elect to interactively match the rejects, editing the addresses as required. You can also adjust the geocoding preferences to change the spelling sensitivity or adjust the minimum match score, if desired.

   Additionally, the Show Candidates button can be used to display the feature attributes of the candidates. The highlighted candidate can be located by clicking on the Flash Candidate button.

5. Add the geocoded theme to the view. When the matching is complete, an ArcView shapefile is created that contains point locations for each event in the event table. A theme based on this shapefile can then be added to the view.

As you will discover when adding a geocoded theme, working with address events can be rather tricky. However, given the ubiquitous nature of address data, this is a skill worth mastering. For additional information on fine-tuning address data, see "The Science of Geocoding."

## The Science of Geocoding

Street addresses are the most common form of geographic data. Nearly all of us work with addresses every day. Address geocoding in ArcView allows you to create a point theme based on address locations, which allows for each address point to be mapped. A 100 percent match of street addresses against a geocoded theme occurs only when both the address table and the geocoded street net are 100 percent properly coded. Perfect coding, alas, is quite rare. There are ways, however, to improve your chances of obtaining a match between addresses and a geocoded theme.

Consider the following address in Phoenix: 1400 N. 16th Ave. Similar to many cities, Phoenix is laid out on a grid. Numbered streets run north and south, Avenues and Drives are located on the west side, and Streets and Places are located on the east side. Accordingly, it is vital that both the locational prefix and the categorical suffix be identified correctly in order to locate an address along a numbered street. Applying the grid rules above, the 1400 N. 16th Ave. address falls in the northwest quadrant of the city.

Let's examine the results when we locate this address against a geocoded street theme in ArcView. (We have instructed ArcView to display all matches by setting Locate preferences to review all candidates when multiple candidates score higher than the minimum match score. The minimum score to be a candidate has been set at 30.) Entering *1400 N 16th Ave* in the Locate dialog window causes the following matches to be displayed:

```
100 - N 16th Ave
77 - S 16th Ave
75 - N 16th Dr
75 - N 16th St
52 - S 16th St
```

With all address fields present, the address will be properly located. Note, however, that if the 1400 block of N. 16th Ave. were not found in the street theme, the corresponding block on S. 16th Ave.—located many miles away in the southwest quadrant—would be preferentially matched over the corresponding block on N. 16th Dr., which is located only one block to the west.

If the field of street type is missing, then matching the remaining address, *1400 N 16th*, produces the following matches:

```
75 - N 16th Dr
75 - N 16th Ave
75 - N 16th St
52 - S 16th Ave
52 - S 16th St
```

Note that as the score decreases for the best candidate (N. 16th Ave.) from 100 to 75, the scores for the secondary candidates (N. 16th Dr. and N. 16th St.) remain the same. Leaving out just one component of your address can dramatically affect your matching success.

The completeness of the address can also affect the match. If you key in *1400 N 16th Av*, you obtain the same matches as if you had entered *Ave*. If, however, you key in *1400 N 16th Ae*, you obtain no matches. The lesson is clear: Take care to see that your address data is coded accurately, and that the address prefixes and suffixes are coded in a fashion that is consistent with the format used in the street network.

The address style used to contain your site addresses also affects the accuracy of the geocoding process. In the example above, we were geocoding using a single field that held the street address. Optionally, we may have additional information to refine the location of the address, such as Zip Code. With Zip Code stored as a separate field in our table, we could then select the address style *US Streets with Zone*. The Zip Code field would be used to refine the outcome of the geocoding process, resulting in a higher accuracy in the subsequent address point theme.

Note that even painstaking quality control over address entry will not provide a match if the corresponding street in the street net is missing or incorrectly coded. To ensure a high percentage of matches, equal attention must be taken to guarantee that your geocoded street net is accurate and up-to-date.

As mentioned in Chapter 2 (see insert, "Cleaning TIGER Street Nets"), the raw TIGER street files from the U.S. Census Bureau are likely to contain errors and omissions. The errors are not sufficient to render TIGER files unusable, but are certainly enough to be aggravating. The Hutchinson-Daniel Law states that the street address missing from your theme is the street address you most need to match. It is possible, however, to edit the TIGER files to improve accuracy. The attribute table for the street net can be displayed and edited in ArcView. (See Chapter 9 for a discussion of table editing.) If you have a few addresses that do not match and an urgent deadline to meet, editing the attribute table can be enough to get you over the hump. Wholesale address changes or the digitizing of new or revised streets may be beyond your resources. In this case, you might enlist the services of a consultant to clean up your street net. Or you might purchase a revised street net from a commercial data provider.

One last note is in order before proceeding to the exercise on importing data into a project. The same strengths that make ArcView desirable as a stand-alone PC-based application make it attractive for use on a computer network in conjunction with ARC/INFO. There are, however, additional considerations when using ArcView in conjunction with ARC/INFO, particularly when accessing the network via ArcView running on a PC.

The primary concern involves file and directory naming conventions. To reliably access all directory and files from the PC, directory and file names should adhere to the DOS 8.3 file naming convention (i.e., eight characters or less for the file name, and three or less for the file type extension, for example *parkways.dbf*). Use this convention for naming ARC/INFO coverages on the network. Coverage names in excess of eight characters may be truncated by the network protocol software, with unpredictable results.

The internal file names for ARC/INFO coverages and INFO data files should also adhere to the DOS 8.3 naming convention. As of ARC/INFO 7.0, this naming convention is the default. There is also a command to convert any existing ARC/INFO workspace names to the 7.0 format. Note, however, that the 7.0 workspace format applies only to the naming of ARC/INFO and INFO internal files. It is up to the user to ensure that ARC/INFO coverage names do not exceed eight characters in length.

A noteworthy data source available to ArcView users who are networked to an ARC/INFO installation is *ARC/INFO Libraries.* At its most basic, this source is merely a formal directory structure allowing ARC/INFO coverages to be divided into tiles, each covering a portion of a total geographic area. A master index and master data templates allow for access of data across tile boundaries and ensure that the internal data format is kept consistent across tiles. As of version 7.0, ARC/INFO has extended the library data model to a feature-based model known as *ArcStorm Libraries.* ArcView accesses standard ARC/INFO Libraries and ArcStorm Libraries in the same manner.

To improve performance when accessing ARC/INFO Libraries, the number of features accessed from the library can be restricted by setting the *Area of Interest* for the library. This can be done interactively using the Area of Interest tool, or by setting the Area of Interest to the extent of another theme, the view extent, the display extent, or an extent set explicitly using X,Y coordinates. These options can be set by selecting the Area of Interest button from the View Properties dialog window.

Finally, note that the PC user who accesses ARC/INFO Libraries across a network needs to set an additional environment variable named *$ARCHOME*, which contains the path to the ARC/INFO install directory. See the *Accessing ARC/INFO Libraries* entry in the ArcView help system for specifics on setting the environment variable.

## Primary Research

To solidify your thinking about how you can use ArcView in a real-world setting, in the chapters ahead you will conduct market research for a restaurateur from Tempe, Arizona. In this ongoing example, we will demonstrate how ArcView can be used as part of a research project.

First, some background: for many years, retailers and others dealing in products that serve the general public have conducted *pin* studies (a reference to the map pins originally used to locate responses) to identify where their customers reside or work. By knowing where customers come from, "trade areas" can be established and subsequently linked to demographic data for further defining customers. This process can help with designing future marketing strategy, such as direct mail or outdoor advertising. Where franchising is involved, pin studies are also used to define "exclusive rights" territories.

The two major steps involved in conducting a pin study are capturing customer addresses and transferring the customer location data to maps.

Data is typically collected via a customer intercept or "fishbowl" survey. Patrons are asked to fill out a form in which they describe themselves and to drop the form into a jar as they leave the business premises.

The data set provided in our exercise, *cust.dat*, was collected through an actual fishbowl survey at a family restaurant in Tempe. The address-matching capability of ArcView will be used to expedite the traditionally tedious task of locating and analyzing customer records.

Once customers are located, their distribution in space can be mapped. Response to research issues can be mapped and examined in a spatial context. By linking this data to commercial "profiling" data, you can also study what type of customers the restaurant or other business is reaching. Profiling data consists of a wide range of parameters that identify who the customers are and how best to reach them. Examples of these parameters are the TV or cable channels they watch regularly, and magazines they subscribe to.

The GIS approach—through ArcView—dramatically improves the analysis of pin studies. Data is processed much more efficiently, and the modeling and analytical tools allow the user to reach much more powerful conclusions.

## Exercise 3: Extending Project Data

In Exercise 2, you imported the raw data (spatial themes and attribute tables) planned for use in subsequent exercises. In this exercise you will establish the linkages and create the event themes to make the data more usable, and begin to import primary research data.

Begin by opening the project saved in Chapter 3. If you did not complete Exercise 2, open the *ch4.apr* project file in the *$IAPATH\projects* directory. The following steps will establish a link between the demographic data and block groups.

1. Click on *Blkgrp* in the view's Table of Contents to make *Blkgrp* the active theme.
2. Pull down the Edit menu and select Copy Themes. This places a copy of the *Blkgrp* theme on the Windows clipboard.
3. Pull down the Edit menu again and select Paste. A copy of the theme appears at the head of the view's Table of Contents.

**NOTE:** *This two-step copy/paste technique is similar to other Microsoft Windows applications, and it also allows you to copy themes between views.*

4. Click on the copied theme to make it the sole active theme. Select Properties from the Theme pulldown menu and click on the Definition icon in the Theme Properties dialog window.
5. In the Theme Name box, remove the file name *Blkgrp* and key in this new name: *Demographics*. Click on OK to accept the entry and close the Theme Properties dialog window.

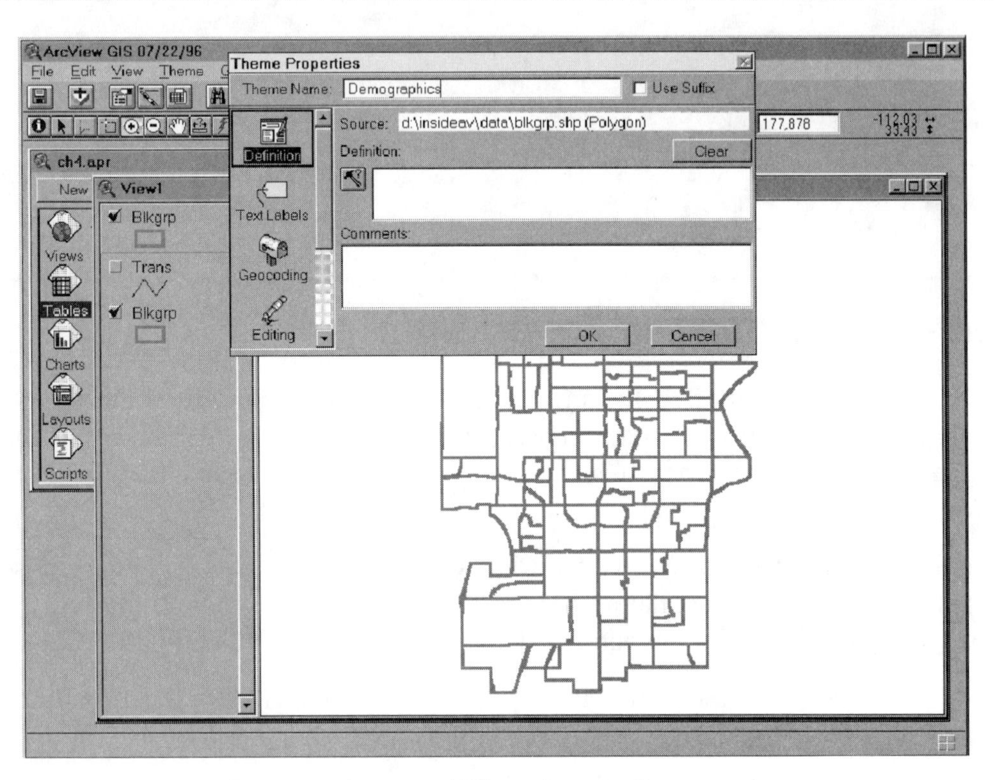


*Renaming the copied Blkgrp theme to Demographics.*

Now you are ready to link the demographic data to the *Demographics* theme. This will be accomplished by joining the demographic table to the *Demographics* theme attribute table.

1. With the *Demographics* theme active, click on the Open Tables of the Active Themes icon on the button bar. The *Attributes of Demographics* table is opened in a window.
2. Click on the Tables icon in the project window, and open the *demog.dbf* table.
3. With both tables now visible, select a common field that will be used to join the tables; that is, a field containing the same attributes in the same format in both tables. In this example, the field is the block group, which is labeled *Blkgrp*.

4. Click on the *Blkgrp* field name below the title bar for the *demog.dbf* table. Note that the field is highlighted.

5. Click on the *Blkgrp* field name below the title bar for the *Attributes of Demographics* table. This field name is highlighted as well.

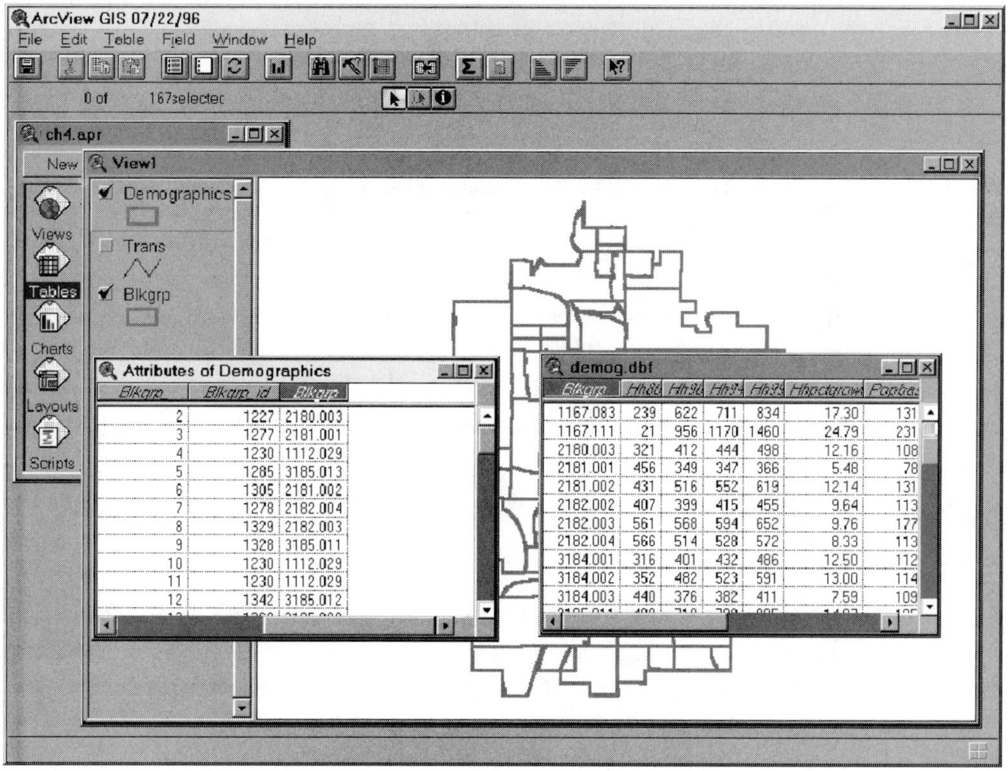

ArcView GIS 07/22/96

File Edit Table Field Window Help

0 of 167 selected

ch4.apr

New View1

Demographics

Trans

Blkgrp

Views Tables Charts Layouts Scripts

Attributes of Demographics

| Blkgrp | Blkgrp_id | Blkgrp |
|---|---|---|
| 2 | 1227 | 2180.003 |
| 3 | 1277 | 2181.001 |
| 4 | 1230 | 1112.029 |
| 5 | 1285 | 3185.013 |
| 6 | 1305 | 2181.002 |
| 7 | 1278 | 2182.004 |
| 8 | 1329 | 2182.003 |
| 9 | 1328 | 3185.011 |
| 10 | 1230 | 1112.029 |
| 11 | 1230 | 1112.029 |
| 12 | 1342 | 3185.012 |

demog.dbf

| Blkgrp | Hh8 | Hh9 | Hh94 | Hh9 | Hhpctgrow | Popbas |
|---|---|---|---|---|---|---|
| 1167.083 | 239 | 622 | 711 | 834 | 17.30 | 131 |
| 1167.111 | 21 | 956 | 1170 | 1460 | 24.79 | 231 |
| 2180.003 | 321 | 412 | 444 | 498 | 12.16 | 108 |
| 2181.001 | 456 | 349 | 347 | 366 | 5.48 | 78 |
| 2181.002 | 431 | 516 | 552 | 619 | 12.14 | 131 |
| 2182.002 | 407 | 399 | 415 | 455 | 9.64 | 113 |
| 2182.003 | 561 | 568 | 594 | 652 | 9.76 | 177 |
| 2182.004 | 566 | 514 | 528 | 572 | 8.33 | 113 |
| 3184.001 | 316 | 401 | 432 | 486 | 12.50 | 112 |
| 3184.002 | 352 | 482 | 523 | 591 | 13.00 | 114 |
| 3184.003 | 440 | 376 | 382 | 411 | 7.59 | 109 |

*The Attributes of Demographics and demog.dbf tables with the highlighted join field, Blkgrp.*

6. With both fields highlighted, select Join from the Table pulldown menu on the menu bar. The two tables are now joined at the common field. A status bar at the bottom of the application window displays progress.

7. When the join is complete, the *demog.dbf* table is closed and the *Attributes of Demographics* table contains the attributes from the *demog.dbf* table. Scroll through the *Attributes* table to view the results of the join. When you have finished examining the joined table, close the window displaying the table.

➾ **NOTE:** *Order is important when joining tables. As mentioned earlier in this chapter, a join requires a one-to-one or a many-to-one relationship. When working with theme attribute tables, the theme attribute is the primary feature of interest. Accordingly, the theme attribute table should always be the destination table. In selecting the fields at which to join, you selected the field from the text file first and the attribute table for the theme second, to ensure that the attribute table window was the active window. When active, the window for the active table is highlighted. Making this window active ensures that the* Attributes of Demographics *table will be the primary table during the join. After the join, the fields from the secondary table are appended to the primary table. Joining these fields to the attribute table for the theme makes the fields available for subsequent analysis and mapping.*

At this point, you need to perform the same operation for a second table titled *cable.dbf.* (The resultant theme will be used in a later exercise.) The steps for the second join follow.

1. Copy the *Blkgrp* theme using the Copy/Paste procedure described earlier. Rename the theme *Cable.*
2. Open the attribute table for the *Cable* theme, and open the *cable.dbf* table.
3. Join the attribute and *cable.dbf* tables at the *Blkgrp* item. (The incremental project has been saved as *ch4a.apr.*)

You have now linked two of our three attribute tables (*demographic* and *cable*) to a spatial theme (*Blkgrp*) so that the data can be mapped. The third file, *restrnt.dbf,* is a little different. In addition to containing business name and SIC Code fields for each restaurant, *restrnt.dbf* contains two fields, X and Y coordinates, which are latitude/longitude points that identify restaurant locations. The X and Y coordinates allow for rapid geocoding.

In order to map the restaurant locations in *restrnt.dbf* using the coordinate values, you need to transform the table into an event theme. The following steps outline this task:

1. Select Add Event Theme from the View pulldown menu on the menu bar.
2. In the Add Event Theme dialog window, click on the XY icon to add an X,Y event theme.
3. Select the *restrnt.dbf* table from the table scroll list.
4. Specify the fields in the table to use for the X and Y coordinates in the X Field and Y Field scroll lists. Select Long and Lat. Click on OK.

*Add Event Theme dialog window with the X and Y fields selected from restrnt.dbf.*

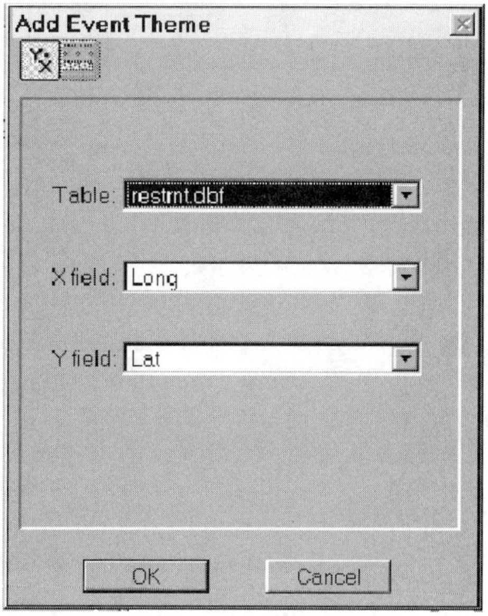

**5.** The completed event theme is added to the legend. A default point symbol and color is assigned, which you can subsequently change using the Legend Editor. Click on the box associated with this theme to display it in the view. You will see a number of points, most located along a grid corresponding to the arterials in our study area. (The incremental project has been saved as *ch4b.apr.*)

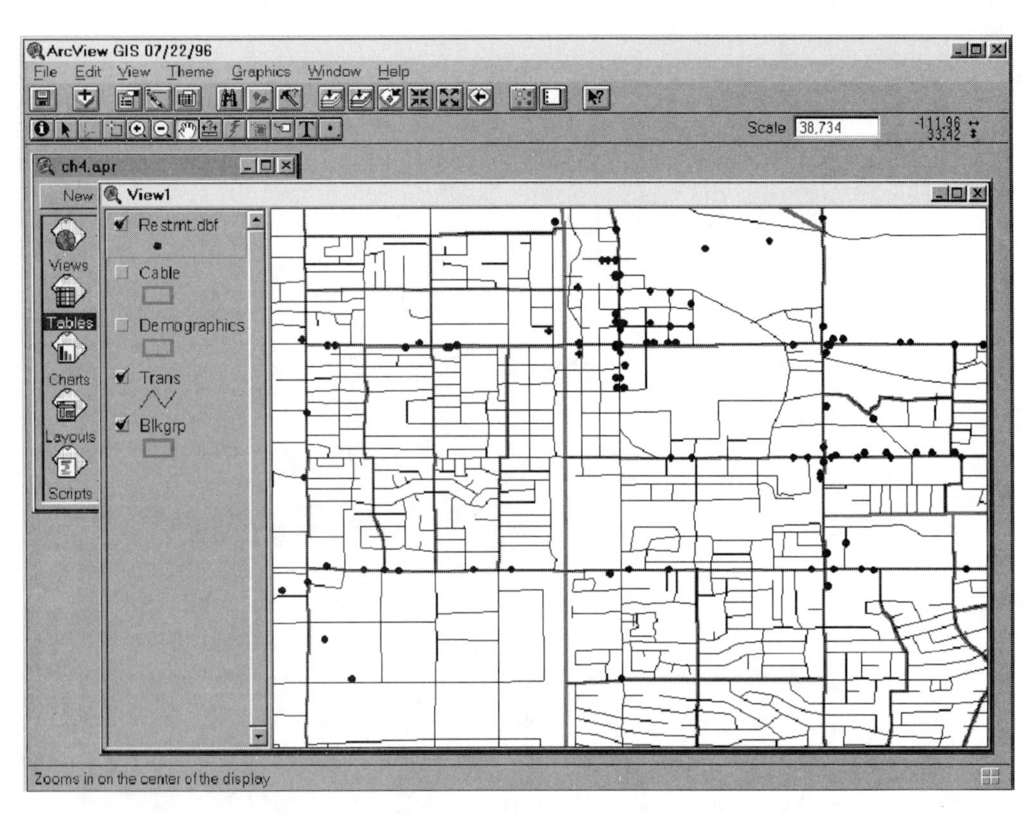


*Restrnt.dbf event theme displayed against the Blkgrp and Trans themes.*

With the demographic data and restaurant locations as background, you now want to direct attention to your client base. As you may recall, the purpose of this exercise is to demonstrate how ArcView can support market research. (See insert titled "Primary Research.") Our overall program is described below.

We worked with a local restaurateur to design a snapshot survey that would capture information about restaurant clientele. In addition to customer home addresses, we asked for additional information on preferences. Two short questionnaires were developed: one focused on potential advertising strategies to reach the clientele, and the other on assessing the client restaurant versus the competition.

We received 67 responses to each form for a total of 134. It should be noted that the number of usable responses (111) was less than the sample size originally targeted when the survey was designed. Statistical formulas indicated that a sample size of at least 300 was needed to gain the level of accuracy desired for evaluating our customer base. While the conclusions you derive from this data will not be statistically significant, the data serve to illustrate the overall process.

To add survey information to the project, take the following steps:

1. From the project window, select Add Table from the Project pulldown menu.
2. Switch to the *$IAPATH\data* directory and add the *ihopad.dbf* and *ihopcomp.dbf* tables. These tables contain the responses from the two customer surveys.

   A quick browse of the records in these tables will reveal that although the restaurant is located in Tempe, customers come from all over the valley. For the purposes of this exercise, you will confine your analysis to respondents located in Tempe.
3. Make one of the tables active by clicking on its title bar.
4. For each table, select Query from the Table menu (or click on the Query Builder tool). For *ihopad.dbf*, select records for which City = Tempe, and for *ihopcomp.dbf*, select records for which City = Tem. Select New Set, and close the dialog window to see the results of the query. The resulting selected set will be 30 of 67 for *ihopad.dbf*, and 19 of 67 for *ihopcomp.dbf*. After viewing the query results, close or minimize the table windows and return to the view.

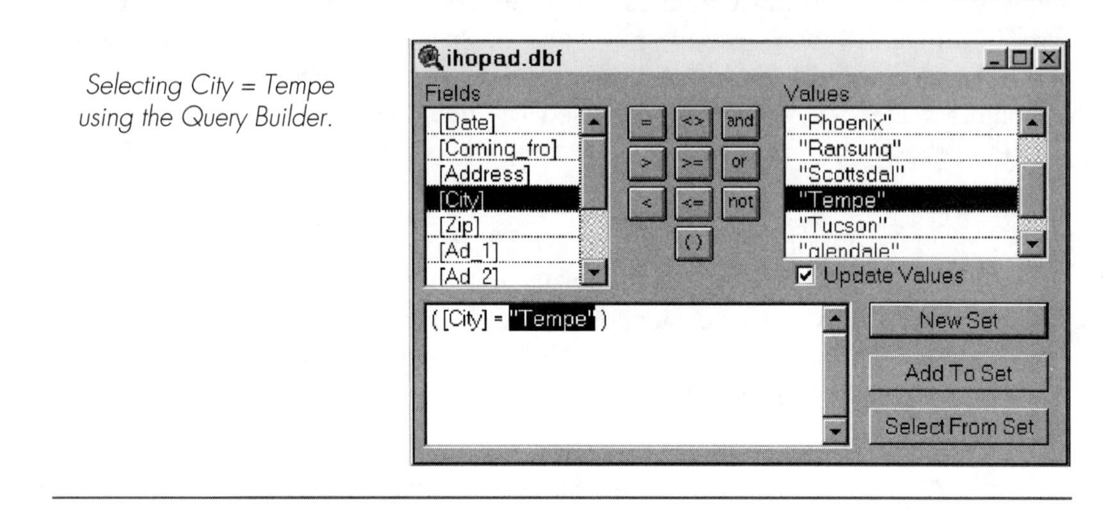


*Selecting City = Tempe using the Query Builder.*

The next task is to create a theme from each *dbf* table with a point located at each record's street address. In ArcView terms, you will create a *geocoded event theme.* The following steps are the same ones used to locate the daycare centers in Exercise 1 (Chapter 2).

1. Click on the *Trans* theme in the Table of Contents to make it the active theme.

2. From the Theme Properties dialog window, select the Geocoding icon from the scroll list on the left. Accept all default address field choices. Click on OK to build the geocoding street index on the *Trans* theme, answering *Yes* to the question, "Build geocoding indexes using Address Style US Streets?"

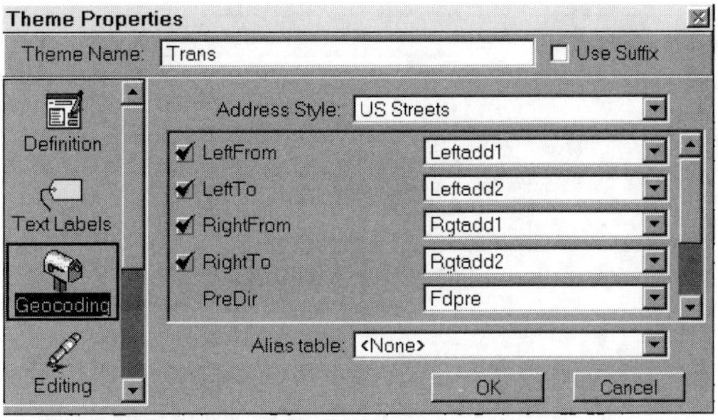

*Geocoding Theme Properties window for the Trans theme.*

3. When the geocoding street index is complete, you are ready to add the event themes. From the View pulldown menu, select Geocode Addresses.

4. In the Geocode Addresses dialog window, select *ihopad.dbf* from the Address Table pulldown list, and Address from the Address Field pulldown list. For Geocoded Theme, click on the browse folder to the right, and navigate to your working directory. For file name, enter *ihop1a.shp*. Click on OK to accept this entry. Click Batch Match to begin geocoding.

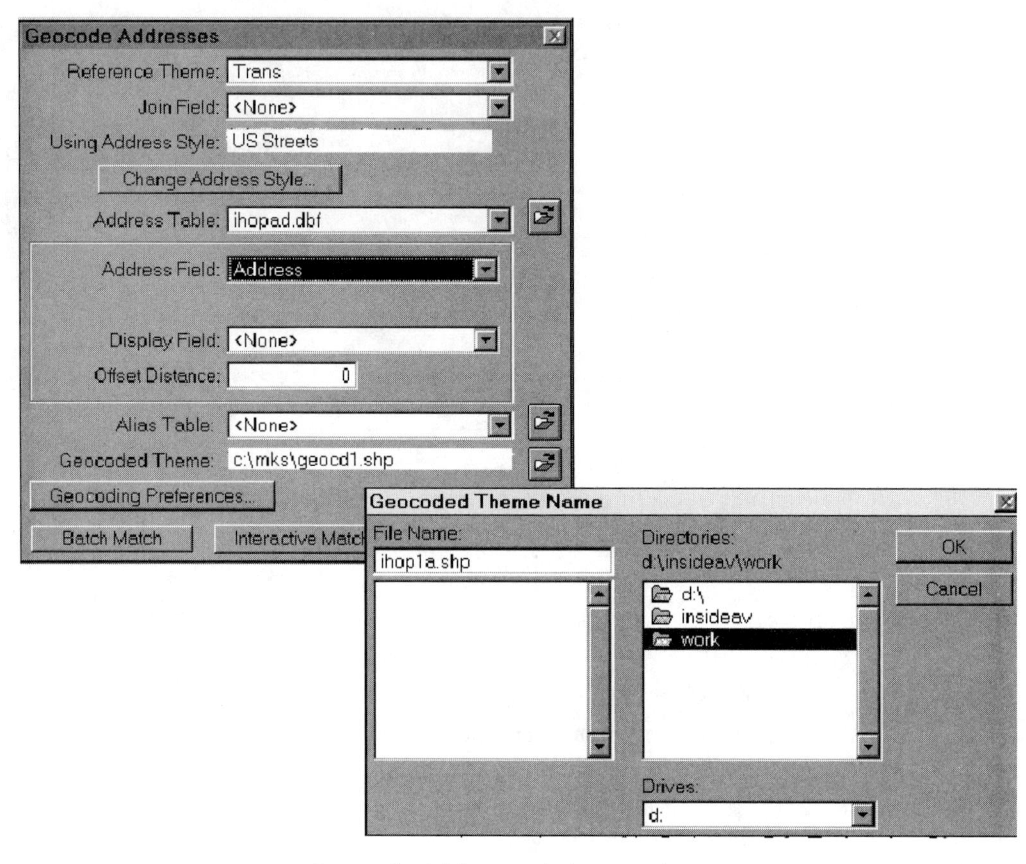


*Geocode Addresses dialog window,*
*with the Geocoded Theme Name window displayed.*

**5.** At the completion of address matching, the Re-match Addresses dialog window is displayed, showing that 17 addresses were geocoded with a good match (score 75–100), 1 address had a partial match (score 50–75), and 12 addresses were not matched. Click on Done to accept the results. The *Ihop1a.shp* point theme, containing the geocoded address locations, is added to the view.

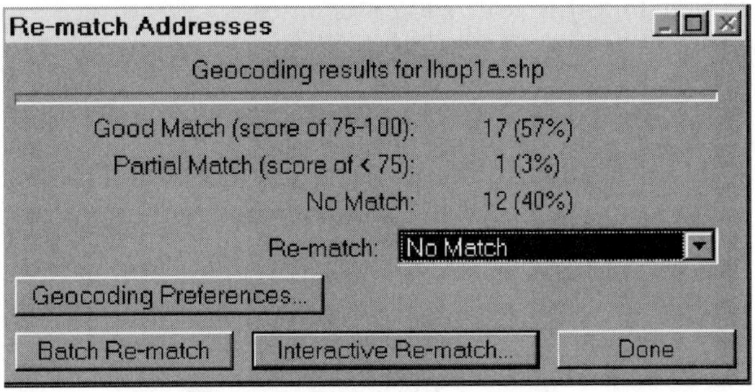

*Re-match Addresses dialog window.*

Re-match Addresses
Geocoding results for Ihop1a.shp
Good Match (score of 75-100): 17 (57%)
Partial Match (score of < 75): 1 (3%)
No Match: 12 (40%)
Re-match: No Match
Geocoding Preferences...
Batch Re-match | Interactive Re-match... | Done

6. Follow the same procedure to create a geocoded event theme from *ihopcomp.dbf.* For the output file name, use *ihop1c.shp.* Upon completion of address matching, 11 records will be matched and 8 unmatched.

As you may have noticed, the match rates for these tables were not particularly high. Because you are working with a small sample, it is important that you match as many records as possible. A quick check of the *ihopad.dbf* and *ihopcomp.dbf* tables reveals several records with incorrect formatting, such as *ASU Registrar's Office* and *Priest // University.* With the aid of a street atlas, we are confident you can obtain a higher match rate. You could match the table again by clicking on the *Ihop1a.shp* theme in the Table of Contents to make it active, and then selecting Rematch from the Theme pulldown menu.

This time you can elect to match the records one by one by stepping through them using the Interactive Re-match button. In this manner you can interactively edit any non-matching record to correct format, spelling or address errors, and then resubmit the address match process following the edit.

**NOTE:** *Interactive address editing using the Geocoding Editor only affects the address used for matching in creation of the event theme. It does NOT change the original address value in the associated table. The address used by ArcView for geocoding the event theme is stored in a separate field,* Av_add, *in the theme's attribute table.*

For this exercise, we have carried out the editing for you. To geocode the edited tables, take the following steps:

1. Make the *Ihop1a.shp* and *Ihop1c.shp* themes active, and select Delete Themes from the Edit menu to delete them from the view.
2. From the project window, switch to the *$IAPATH\data* directory and add the *ihopad2.dbf* and *ihopcmp2.dbf* tables.
3. Geocode the tables as demonstrated above. Remember to select out the Tempe records. For output names, use *Ihop2a.shp* and *Ihop2c.shp*. This time, when the geocoding is complete, you will obtain a 100 percent match.
4. Display the new geocoded event themes so that you can view the newly added points. Select and delete the old event themes because you will no longer need them. Rename the new event themes by accessing the Theme Properties dialog window for the active theme. Rename the *Ihop2a. shp* theme to *IHOP - AD*, and the *Ihop2c.shp* theme to *IHOP - COMP.*

You are now at the point where analysis begins. In Chapter 5, you will learn about querying and classifying data, the first phase toward obtaining an initial overview of what the data represents.

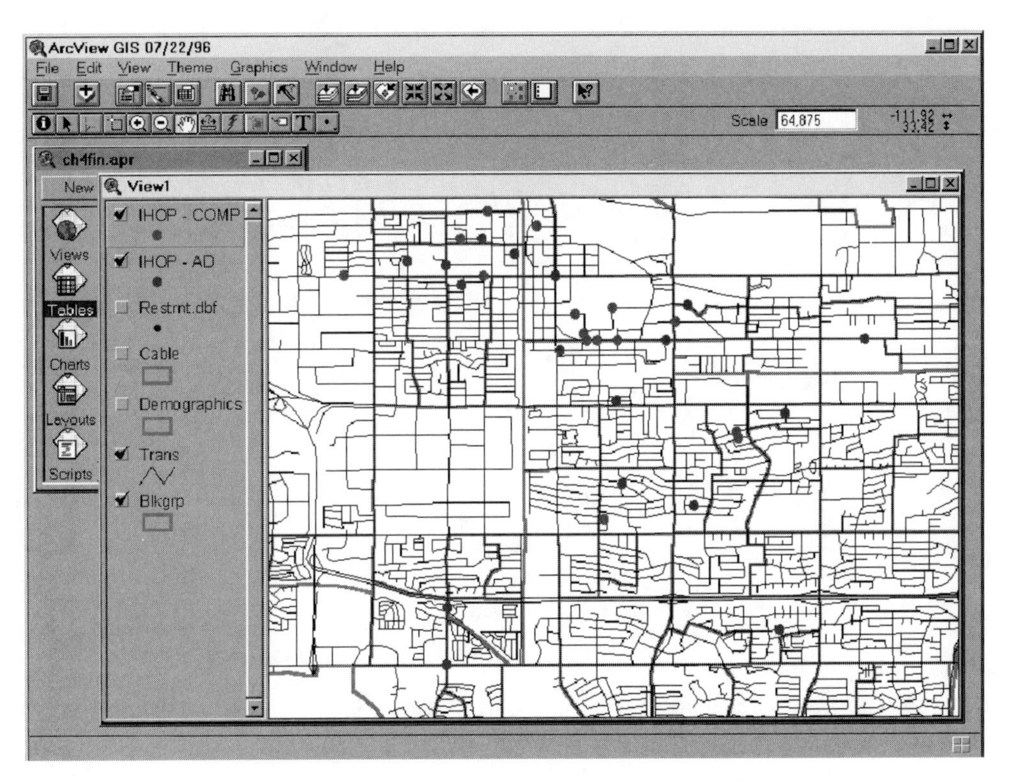


*Completed IHOP - AD and IHOP - COMP event themes.*

# Chapter 5

# Displaying Data

In Chapter 3 you imported tables and spatial data, and in Chapter 4 the data were extended by joining tables and creating event themes. At this juncture, the data are ready for display. Although we have touched upon some display issues in previous exercises, in this chapter you will explore the concepts of data display in greater depth.

## Defining Symbology

When a theme is added to a view, ArcView assigns it a default symbol and the symbol color is randomly selected. Because you will often wish to override the system-assigned choice, ArcView makes it easy to change symbols.

The first step to changing a symbol is to access the Legend Editor window by double-clicking on a theme's entry in the Table of Contents. If the theme is already active, you can call up the Legend Editor by selecting Edit Legend from the Theme menu or by clicking on the Legend Editor icon on the button bar.

> ✓ **TIP:** *If the Legend Editor window is already open, double-clicking on the theme entry in the Table of Contents will reinitialize the Legend Editor for the current theme.*

There are several display options in the Legend Editor. With the editor you can change a theme's symbology or classify a theme. Classification will be discussed later.

When a theme is initially added to a view, one symbol is assigned to draw the entire theme. When brought into the Legend Editor, the symbol for the theme is displayed, and the area for display of legend text is blank. To change the symbol for theme display, double-click on the symbol; the Symbol Palette window will be displayed.

The functional areas accessed through the Symbol Palette window are indicated by six icons displayed across the top of the window. From left to right, these icons access the Fill Palette, Pen Palette, Marker Palette, Font Palette, Color Palette, and Palette Manager windows.

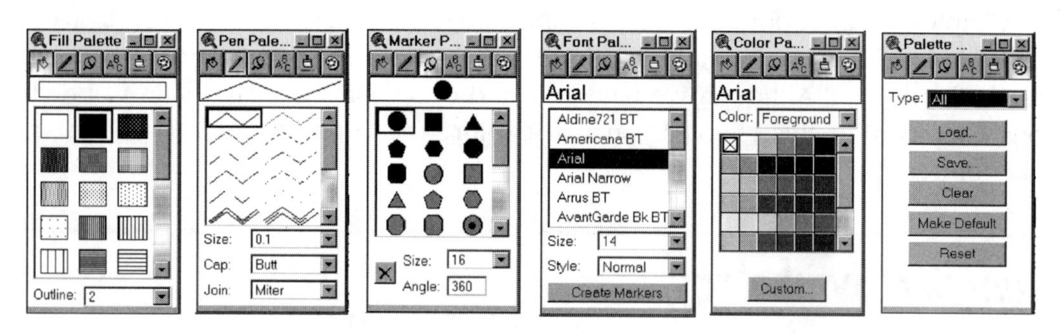


*Fill Palette, Pen Palette, Marker Palette, Font Palette, Color Palette, and Palette Manager windows.*

The Fill Palette controls how areas are shaded. Through this window you can indicate if the corresponding polygons should be solid, shaded in full with hatchings, or not filled at all. In addition you can control the polygon outlines by setting border width and appearance.

> **NOTE:** *Polygon themes that consist of relatively small polygons covering a large area, such as those representing land use or major land ownership, are better shaded without outlines, particularly if those themes will be displayed at large scales.*

The Pen Palette controls the appearance of line features and line themes. You can adjust the pattern and width of lines. You can also determine how ArcView draws ends of lines and line vertices by setting the Cap and Join options.

The Marker Palette controls the appearance of point features and point themes, allowing you to change a point's symbology and size.

The Font Palette controls how text and labeling appears. Available options include font type, size, and style (normal, bold, italic, and bold italic). On the Windows platform, all TrueType fonts installed on your system are available.

The Color Palette controls color. Whether you are working with areas, pens, markers, or fonts, this palette allows you to control a feature's foreground color, background color, outline, and text. In addition to the standard color palette, a custom option is available that allows color mixing by specifying Hue, Saturation, and Value.

Finally, the Palette Manager allows you to import new symbol palettes or revert to the default palette. ArcView ships with a number of additional color and marker palettes designed for special applications. Marker palettes are available with special cartographic, mineral, transportation, USGS, and weather symbols, among others. In addition, ArcView supports the conversion of any TrueType font into marker symbols, as well as the creation of custom marker symbols from bitmaps. ARC/INFO linesets and shadesets can be loaded into ArcView as well.

ArcView palettes enable you to control every aspect of how your features appear. As you gain skill in map customization, using them will become second nature.

## Classification

Thus far we have addressed changes to theme symbology in which one symbol is used to represent the entire theme. The next step is to *classify* the theme based on the values of an attribute field associated with the theme. Classification by attribute value is the cornerstone of thematic mapping, and of GIS as well. Through this single function you will open up a wealth of information pertaining to *thematic data*. By giving you easy access to the tools for thematic classification, ArcView allows you to quickly analyze the patterns underlying thematic data.

The first step in thematic classification is to access the Legend Editor by double-clicking on a theme entry in the Table of Contents. You must first select the legend type for the classification. Six options are possible from the Legend Type pulldown list—Single Symbol, Graduated Color, Graduated Symbol, Unique Value, Dot, and Chart. By default, the legend type defaults to Single Symbol.

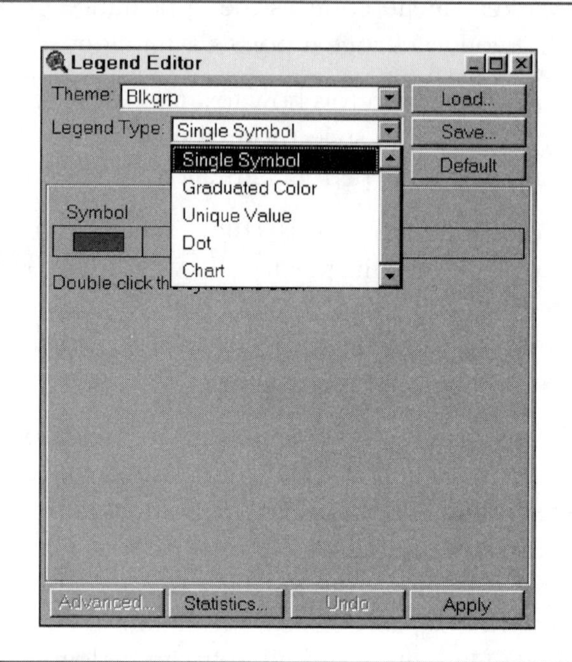


*Legend Editor dialog window, with available legend types for polygon themes.*

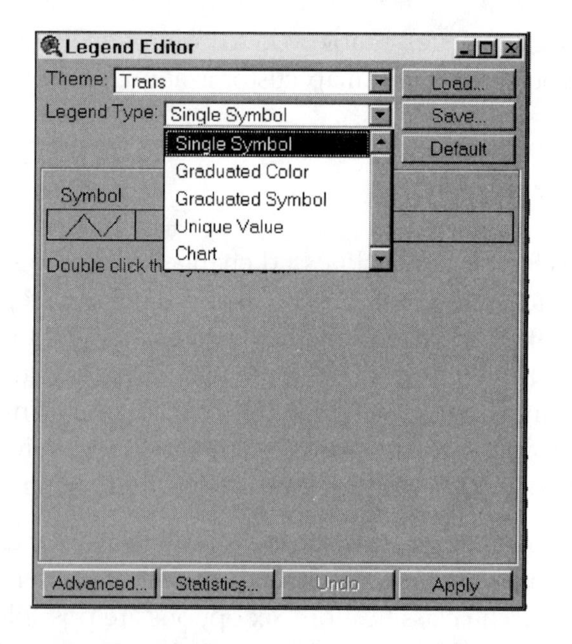


*Legend Editor dialog window, with available legend types for line and point themes.*

Based on the legend type selected, the Legend Editor initializes the options appropriate to that choice; this includes options such as classification field, symbol, color scheme, and chart type. (See the illustrations below for the Legend Editor options for each legend type.)

The next step is to select the field by which to classify. The fields available from the Classification/Values Field pulldown list are determined by the legend type selected. Unique Value legends can be applied to character and integer data, whereas Graduated Symbol, Graduated Color, Dot Density, and Chart legends can be applied only to numeric field types.

The final step is to select the classification type, define the symbology (if applicable), and apply it to the theme.

## Legend Type

To select the appropriate legend type, you need to be familiar with your data. Each legend type has a corresponding data type for which it is most suited. Categorical data, such as zip codes, are suited for display using Unique Values. Displaying zip codes with Graduated Color based on a quantile classification, while technically feasible, would produce a totally meaningless map.

Numeric data that exhibits a continuous range of values, such as household income or daily traffic volume, are well suited to display using Graduated Colors or Graduated Symbols. These legend types further allow you to select the appropriate classification type to be applied, such as quantile or equal interval.

Raw counts, such as population data, are well suited to display using a Dot Density legend. A Dot Density map provides a visual correlation between the raw data and the area for which the data was collected. Alternatively, raw count data can be converted to a per area basis, such as population per square mile, and subsequently displayed using Graduated Colors or Graduated Symbols. Converting data to a per area basis is an example of normalizing your data. ArcView gives you the ability to normalize your data from within the Legend Editor.

If you have more than one field of normalized data you wish to map, you may find that using a Chart legend to display the data in a pie or bar chart is appropriate.

*Legend Editor dialog windows for Graduated Color, Dot, and Chart legend types on polygon themes.*

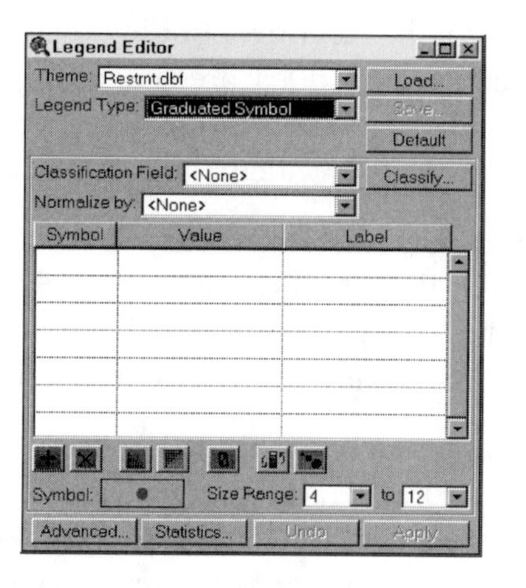


*Legend Editor dialog window for the Graduated Symbol legend type on point themes.*

## Classification Type

When mapping continuously ranging numeric data, selecting the Graduated Color or Graduated Symbol legend type allows you to select the appropriate classification type. These are the supported classification types:

- *Equal area.* Class breaks are determined by dividing the summed area for all features by the number of classes selected, then setting class intervals such that the summed area for all features in each class is made equal.
- *Equal interval.* Each class contains an equal range of values.
- *Manual equal interval.* This is similar to equal interval; however, it also allows you to specify the minimum and maximum values to classify. (Values outside this range are displayed with the null symbol).
- *Natural breaks.* This optimizes the breakpoints between classes by using an algorithm that minimizes the variance within classes. This is the default classification applied when the Graduated Color or Graduated Symbol legend types are selected.
- *Quantile.* Each class contains an equal number of records.
- *Standard deviation.* Class breaks are determined above and below the mean, based on intervals of standard deviation from that mean.

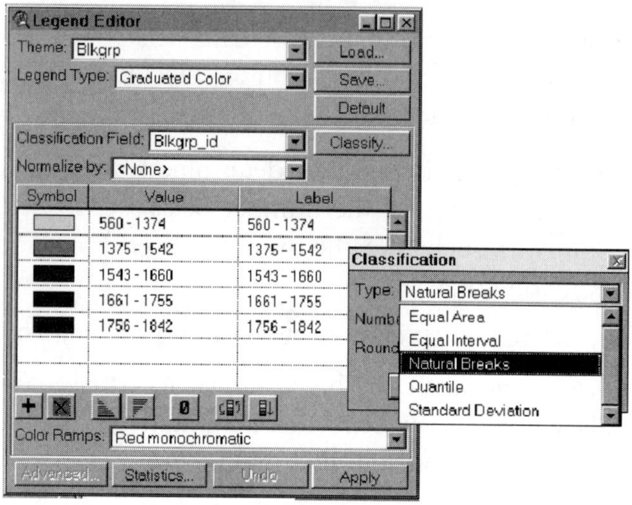

*Available classification types for the Graduated Color legend type on polygon themes.*

A Unique Values legend type can be applied to any field, numeric or character. It is, however, best suited to data comprised of a few discrete values, such as soil type or zoning. Although typically each category in a Unique Values legend corresponds to a specific value, classes can be altered to contain more than one discrete value. You can do so by typing the values in the Legend Editor, separated by commas, for example, *R-1, R-2, R-5.*

> **NOTE:** *As of ArcView 3.0, the 64-class limitation for interval and unique value classifications has been removed.*

## Class Values

A class range can be modified by opening the Legend Editor, clicking on a Values cell that you wish to modify, and typing in a new value. Updating the Value cell will automatically update the Label cell as well.

> **NOTE:** *ArcView does NOT verify that you have entered valid class ranges. If class ranges overlap, ArcView will assign the feature to the*

*first class into which the feature's attribute falls. In the event of gaps between classes, features falling in this range will not be drawn.*

Typing in new values for class ranges.

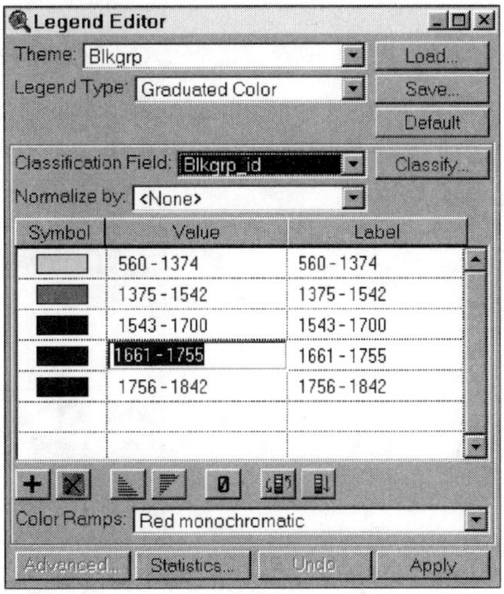

Additionally, the Legend Editor provides you with the ability to assign a value and symbol to Null or No Data values. Values declared to be null are ignored in the theme classification, and by default do not appear in the theme legend.

> **NOTE:** *You may set a null value for more than one field in your data. Empty strings in character fields are automatically ignored by ArcView during theme classification.*

Classification may be further customized by adding or deleting classes. To delete a class, highlight the class in the Legend Editor and click on the Cut Class icon. To add a class, click on the Add Class icon. (See figure below.) The new class range can then be typed into the Value cell. Additionally, classes may be reordered in the legend by highlighting and dragging them to new positions in the Legend Editor.

To help you determine any custom class values, use the Statistics button to display the minimum, maximum, count, sum, mean, and standard deviation for any numeric field.

*The classification customization buttons for the Graduated Color legend type.*

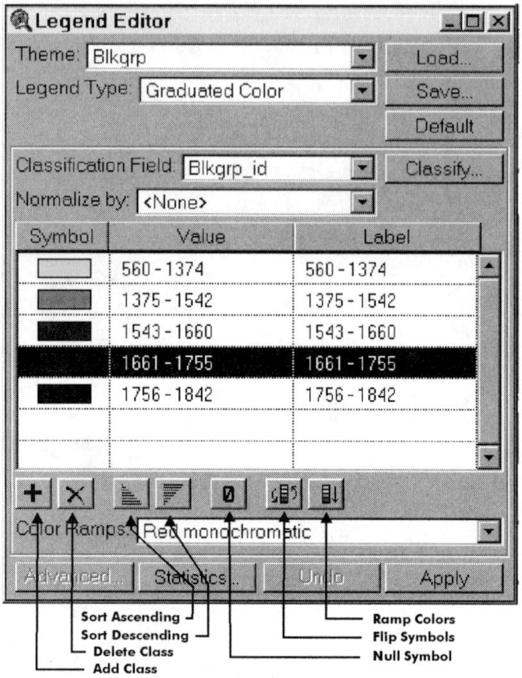

## Classification Revisited

Classification methods can have a profound effect on how data will be displayed on a map. The selection of *quantile*, *equal interval*, or *unique values*, combined with different total class numbers, may produce strikingly different results.

Consider a small data set of 12 values: 24, 25, 26, 29, 32, 43, 44, 49, 51, 69, 78, and 113.

Nearly any method of automatic ranging (quantile or equal interval) will skew the results toward some type of bias. Using four classes, the quantile approach will set breakpoints at 26, 43, and 51 in a manner that arbitrarily splits and artificially communicates breaks in the natural progression between 26 and 32 and 43 and 44. Using the equal interval approach, the breakpoints will be set at 46, 69, and

92. Again, the breaks seem less than optimal: this time, 49 and 51 are unnecessarily separated from the lower group, and 69 and 78 are split apart.

Could this situation be resolved with fewer classes? Experimenting in this manner is healthy, but not guaranteed to be productive. Using the same numbers, three quantiles produce breakpoints at 29 and 49, again resulting in seemingly artificial splits between values in very close proximity. However, three classes, with breaks at 54 and 84, might work. This group, more than any of the others above, represents the actual distribution of the data.

Is such careful work with ranges typical? Actually, it is. Data rarely fall into a distribution for which a quantile or equal interval classification is optimal. For many data sets, automatic classifications might not be the answer. In these situations, you should feel free to explore custom breakpoints. One way to proceed is to sort the table and visually determine the natural breaks. Another is to use ArcView's statistical capabilities.

ArcView can provide summary statistics for a table based on the active field. Available statistics include average, sum, minimum, maximum, standard deviation, first, last, and count. These statistics can be generated for any numeric field, and they are written to a new table. The new table can subsequently be joined to the Theme Attribute table and used for additional analysis.

Why is the careful choice of ranges so important? It is because ranges will be used to determine the color breaks in your map, so users will ultimately remember the colors better than the specific ranges that the colors represent.

In *How to Lie with Maps* (The University of Chicago Press, 1991), Mark Monmonier examines how reality can be distorted by the way maps are designed. We recommend that you seriously consider the issues mentioned above; otherwise, your maps could end up being more misleading than helpful.

## Symbology

Once you have arrived at a classification that is appropriate for your data, you need to determine a symbolization schema appropriate for displaying your data on the map. Properties for each class, such as symbol, size, and color, can be altered by double-clicking on any symbol in the Legend Editor to access the Symbol Palette. Selecting the Graduated Color legend type allows classes to be displayed using predefined or custom color ramps.

## Undoing the Classification

As of ArcView 3.0, you now have the ability to undo a previously applied classification. The Undo button at the bottom of the Legend Editor allows you to revert successively to the last five applied classifications. This feature extends the functionality of ArcView as a data exploration tool by allowing you to examine alternative classification options until you arrive at a schema that reveals the underlying patterns in the data.

## Cleaning Up the Legend

Once the symbology has been set, the only task remaining is to edit the legend text. This, too, is accomplished from the Legend Editor window.

To edit the legend text, click on the appropriate cell in the Values column and type in the new value. When all desired changes have been made, click on the Apply button to apply these changes to the Table of Contents. As with other changes in theme classification, click on the Undo button to revert the theme to its previous state.

## Hiding the Legend

By default, when a classification is applied to a theme, the theme legend is displayed in the Table of Contents. However, legends can be quite extensive, particularly those resulting from a classification using unique values. As themes are added to a view, it does not take long before the Table of Contents is too long to be displayed on the screen, requiring you to scroll to view all theme entries. To reclaim space in the Table of Contents, you have the option of hiding or showing the legend for each theme. When a legend is hidden, its theme will be represented with just the theme title and check box in the Table of Contents. You can access this feature by selecting the Hide/Show Legend choice from the Theme pulldown menu on the View menu bar.

## Advanced Features

Our discussion to this point has focused on the basic principles pertaining to thematic classification and display. ArcView version 3.0 significantly extends the theme classification functionality: New features include the ability to normalize your data, support for dichromatic color ramps to better represent bi-modal data, scaleable marker and line symbols, support for marker symbol rotation and line symbol offset, and enhanced class editing tools. These functions, among others, will be explored further in Chapter 11, "Beyond the Basics."

# *Identifying Features*

At this point, and continuing through Chapter 6, you will look more closely at the data associated with spatial themes. One straightforward way to examine the attributes of a theme is with the Identify tool, which is located at the far left side of the ArcView tool bar. The Identify tool allows you to display the attributes of a feature in an active theme. To make a theme active, click on its entry in the Table of Contents.

✓ **TIP:** *If you click on a theme name and simultaneously press the <Shift> key, that theme becomes active, along with all previously active themes.*

When the Identify tool is active, clicking on a feature in the view brings up a window that displays the feature's attributes. The Identify Results window displays each column and the corresponding values for the identified feature.

*Identify tool icon.*

A list is maintained on the left side of the window of all identified features. If more than one feature was found at that location, multiple records will be displayed in the list. Clicking on a record from the list will refresh the display of the record's fields. As additional features are identified, they are added to the list. Consequently, current and previous Identify results can be compared.

If two or more themes are active, a record will be added to the list for each feature found from each active theme. This is a simple way to compare values for overlapping themes at a specific location.

Within the Identify Results window, the Clear button clears the record for the current Identify, whereas Clear All removes all records from the Identify report.

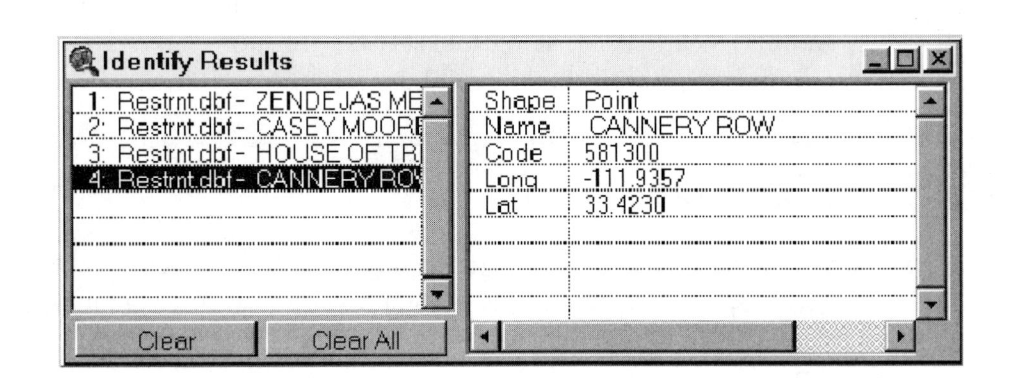


*Results of an Identify on the restaurant theme.*

# Labeling Features

## Label Tool

Another way to identify the features in a theme is with the Label tool. The Label tool, also located on the ArcView tool bar, allows you to select a feature from the active theme. This feature is subsequently labeled with the value from a designated field.

> **NOTE:** *Unlike the Identify process, Label displays the value for only one field from a single active theme.*

To specify the field for feature labeling, access the Theme Properties dialog window from the Theme pulldown menu. Click on the Text Labels icon to access the text-labeling options. A pulldown list displays the available choices for Label Field, as well as options for how to orient the text.

*Text Labels portion of the Theme Properties dialog window.*

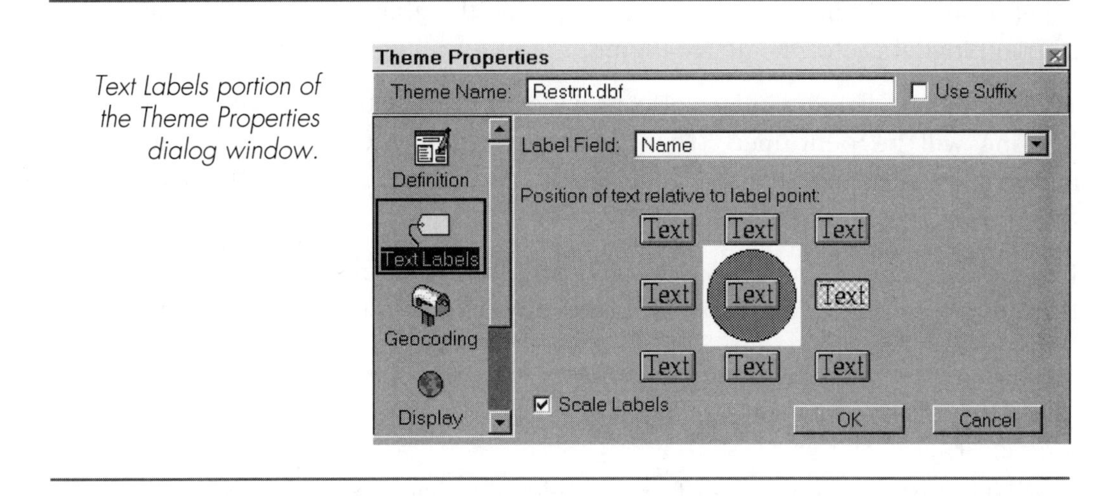


To label features, the correct theme must be active and currently displayed. With the Label tool active, click on a feature in the active theme. As you select features, a text label is placed adjacent to the feature. This text label contains the selected attribute from the Label Field (e.g., Name).

The resulting text labels are ArcView graphic elements (see the next page), and can be moved, resized, and edited in the same manner as any other graphic element. In addition, by default, the text labels are attached to their themes. The elements are turned off when the theme is off, and if the theme is subsequently deleted, the attached text labels will be deleted as well.

It is also possible to detach text labels from a theme. To do so, make the appropriate theme active and select Detach Graphics from the Graphics pulldown menu.

## Auto-label

The Auto-label tool allows you to label many features at once. Like the Label tool, Auto-label works on the active theme and inherits the properties that have been set for Text Labels in the Theme Properties window. Only one theme should be active when using the Auto-label tool. To automatically label features, make a theme active and select Auto-label from the Theme pulldown menu.

Additional choices in the Theme menu provide the ability to remove all labels or overlapping labels for the active theme.

The Auto-label dialog window also provides several options for controlling how labeling will be performed. These include finding the best label placement, allowing overlapping labels, scaling labels, and labeling only features visible in the view extent.

If a set of features is currently selected, only those features will be labeled. If no features are selected, Auto-label will label all features in the active theme.

---

# *Adding Graphics to a View*

Text labels are only one type of graphic element that can be created in ArcView. Drawing tools enable you to create points, lines, polylines, rectangles, circles, and polygons. These graphic elements can be added to a view or a map layout. When added to a view they can be linked to a theme, or they can be used to spatially select features from the theme.

## Adding Graphics

The Draw tool is used to add all graphic elements, with the exception of text. When you position your mouse cursor over the Draw tool icon and click the left mouse button, a series of icons pop up that correspond to the graphic types listed above. Graphic elements are added by clicking with the mouse as described below.

- *Points* are located with a single mouse click.
- *Lines* are graphic elements that contain a start and an end point. The start point is identified by clicking; the mouse button is held down

while the mouse is dragged to the end point, where the button is released. The length of the line is displayed as the mouse is dragged.

- *Polylines* are lines that contain more than the two start and end points. To add a polyline, click at the start point and again at the location of all intermediate points. *Double-click* at the end point to end the polyline. The length of the most recently drawn line segment and the cumulative line length are both displayed as the polyline is drawn.
- *Rectangles* are added by clicking at the start point, holding down the mouse button, and dragging the mouse to define the box. The display shows the current area of the rectangle as the box is dragged.
- *Circles* are added by clicking at what will be the center and then holding down the mouse button while dragging to define the radius. The display shows the current radius as the circle is dragged.
- *Polygons* are added by clicking at the start point and subsequent vertices. Double-click at the last vertex to close up the polygon. The display shows the last segment length, perimeter, and area as the polygon is being defined.

**NOTE:** *The units of length or area displayed as graphic features are set via the Distance Units selector in the View Properties dialog window. In addition, the Measure tool on the View tool bar can be used to measure distance without adding a polyline to the view.*

Text is added to a view by using the Text tool from the ArcView tool bar. With the Text tool active, click on a point in the view. This anchors the text. The Text Properties dialog window pops up. The window includes an area for text input, as well as options for line justification, vertical spacing between text lines, and rotation angle for the text. After the text is entered and properties set, click on OK to add the text to the view.

*Text Properties dialog window.*

Text Properties
ZENDEJAS MEXICAN RESTAURANT
Horizontal Alignment:
Vertical Spacing: 1.0 lines
Rotation Angle: 0 degrees
OK Cancel

When graphic elements are added to a view, they take on the current settings from the Symbol Palette. To access the Symbol Palette, choose Show Symbol Window from the Window pulldown menu, or press <Ctrl>+p.

# Editing Graphics

Graphic elements can also be moved or edited after they have been added to a view. To select a graphic element, use the Pointer tool on the ArcView tool bar.

*Pointer tool icon.*

Individual graphic elements are selected by clicking on them with the mouse while the Pointer tool is active. Several elements may be simultaneously selected by dragging a selection box around them with the mouse. To add additional elements to a selected set, hold down the <Shift> key while clicking on additional elements.

To delete selected elements, choose Delete Graphics from the Edit pulldown menu, or press the <Delete> key after the elements are selected. In a similar manner, selected elements can be copied, grouped, or moved between views using the Edit pulldown menu.

When a graphic element is currently selected, selection handles for the element are displayed. To move an element, place the mouse pointer inside the element's boundary box. The appearance of the cursor will change to a move symbol (two arrows that form a cross shape). Hold down the left mouse button and drag the element to a new location. If more than one element is selected, all elements are moved as a group.

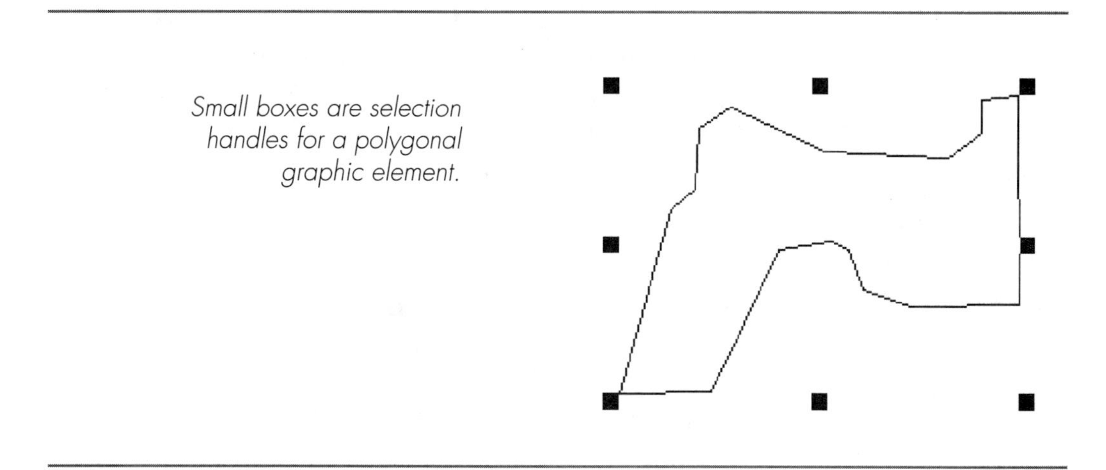

*Small boxes are selection handles for a polygonal graphic element.*

To resize a graphic element, click on and drag one of its selection handles. Dragging on a corner selection handle will resize the graphic element proportionally. Dragging on a side element will stretch the element in that direction.

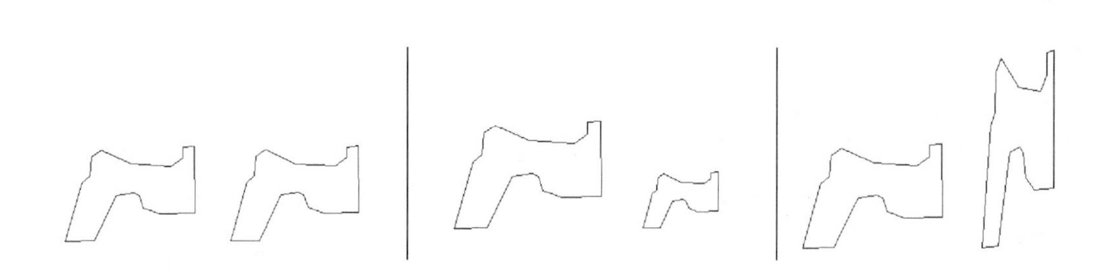

*Original polygonal graphic element duplicated (left), proportionally resized (center), and stretched (right).*

To change the display properties of an element after it has been added, select the element or elements and access the Symbol Palette. Symbol and color choices will be applied as they are selected from the palette.

## Reshaping

In addition to moving graphics, you may wish to occasionally edit their shapes. To reshape polylines and polygons, click on the Vertex Edit tool icon on the ArcView tool bar. If you have previously selected a graphic element, its vertex handles will be revealed. Clicking on a graphic element when the Vertex Edit tool is active will also reveal its vertex handles. Click and drag a vertex handle to a new location, thereby moving the vertex and reshaping the graphic.

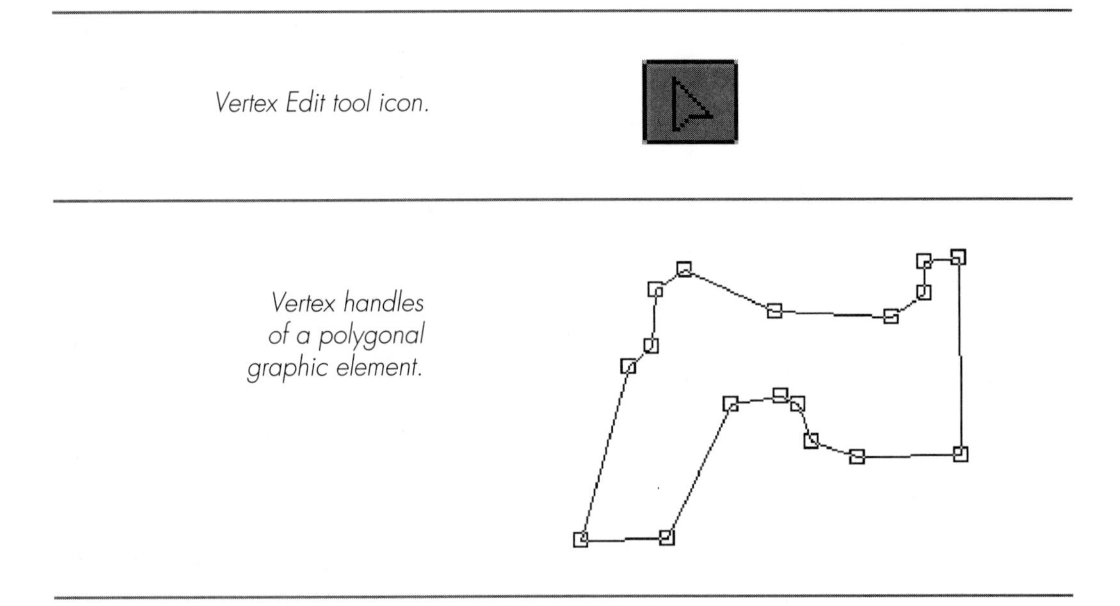

*Vertex Edit tool icon.*

*Vertex handles of a polygonal graphic element.*

## Text Editing

Text labels may be resized and repositioned using the Pointer tool. In addition, text labels can be edited by clicking on a text label while the Text tool is active. The text element will be displayed in the Text Properties window; the text can then be edited and additional properties changed as needed.

# Exercise 4: Thematic Classification and Graphics

1. Open the project you saved at the end of Exercise 3. If you did not save the project, open the *ch5.apr* project file in your working directory.

2. Turn off the *IHOP* point themes, and the *Blkgrp* and *Trans* themes. Turn on the *Demographics* theme and click on its theme entry in the Table of Contents to make it active.

3. From the Edit pulldown menu, select Copy Theme and then Paste Theme from the same menu. A copy of the *Demographics* theme will appear at the top of the Table of Contents. Click on this new theme to make it active.

4. The *Demographics* theme contains a variety of demographic data, including information about population and income. We will examine some of these attributes in greater detail, starting with the field that contains the projected growth (as a percentage) in households from 1994 to 1999.

5. Access the Theme Properties dialog window and change the name of this new theme from *Demographics* to *HH Pct Growth*. Next, double-click on the theme in the Table of Contents to bring up the Legend Editor. For legend type, select Graduated Color. From the Classification Field pulldown list, select *Hhpctgrowt*. A five-class classification, using natural breaks, is immediately generated. Apply this classification, and examine the results.

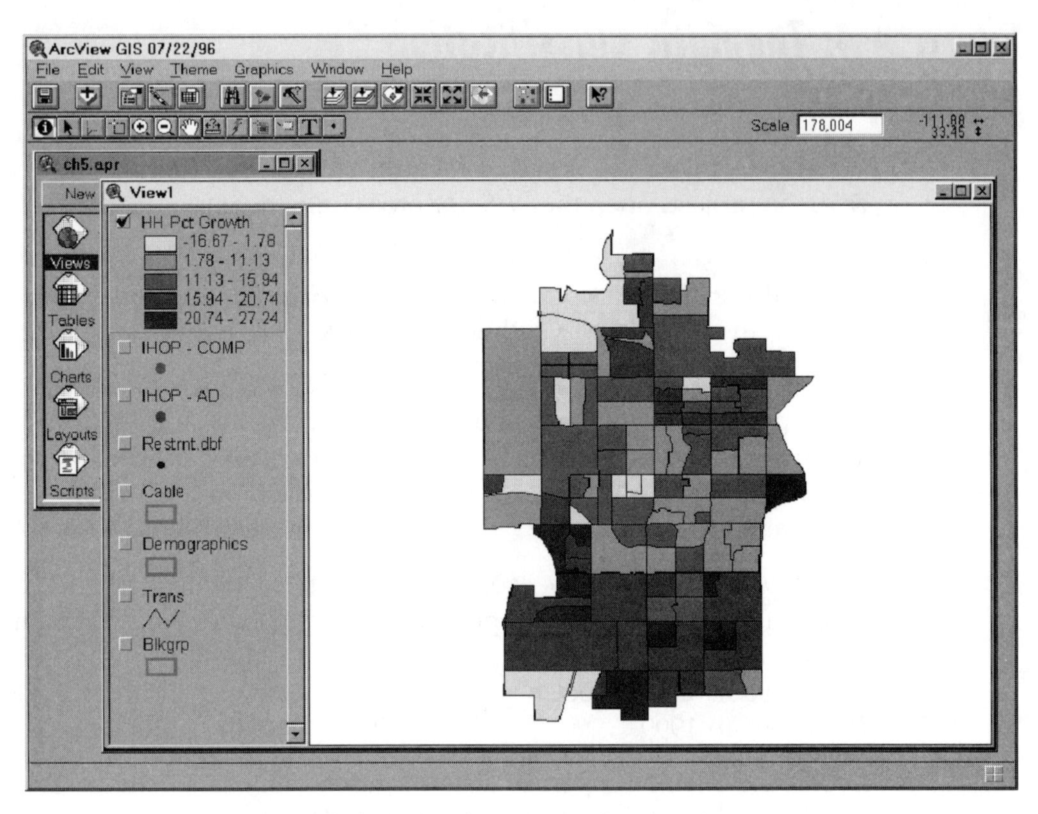


*Default classification on Hhpctgrowt.*

6. The default symbolization uses the Red monochromatic color ramp. Because you are working with continuous data, the color ramp is appropriate. Experiment with the available color ramps from the Color Ramps pulldown list until you find one that works well.

7. To create a color ramp manually, double-click on the symbol for the first class. This brings up the Symbol Palette window. Switch to the Color Palette and select a pale color. Now, click on the highest class and select a color with strong saturation. Click on the Ramp icon to create the color ramp.

8. ArcView constructs a color ramp grading between the two colors. In some cases, all the intermediate colors are readily distinguishable and easily interpreted as representing classes of continuous data. In other cases, one or more colors may be difficult to distinguish. Click on Apply to apply the symbols to the classified theme. If you cannot find a color selection that *ramps* well, select the Orange monochromatic ramp from the Color Ramp pulldown list.

9. With the color ramp applied, the results of the classification should be evident. A pattern is discernible, but suppose you want to modify the class breaks to further explore the data. To change class breaks, go to the Legend Editor window. Click on the cell for the first class in the Value column. The text will be highlighted. For the first class, key in this range: *-17 - 0* (negative 17 to zero). Highlight each cell in the column in turn, and key in the following class ranges: *0 - 10, 10 - 15, 15 - 20,* and *20 - 28.* When the class range changes are complete, click on Apply. The new classes are immediately applied, and the view display is updated.

**NOTE:** *It is customary to repeat the value that was used for the upper range of a class as the lower range for the next class. If a field value falls on the break, it is assigned to the first class containing that value. Thus, the value 10 would be assigned to the 0 - 10 class.*

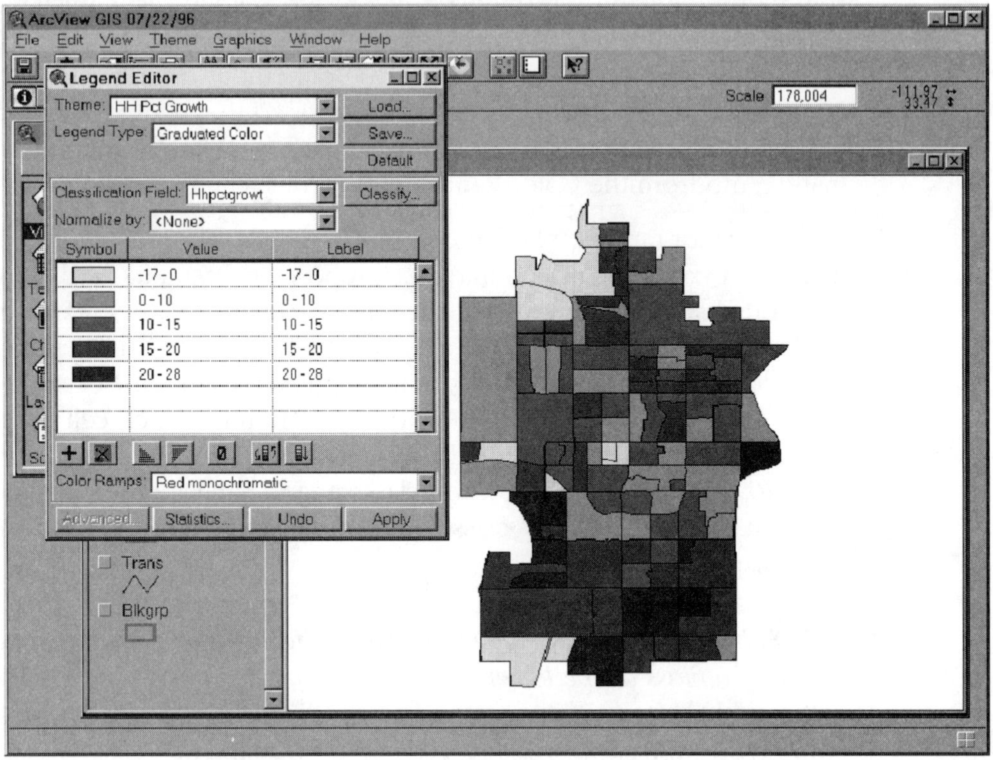

*Adjusted class ranges for the HH Pct Growth theme.*

**10.** You can likely obtain an even better pattern if you reduce the number of classes. In the Legend Editor dialog window, click on the Classify button. Change the number of classes from five to four. A new four-part classification is generated. Note that your previous color choices are preserved.

**11.** Edit the class ranges again, this time using the following ranges: *-17 - 0, 0 - 12, 12 - 18,* and *18 - 28.* Apply the classification and examine the results. The four-part breakdown appears to adequately portray the data breaks. (The incremental project has been saved for you as *ch5a.apr.*)

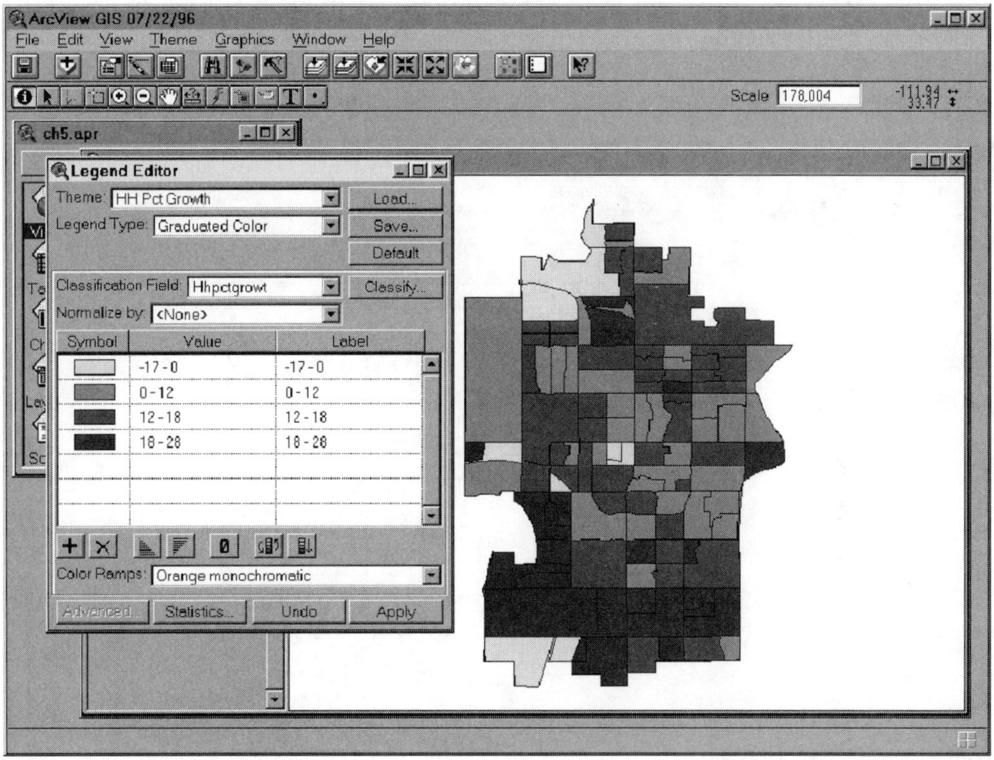

ArcView GIS 07/22/96
File Edit View Theme Graphics Window Help
Scale 178,004
ch5.apr
Legend Editor
Theme: HH Pct Growth
Load...
Legend Type: Graduated Color
Save...
Default
Classification Field: Hhpctgrowt
Classify...
Normalize by: <None>

| Symbol | Value | Label |
|---|---|---|
| | -17-0 | -17-0 |
| | 0-12 | 0-12 |
| | 12-18 | 12-18 |
| | 18-28 | 18-28 |

Color Ramps: Orange monochromatic
Advanced... Statistics... Undo Apply

Adjusted four-class classification on the HH Pct Growth theme.

Next, you will examine the *Spent* item in the restaurant customer survey. (The *Spent* item reflects the amount of money spent.) You will map this item in an attempt to identify patterns.

1. Double-click on the *IHOP - AD* theme to access the Legend Editor. For Legend Type, select *Graduated Color*, and for Classification Field, select *Spent*.

2. The default of natural breaks classification is applied. A cursory look at the class breaks suggests that five classes may be excessive. Click on the Classify button, and change the number of classes from five to three, and the classification type to *Quantile*. The class breaks are

now *0 - 8, 9 - 18,* and *19 - 30.* Adjust the values for class breaks to *0 - 7, 7 - 15,* and *15 - 30.* Change the marker colors as desired to better distinguish the three classes. Then click on Apply.

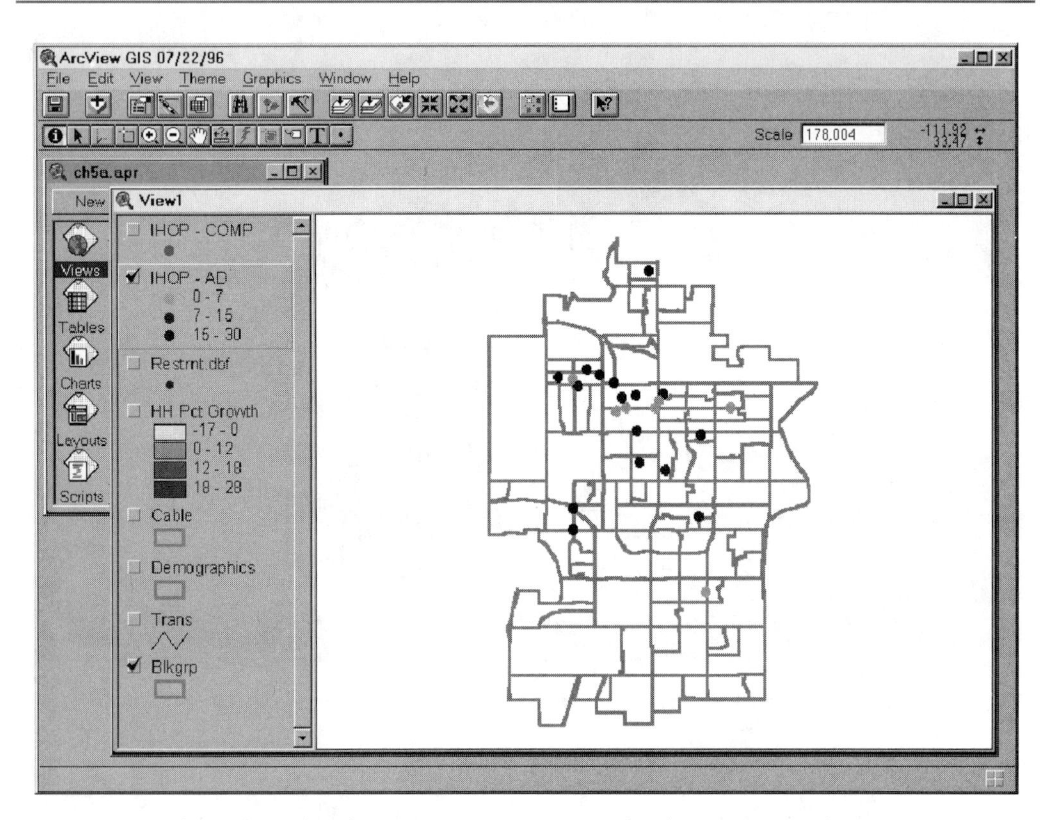


*Three-class classification on amount spent, for the IHOP - AD theme.*

3. Perform a similar type of adjustment to the *IHOP - COMP* theme. Adjust the class breaks so that they correspond to those in the *IHOP - AD* theme. Assign the same symbol colors, and click on Apply.

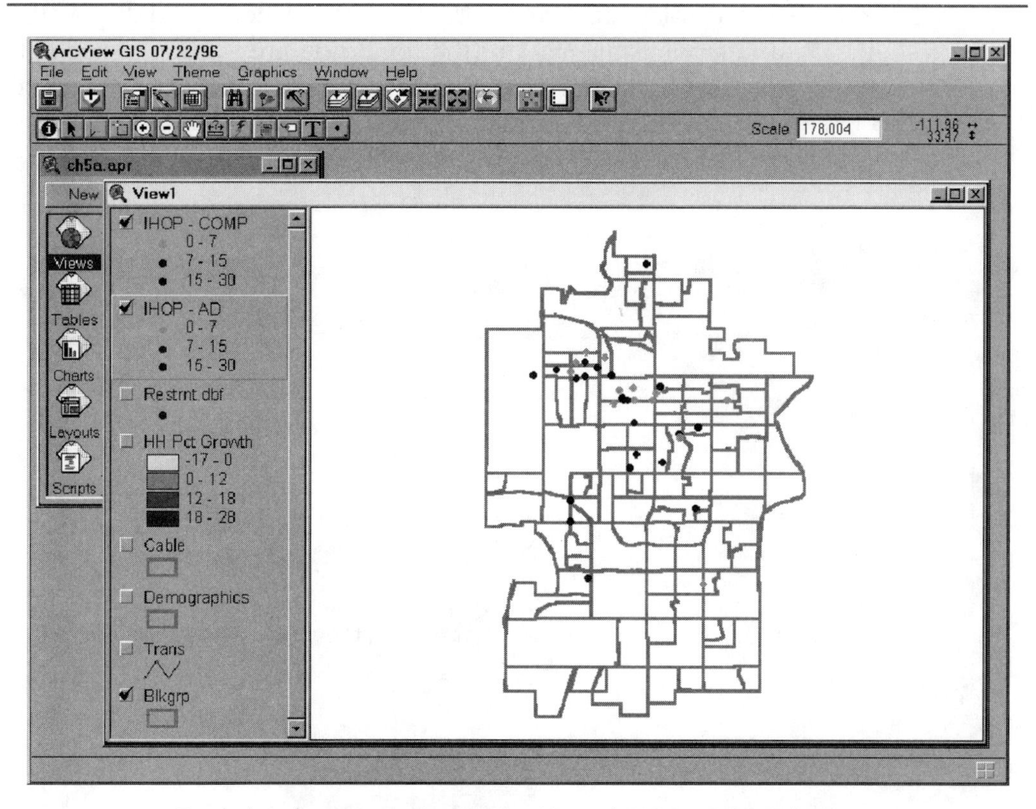


*Final classification on the IHOP - AD and IHOP - COMP themes.*

**4.** In the display, the dot size was made smaller by changing the size from 8 to 6 via the Marker Palette. To change dot size after classifying the theme, double-click on the symbol for the first class to re-initialize the Palette window. Select the Marker icon (third from the left) to access the Marker Palette. While holding down the <Shift> key, click on the remaining class symbols until all are highlighted. Click on the arrow to the right of the size input box to access the scrolling list of available marker sizes. (Although the scrolling list jumps from 4 to 8, it is possible to backspace over a number and enter a custom value for size.) In this instance, enter *6* and press <Enter>. Apply the new marker size to the classified theme.

5. For the final display, turn on the *HH Pct Growth* theme and display the IHOP survey points against this theme. If you are starting to see patterns emerging, and are considering the application of additional classifications, you are beginning to appreciate the power of classification in GIS analysis. (This project has been saved as *ch5fin.apr.*)

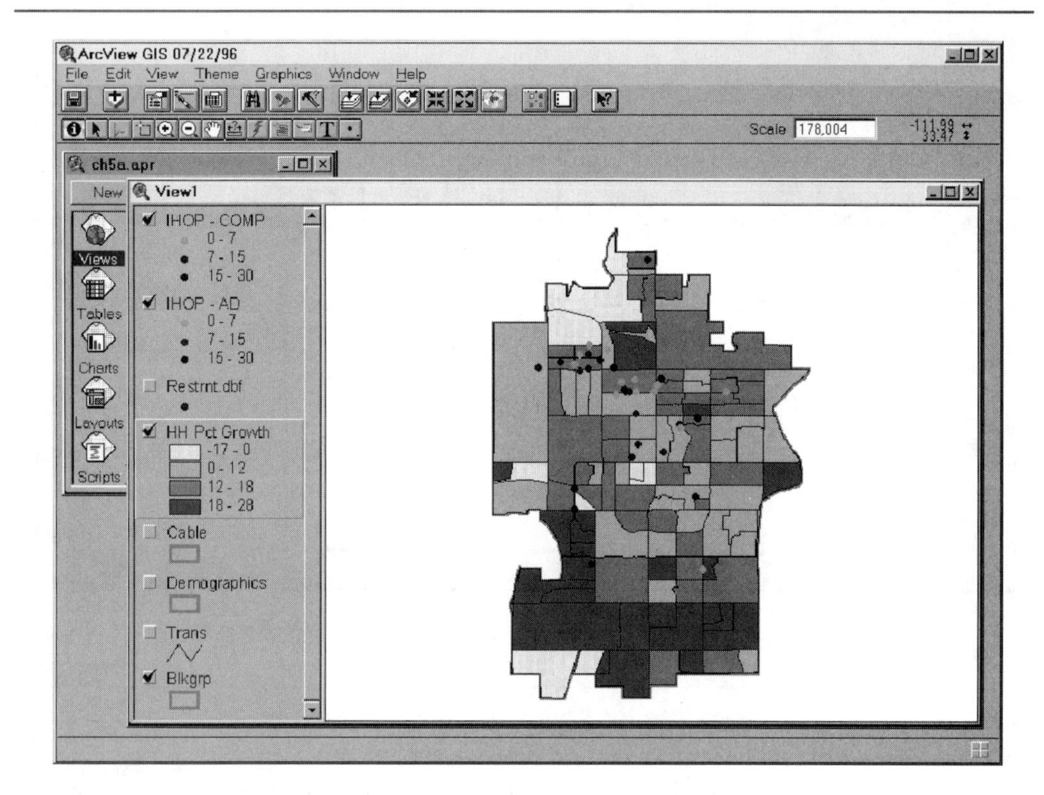


*IHOP - AD and IHOP - COMP themes displayed against the HH Pct Growth theme.*

The Identify and Labels tools are fairly straightforward. We suggest that you turn on an active theme and identify and label features until you feel comfortable. The same applies to the graphic tools. Turn off enough themes so that your display is relatively uncluttered and experiment with adding different types of graphics. Switch to the Pointer tool to select graphics, then move, resize, stretch, and reshape them until you feel comfortable with these manipulations. Select and manipulate a few text labels as well. Mastering ArcView graphics is particularly useful for performing spatial manipulations on themes.

# Chapter 6

# Data Queries

This chapter focuses on the manipulation of tables as an analytical tool. In preceding chapters, we discussed attribute tables associated with X,Y and addres-geocoded event themes. Tables were imported from delimited text files and joined to theme attribute tables. Through all of these operations, table appearance did not change.

From the ArcView user's perspective, all tables *are* the same, regardless of table source (INFO, dBase, or delimited text file). ArcView defines a standard template to reference the tables you access. The tabular data itself is *not* imported, but rather continues to be stored in its source file in native format. The ArcView link to the data is *dynamic,* meaning that changes to your data outside ArcView will be reflected in ArcView projects that reference the data. Consequently, ArcView sees the same snapshot of your data as other application packages. There is no redundancy because you do not manage multiple databases. Your database software handles your tabular data, freeing ArcView to focus on managing and organizing spatial data. (ArcView has its own built-in database manager that works quite well for simple mapping applications.)

## Basic Table Operations

When you open a table, all fields are displayed, along with the field names from the initial table definition. This may or may not be satisfactory, depending on the size of your table and how the fields are named. Tables containing many fields require scrolling of the display to examine widely separated fields. Field names may be incomprehensible, making interpretation difficult. Fortunately, these properties can be changed.

The Table Properties window controls which fields are displayed as well as an *alias* for each field. An alias allows you to substitute a more easily recognizable name for a cryptic field name. For example, a field defined as *VLPCT* can be aliased as "% Vacant Land." The new name will then be used in future queries and output. To access the Table Properties window, click on the title bar of a table to make it active, and select Table Properties from the Table pulldown menu.

*Table Properties window.*

Table Properties

Title: Attributes of Demographics — OK

Creator: — Cancel

Creation Date: Saturday, August 31, 1996 08:13:47

Comments

| Visible | Field | Alias |
|---|---|---|
| ✓ | Hh90 | |
| ✓ | Hh94 | |
| ✓ | Hh99 | |
| ✓ | Hhpctgrowt | |
| ✓ | Popbase | |
| ✓ | Hhincavg | |
| ✓ | Hhincmedia | |

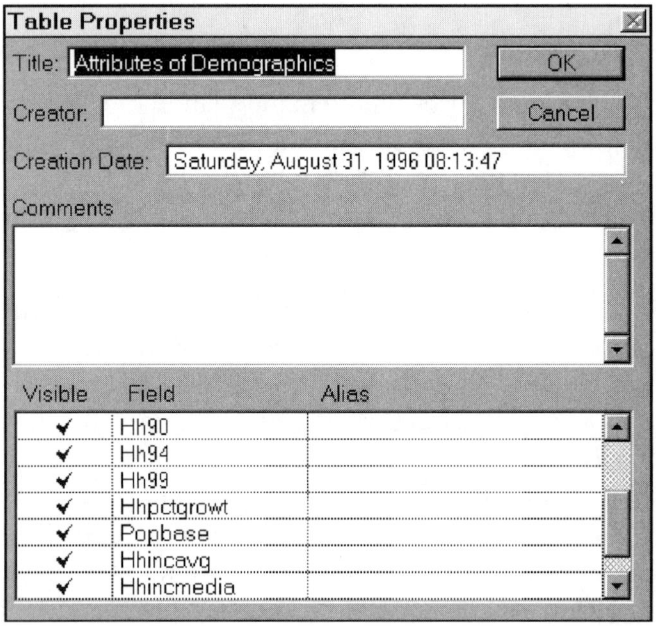

Within the Table Properties window are three columns: one that indicates which fields will be visible in the table, one that lists all fields in the table, and one in which you can specify an alias for each field.

To change an alias in the table, click in the Alias cell for a field. Type in the new name. This name will now be associated with this field in all subsequent operations.

The Visible column controls which fields are displayed in the table. Initially, all fields are set to be visible. To make a field invisible, click in the Visible cell for that field. This will turn off the check mark and render the field invisible.

**NOTE:** *Invisible (hidden) fields are not available for query or classification. They will not be shown in the results of the Identify process or in feature labeling, and will not be included when you print or export the table.*

In addition to making a field visible or invisible, you can resize or reorder the fields on your display. You can make these changes directly to the table.

To resize a field, place the mouse pointer on the border of a cell at the top of a column. Click and drag the border to increase or decrease the display width for the field. If you wish, the display width can be collapsed so that the field does not display at all; the field will still be available for all other operations.

To change the order of the fields in a table, click on and begin to drag the name of the field. As you do so, the outline of the cell will be displayed. When you have positioned the field where you want it, release the mouse button. The table will be redisplayed with the field in the new location.

Sorting a table may also make it easier to locate specific records in that table. To sort a table, select the field name on which to sort, and then click on either the Sort Ascending or Sort Descending button from the button bar. The table will be redisplayed in sorted order.

*Sort Ascending and Sort Descending button icons.*

## Using Tables with Views

Once you have added themes to a view, opened the attribute tables associated with these themes, and joined tables as needed, the next step is to associate records from these tables with features of the view. These operations can range from simple selection and identification of features to robust logical queries.

## Selecting Features

Feature selection is one of ArcView's basic and most frequently used functions. The Select Feature tool, available from the View tool bar, is used to choose features from the active theme by pointing or by dragging a box. The selected features are then displayed in the view using the current selection color (the default color is yellow). The color can be changed by accessing Properties from the File menu (with a project window active).

The ability to select features is also present when working with tables. The Select tool, available from the Table tool bar, allows records to be selected from the active table; these selected records are highlighted using the current selection color.

*Select Feature tool icon.*

When you access the feature attribute table for an active theme, you gain additional functionality from the integral link that is maintained between the records in the table and the corresponding features that are displayed for the theme in the view. When you select features in a theme using the view's Select Feature tool, the corresponding records are highlighted in the theme's attribute table as well. Conversely, when you select records using the table Select tool, the theme's corresponding features are highlighted in the view. As you will soon see, this integral link between features in a theme and their corresponding records in the theme's attribute table is at the heart of much of the analytical functionality of ArcView.

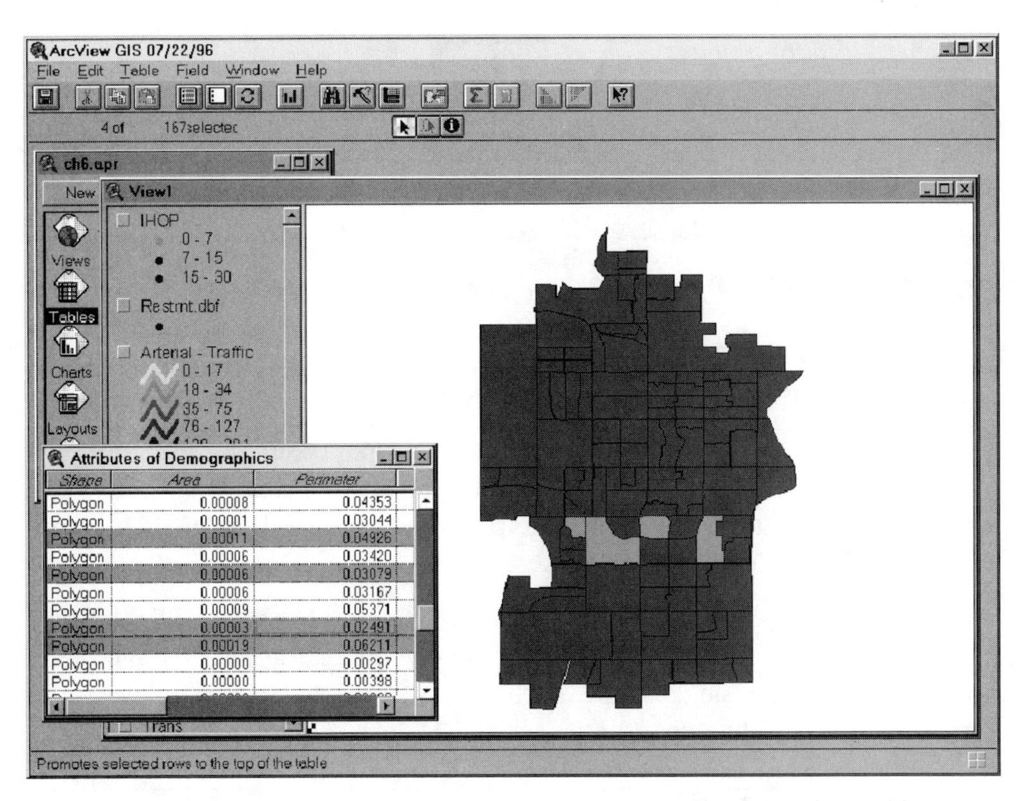


*Selected features highlighted in both the view and the theme's attribute table.*

When using the Select tool, clicking on a feature causes it to be selected. Clicking on a new feature causes the new feature to replace the old one as the selected feature. To add to the selected set, hold down the <Shift> key while selecting additional features. Clicking on a previously selected feature with the <Shift> key depressed will remove only that feature from the selected set.

Additionally, features within the active theme can be selected using the view's Select tool by dragging a box shape on the display while holding down the mouse button.

You can easily view the records you have selected by using the Promote tool on the Table button bar. This tool reorders the table records, displaying the selected records at the top of the table.

Promote tool icon.

# Selecting Features with Shapes

The ability to select features by dragging out a box shape with the mouse (as described earlier) is only a sample of how you can select features by shape.

Selecting features by shape is a two-step process. First, create the shape using the Draw tool on the View tool bar. Then, use the resultant shape or shapes to select features from active themes. Existing graphics may be used, as well as graphics created specifically for selecting features.

The Select Features Using Shape tool on the View button bar selects all the features in the active themes that fall partially or totally under the selected graphic or graphics. This can be a single shape, such as a line or polygon, or a mix of shapes, such as circles and polylines.

Select Features Using Shape tool icon.

The ability to select features using shapes lends itself to basic analytical techniques. A circle could be created with a specific radius (e.g., one mile) to delineate areas of influence for point locations. Polygons could be drawn to represent service areas and used to select businesses or homes lying within this area.

Circle drawn on a view (left); features selected with the circle (right).

## Selecting Features by Query

Occasionally, you will want to identify a feature by its value. In this case, you do not know the feature's location or exact position, but you are aware of its address or ID or some other attribute. In this instance, you can query the associated attributes to locate specific features.

The simplest way to query is by using the Find button. The Find function allows you to search all character fields in the active theme for the first occurrence of a specified string. The Find function is not case-sensitive, and it can match on a partial string (e.g., *St. Louis* will match the entered string of *Louis*). If a match is found, the display is redrawn with the selected feature in the center of the display.

Find button icon.

If the feature that is matched is not the desired feature, you can select Find again to locate the next matching feature for the same string.

Except for the most specific and simple queries, selecting features by attribute is best accomplished with a logical query. A logical query is used to select features based on how their columnar data matches search criteria.

To select features via logical query, click on the Query Builder button on the button bar. This brings up the Query Builder dialog window, which displays a selection of fields, operators, and values from which to construct a query. A query can be used to construct a new selected set or to add to or remove items from an existing set. When a query is executed, the selected elements are highlighted in the table and on the view.

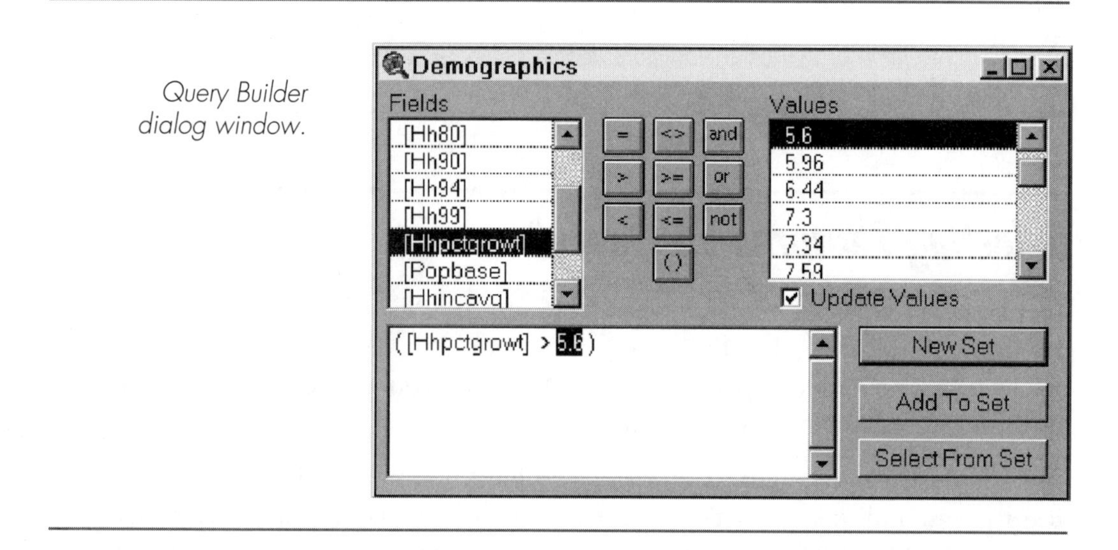


*Query Builder dialog window.*

It is often easier to define features you do *not* want than those you do. The Switch Selection button on the Table button bar allows you to move all features from those not selected to the selected set.

The Clear Selected Features button is used to deselect the selected features in all active themes.

*Select All, Clear Selected Features, and Switch Selection button icons.*

## The Undervalued GIS Skill of Database Management

As practitioners and promoters of a graphical system, we often focus on and extol the virtues of visual analysis and the cartographic skills that advance the power of its usage. It is easy to overlook the fact that many important GIS operations revolve not around graphics, but around tabular data management.

Familiarity with database concepts, and especially a background in SQL (structured query language), can be a significant asset to the GIS operator. With that knowledge, the ease and power of conducting ad hoc logical queries increases immensely, enabling much more rapid exploration of your overall spatial database. In brief, strong tabular database knowledge is the perfect complement to strong cartographic knowledge.

Whether you are the manager of a shop or the newest kid on the block, we strongly encourage you to include database management training in your plans. If you work with Microsoft Access, be sure to train with Access. If you work with dBase, train with dBase. Many GIS training programs assume that you will become familiar with your tabular database. Be sure to put SQL or other database management training on your schedule.

## Locating the Selected Set

Often, the results of a specific query from a table or theme may not be readily apparent, even when the features are highlighted. The Zoom to Selected button on the View button bar zooms to the extent of the selected features on the active themes.

*Zoom to Selected button icon.*

✓ **TIP:** *When executing a specific query against a theme that contains a large number of features, zooming to the extent of the selected features before drawing may speed redraw time. In addition, opening the attribute table for the active theme will cause the number of selected elements to be displayed in the Table tool bar. This number can provide useful feedback about the outcome of a query before the elements are examined or displayed.*

## Logical Queries on Themes

In the examples above, all features present in the original theme were available for query. In many instances, however, restricting a theme to a subset of the available features is desirable. For example, a data set of soils information might be used to extract a theme showing only soils with a high water table.

To define a logical query for extracting features from a theme, make the theme active and select Properties from the Theme pulldown menu to access the Theme Properties dialog window. Click on the Definition icon, then the Query Builder icon to call up the logical query dialog window.

You construct a logical query in the same way you use the Query Builder to select features from an active theme, with one difference. Because this query defines the base set of features comprising the theme, there is no option to add to or remove from the selected set.

When the query is complete, click on OK to accept the query. The query will then appear in the Theme Properties definition box. Click on OK in the Theme Properties dialog window to apply this query to the theme. Now only the features selected in the theme will be available for display and query.

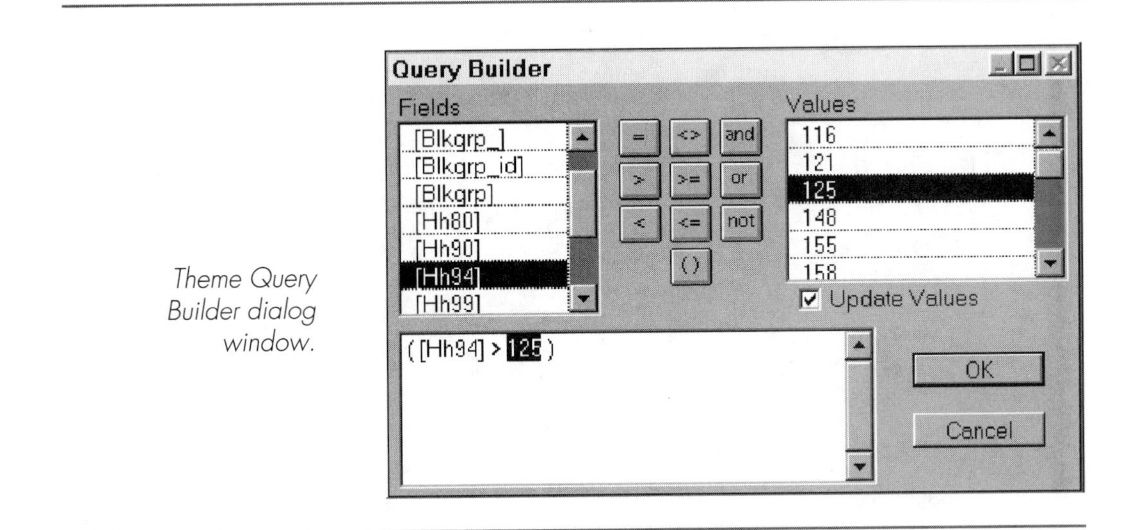


*Theme Query Builder dialog window.*

## A Review of Logical Queries

Selecting features with the Query Builder has been mentioned many times. For those not familiar with logical queries, some elaboration is in order.

All logical queries are comprised of *fields, operators,* and *values.* Although complex queries can be built from these elements, all queries share this basic form. At its most basic, a logical query is comprised of one field, one operator, and one value. For example, the following query,

```
( block_group = 3194.102 )
```

can be used to select any features or records for which the *block_group* is 3194.102.

The query

```
( acres > 640 )
```

can be used to select all features with an area greater than 640 acres.

Complex queries can be constructed by combining simple expressions using the `And` or `Or` operators. For example, the query

```
( acres > 640 and owner = "Jones")
```

can be used to select all parcels greater than 640 acres owned by Jones. Note that character fields must be enclosed in double quotes.

The following operators are available in the Query Builder:

= (equals)

> (greater than)

< (less than)

<> (not equal to)

>= (greater than or equal to)

<= (less than or equal to)

For complex queries, the following joining operators are available:

`and` (both expressions are true)

`or` (either expression is true)

`not` (excluding the following expression)

The following wildcard operators are available:

* (multiple character wildcard)

? (single character wildcard)

Next, the following mathematical operators are available:

+ (addition)

- (subtraction)

* (multiplication)

/ (division)

Note that all logical expressions evaluate from left to right *regardless* of the mathematical operator. Thus, the expression

```
( 3 + 8 * 7 )
```

will yield a value of 77, not 59, as you might expect.

Parentheses are used to force an expression to be evaluated first. Thus, the expression

```
( 3 + ( 8 * 7 ) )
```

will yield the value 59.

Queries can be used to compare the values of two fields. For example, the query

```
( [value80] < [value90] )
```

will return the records for all parcels that have increased in value from 1980 to 1990.

Calculations can be performed on fields as well. For example, the query

```
( [value90] / [acres] > 10000 )
```

can be used to return the records for all parcels valued at greater than $10,000 per acre in 1990.

➻ **_NOTE:_** *In ArcView,* `( [name] = "*Main Street Cafe*" )` *will match "Main Street Cafe" in your table, regardless of how*

> *many blanks the field is padded with. If you are in doubt as to whether a particular field contains padding blanks, the list of values produced by selecting a field in the ArcView Query Builder will contain all blanks from that field. Double-clicking on a value to add it to the query will cause the requisite number of blanks to be included within the quoted string.*
>
> Our best advice to new users is to dive into the Query Builder and experiment. Once the structure and vernacular of queries are understood, the flow becomes easy. Do not worry if seemingly simple queries occasionally fail to fire off as expected. With some practice, soon you will be executing them with ease.

# Exercise 5: Making a Smarter Map

By the end of this exercise, you will have produced visually appealing and powerful maps!

The first step is to open the *ch6.apr* project file in your working directory. Do not open the project you saved at the end of the previous exercise, because *ch6.apr* is a different file. We have made some additions to the project.

As you look at the current version of the project, you will notice three changes: the two IHOP survey themes were combined into a single theme, an *Arterial - Traffic* theme containing traffic count information has been added, and a MicroVision theme containing Equifax MicroVision Segment information was added. In addition, an initial classification of the *IHOP* and *Traffic* themes has been performed.

The market research project has progressed as a result of combining the IHOP surveys into a single theme, and now contains a single point theme that locates customer responses classified by amount spent. The next step is to associate these responses with the block group in which they are located. These responses can then be compared to other data mapped by block group, such as demographic data.

To associate the survey response points with the block group, you need to join the *IHOP* point theme to the *Blkgrp* theme. This spatial association requires a special type of join, known as a *spatial join.*

1. The first step in the spatial join is to open the attribute tables for the *IHOP* and *Blkgrp* themes. Next, highlight the Shape field in each table. Because you want to associate a block group with every IHOP survey point, the *Blkgrp* attribute table will be the source table, and the *IHOP* attribute table the destination table.

2. Making sure the *Attributes of IHOP* table is active, select Join from the Table pulldown menu. Upon completion, the *Blkgrp* attribute is associated with every response point in the *IHOP* theme.

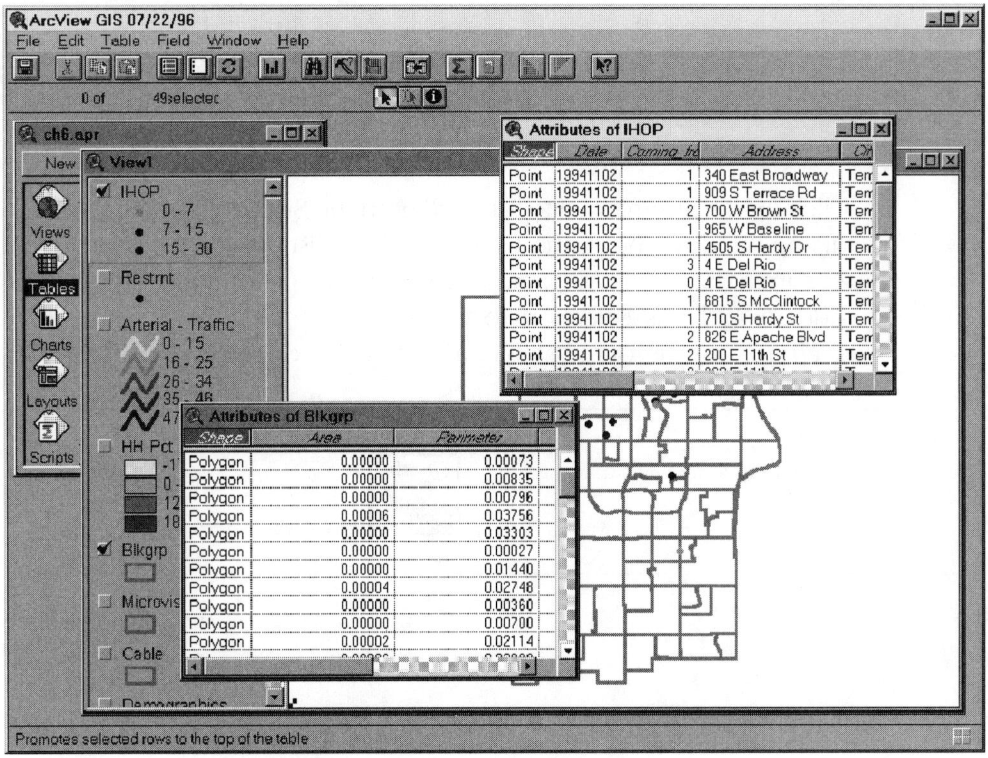

*IHOP and Blkgrp theme attribute tables, ready for the join.*

At this point, you want to summarize the amount spent per respondent for each block group.

1. With the *Attributes of IHOP* table open, highlight the *Blkgrp* field name. Note that when you do so, the table manipulation tools—Sort Ascending, Sort Descending, and Summarize—are no longer grayed-out.

---

*Results of the spatial join to the IHOP theme.*

**Attributes of IHOP**

| Av_add | Av_stat | Av_score | Av_side | Blkgrp_id | Blkgrp |
|---|---|---|---|---|---|
| 340 EAST BROADWAY | M | 83 | L | 1536 | 3190.002 |
| 909 S TERRACE RD | M | 67 | L | 1498 | 3191.003 |
| 700 W BROWN ST | M | 99 | R | 1447 | 3188.004 |
| 965 W BASELINE | M | 83 | L | 1713 | 3200.011 |
| 4505 S HARDY DR | M | 99 | L | 1685 | 3197.022 |
| 4 E DEL RIO | M | 83 | L | 1616 | 3196.003 |
| 4 E DEL RIO | M | 83 | L | 1616 | 3196.003 |
| 6815 S MCCLINTOCK | M | 75 | L | 1755 | 3199.061 |
| 710 S HARDY ST | M | 67 | R | 1452 | 3188.003 |
| 826 E APACHE BLVD | M | 99 | L | 1536 | 3190.002 |
| 200 E 11TH ST | M | 99 | L | 1487 | 3190.001 |

---

2. Click on the Summarize tool to bring up the Summary Table Definition dialog window. From the Field pulldown list, select *Spent* as the field to summarize, and Average as the option to summarize by. Click on Add to add this selection to the summary table. Specify the name *spent.dbf* for the output table, and then click on OK to proceed.

---

*Summarize tool icon.*

---

Upon completion, the output table *spent.dbf* is created. The table contains three fields: *Blkgrp, Count,* and *Ave_spent.*

> ✘ **WARNING:** *Due to the manner in which a saved project is reopened, accessing a theme that results from a spatial join in a restored project can cause a fatal error. To prevent such an error, it is necessary to remove all spatial joins from the project before saving.*

*Summary Table Definition dialog window.*

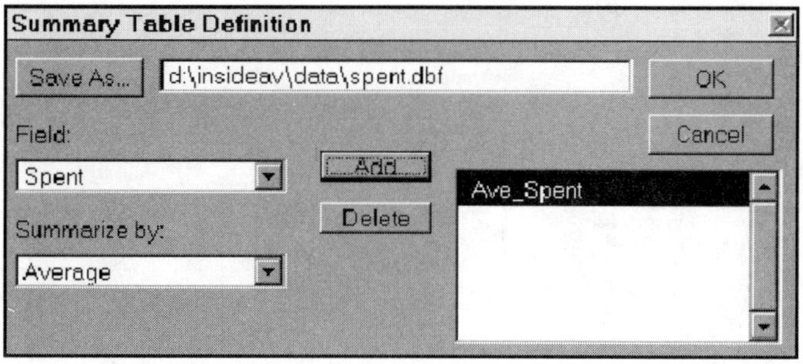

Because you have successfully created the summary table, you can remove the spatial joins now.

1. Click on the title bar of the *Attributes of IHOP* table to make it active.
2. From the Table pulldown menu, select Remove All Joins. The *Blkgrp* field is no longer associated with the *Attributes of IHOP* table.
3. Close the *Attributes of IHOP* table and return to the *spent.dbf* table.

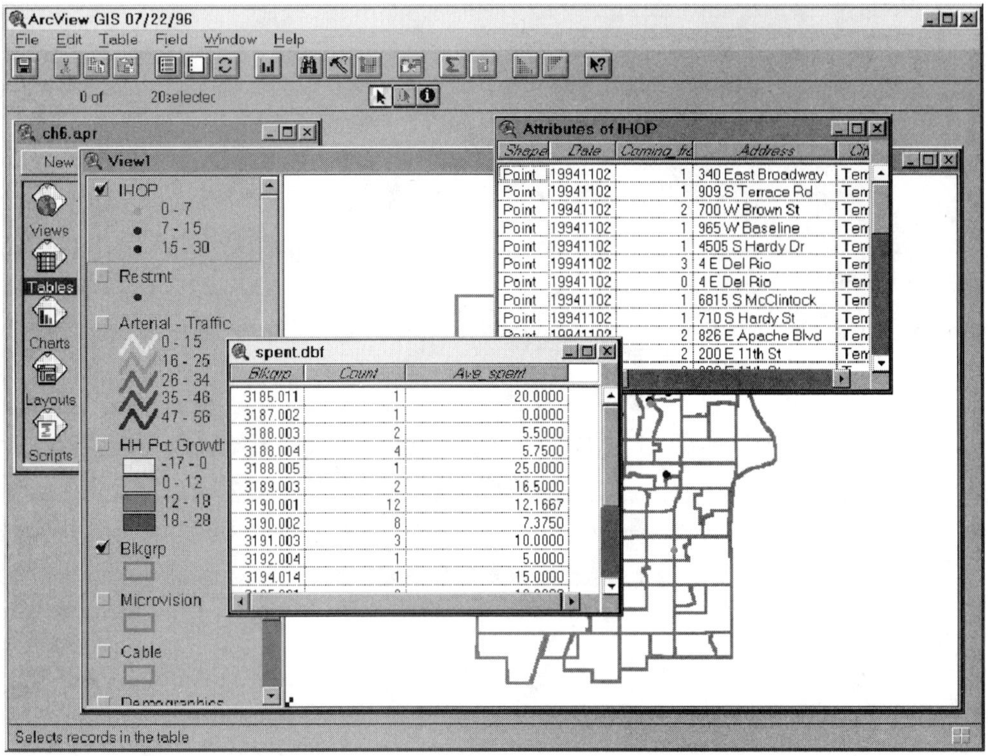

Resultant spent.dbf summary table.

You now have a table that contains the average amount spent by block group, which you can relate back to the *Blkgrp* theme.

1. Make the *Blkgrp* theme active, and make a copy of the theme using Copy Themes and Paste from the Edit pulldown menu.
2. The new theme appears at the top of the Table of Contents. Click on this theme to make it active, access the Theme Properties dialog window, and change the name of the theme to *Avg Amount Spent.*

3. Open the *Attributes of Avg Amount Spent* and *spent.dbf* tables, and join *spent.dbf* to *Attributes of Avg Amount Spent* using the common *Blkgrp* field, making sure that the *Attributes of Avg Amount Spent* table is the active table.

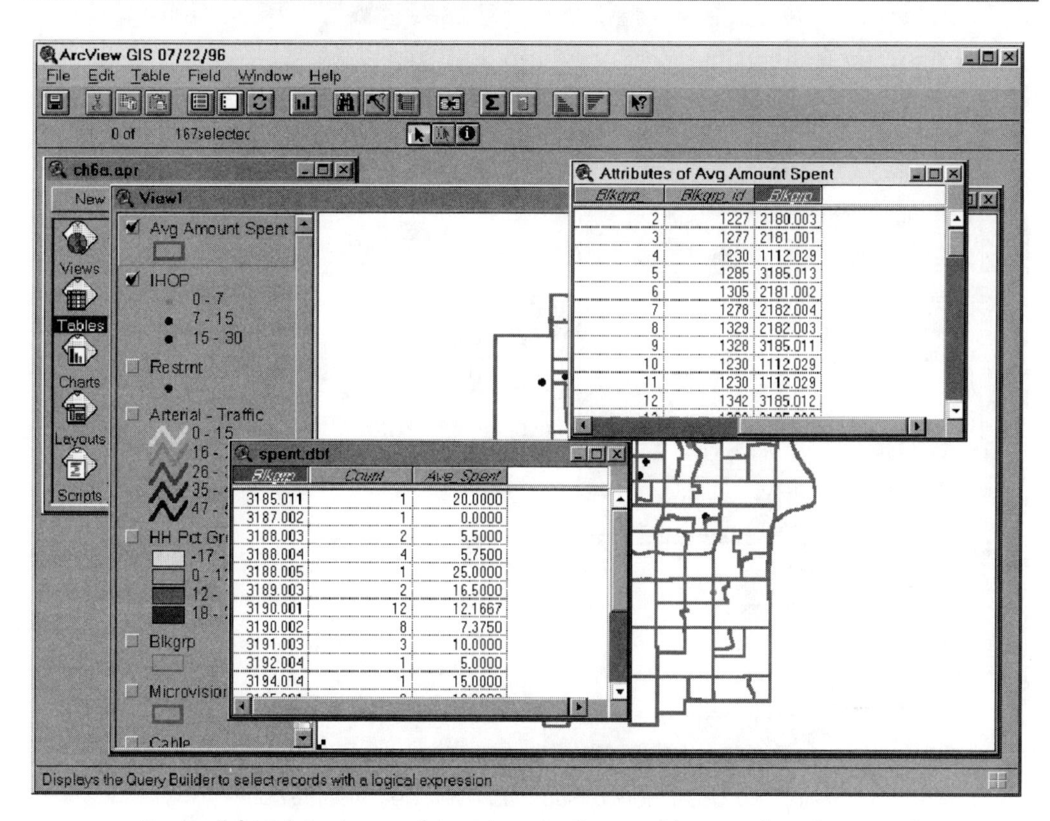


*Spent.dbf and Attributes of Avg Amount Spent tables, ready to be joined.*

Once the tables are joined, you can classify the *Avg Amount Spent* theme on *Ave_Spent*.

1. Double-click on the *Avg Amount Spent* theme to bring up the Legend Editor.

2. Classify the theme on *Ave_Spent* in four classes, using the Graduated Color legend type and Quantile classification, and Apply. (The incremental project has been saved as *ch6a.apr.*)

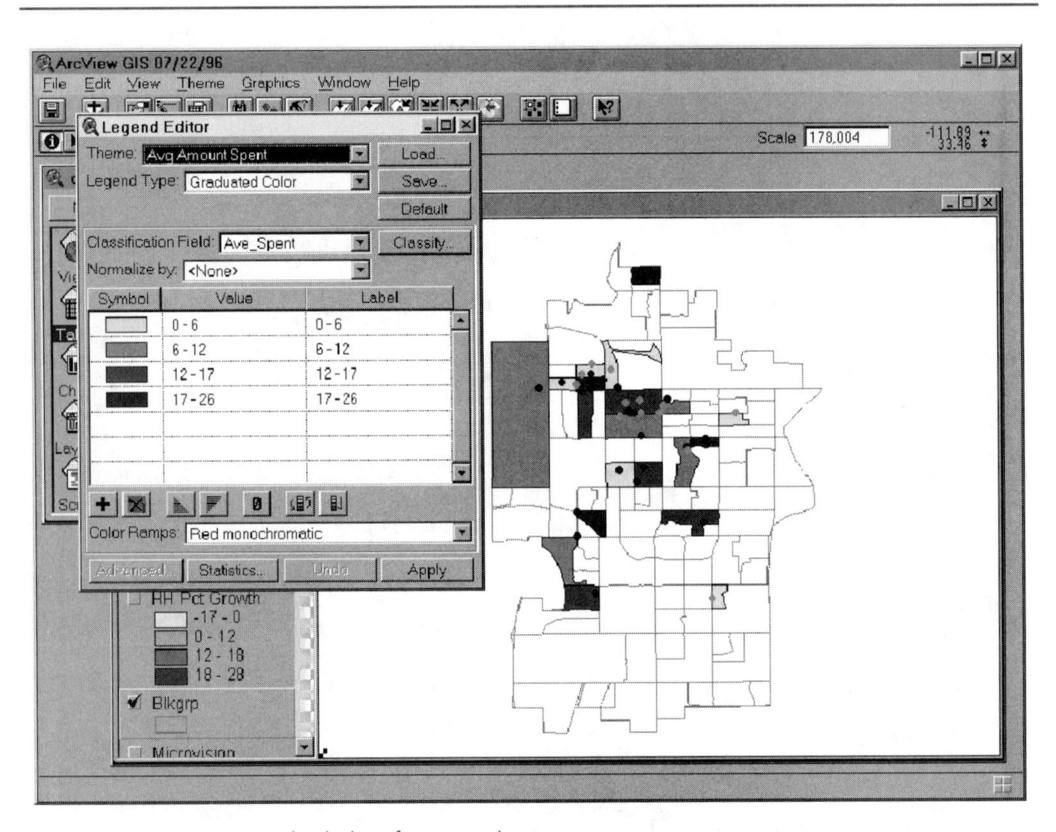


*Applied classification, showing average amount spent, along with the IHOP survey response points.*

You have now successfully generated an indication of how the amount spent varies by neighborhood. For the next series of analyses, you will examine the area immediately adjacent to the client's location. First, you need to determine the location of the client by querying the *Restrnt* point theme.

1. Turn on the *Restrnt* theme, and click on the theme to make it active.

2. To access the Query Builder, click on the Query Builder icon from the button bar, or select Query from the Theme pulldown menu.

3. Double-click on the *Name* field. The Values list at the right will be populated with the values for *Name* from the *Restrnt* theme attribute table.

4. Click on the equals ( = ) operator, then scroll down the list until you locate the name INTERNATIONAL HOUSE OF PANCAKES. Double-click on this entry.

5. When the query statement is complete, click on the New Set button. The location of the client, the International House of Pancakes, will be highlighted. Make sure that the *IHOP* theme is turned off so that you can see the result of this query.

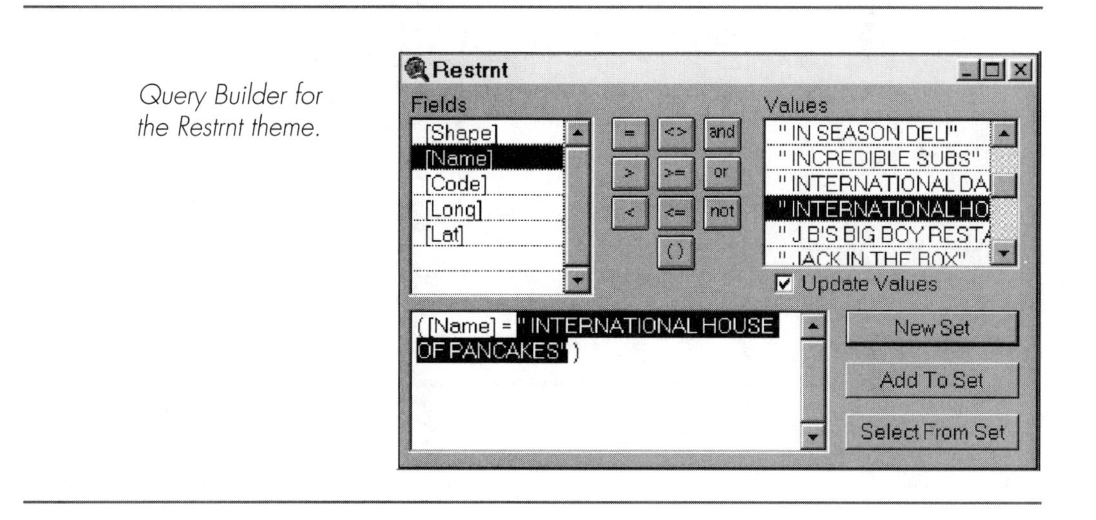


*Query Builder for the Restrnt theme.*

6. To select all demographics and customers within a three-mile radius of the client's location, click on the Graphics tool (on the far right side of the tool bar), and select the Circle tool.

7. Click on the IHOP location and drag a circle until the status bar reports a radius of 3.00.

**NOTE:** *To measure the radius distance in miles, the Distance Units for the View must first be set to miles. If the units have not been set to miles, or you are not sure whether units have been set, access Properties from the View pulldown menu and select* Miles *from the list for Distance Units.*

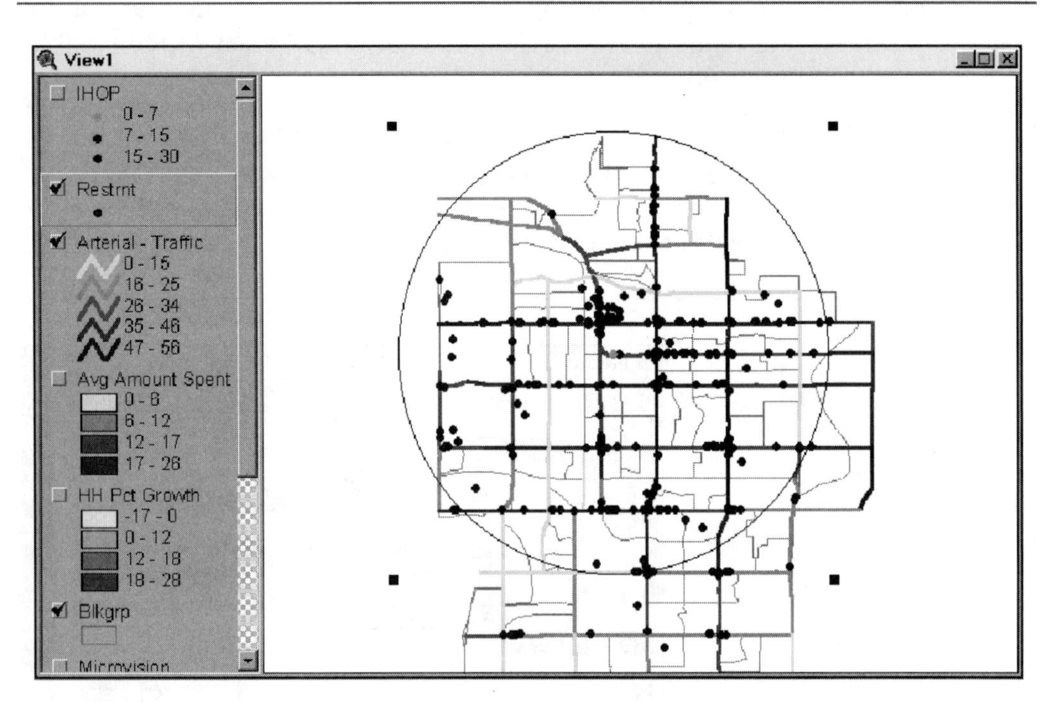


*Three-mile radius circle centered on the IHOP location.*

You are now ready to use this circle to select block groups that fall within the three-mile radius, but look closely. Something is amiss with the circle.

The arterial streets in Tempe are located on a mile grid. The IHOP restaurant is located almost exactly halfway between the mile grid arterials in both the north/south and east/west directions. The circle describing the three-mile radius should also fall between the mile arterials. Note that although this is the case to the east and west, to the north and south the circle falls directly on the mile arterial, a distance of 3.5 miles.

This distortion results from the fact that you are still working in geographic coordinates. Although you have set your map units to miles, geographic coordinates do not constitute a true projection, and distances measured are not uniform in both the X and Y directions. Thus, whereas working in geographic coordinates is adequate when *relative* positions are sufficient, when more precise information is desired, it is necessary to switch to a projection in which accurate area and distance are maintained. For your project area, Arizona State Plane coordinates are suitable.

From the View Properties window, select Projection to access the Projection Properties window. Select *State Plane - 1983* from the *Category* field, and *Arizona, Central* from the *Type* field. Click on OK to apply.

*Projection Properties dialog window.*

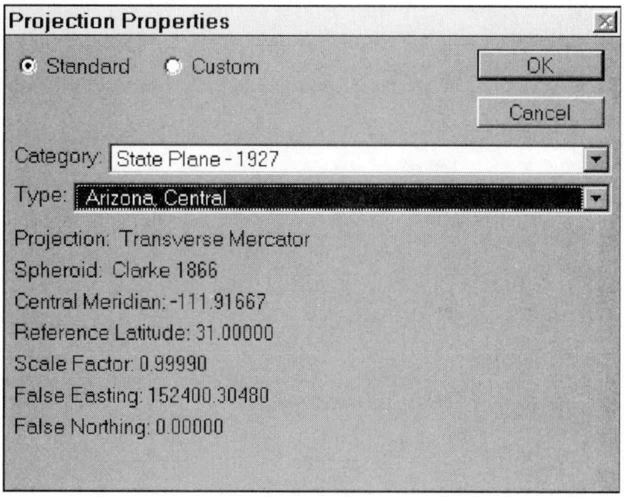

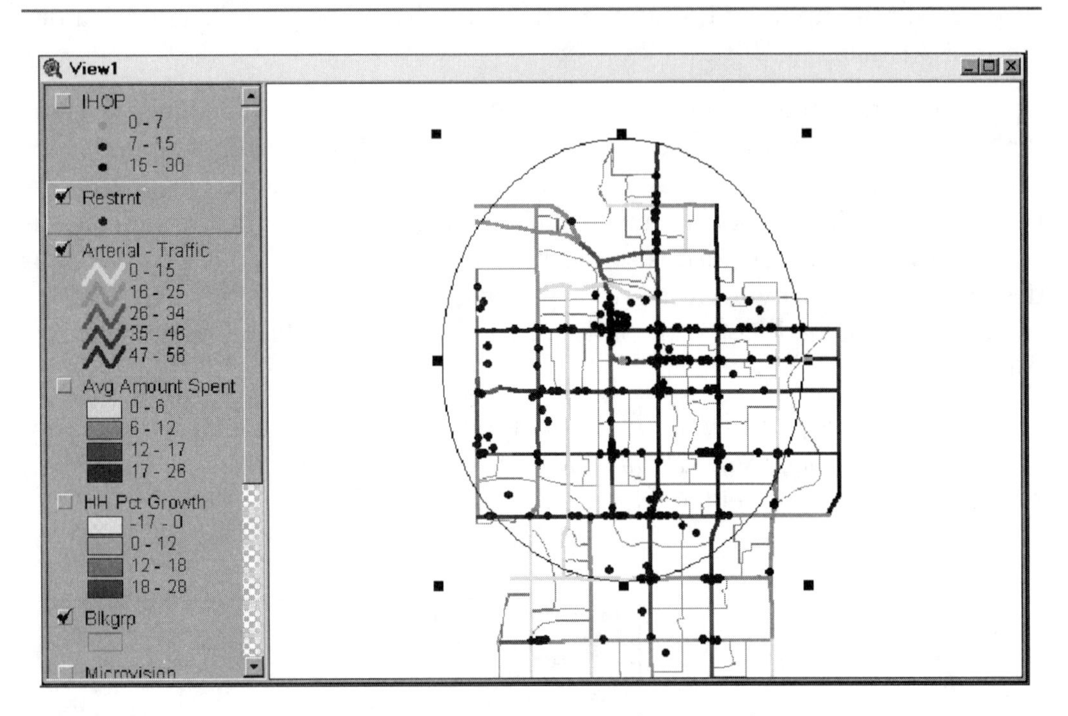


*View1, projected to Arizona State Plane coordinates.*

Note how the circle is now quite clearly an oval. You will need to re-do the three-mile radius. Select and delete the old circle, either with the Delete key or by selecting Delete Graphics from the Edit menu. Add a new circle with the same three-mile radius.

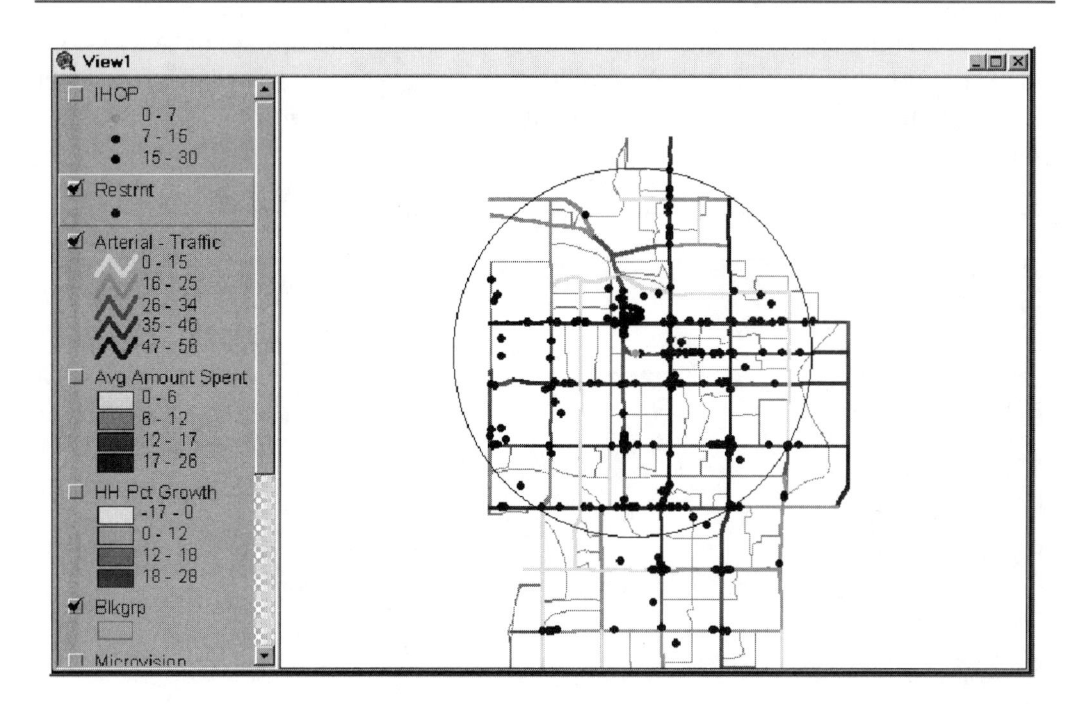


*New three-mile radius circle for View1.*

You can now identify those block groups within the three-mile radius that spent the greatest amount per capita. First, make the *Avg Amount Spent* theme active. Next, use the selection tool (the arrow) to select the circle. At this point, you can use the Select Features by Selected Graphics tool.

*Select Features by Selected Graphics tool icon.*

Based on your initial classification, you determined that $15 was a suitable cut-off point for identifying block groups with a high average amount spent. Access the Query Builder and construct the query *Ave_ spent >= 15*. Choose Select from Set to extract this new set from the currently selected block groups. Seven block groups are now selected.

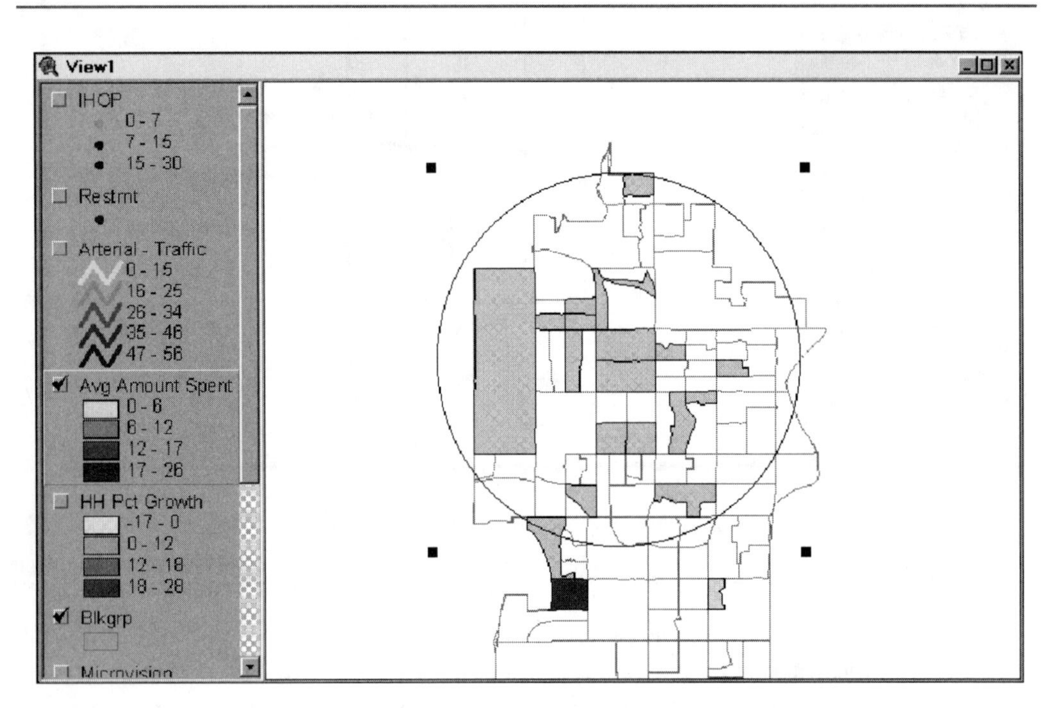


Selected block groups from the Avg Amount Spent theme.

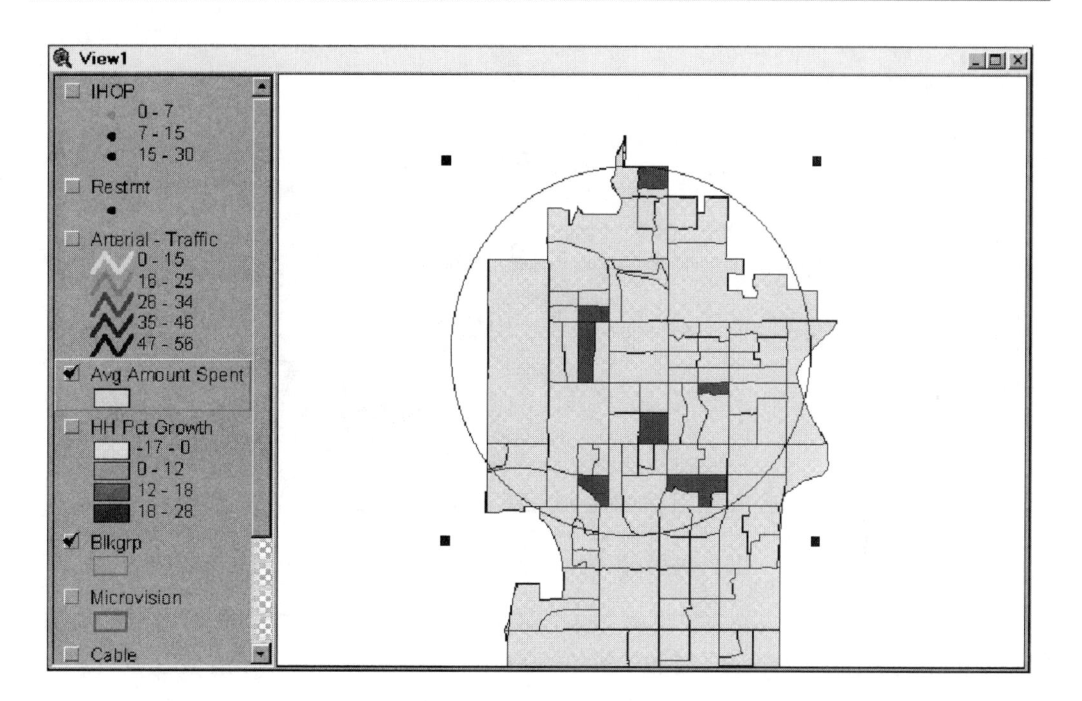


*Seven selected block groups who spent an average amount of at least $15, and who are located within a distance of three miles.*

In the figure above, the polygons are displayed in a uniform color by removing the classification on the *Avg Amount Spent* theme and by selecting Single Symbol for the Legend Type in the Legend Editor. A light shade was assigned to the theme and the color used to draw selected features was subsequently changed to a dark color. This was done by making the project window active, accessing the Project Properties dialog window from the Project pulldown menu, and setting the color for the display of selected features to a dark color with the Specify Color sliders. (The incremental project has been saved as *ch6b.apr.*)

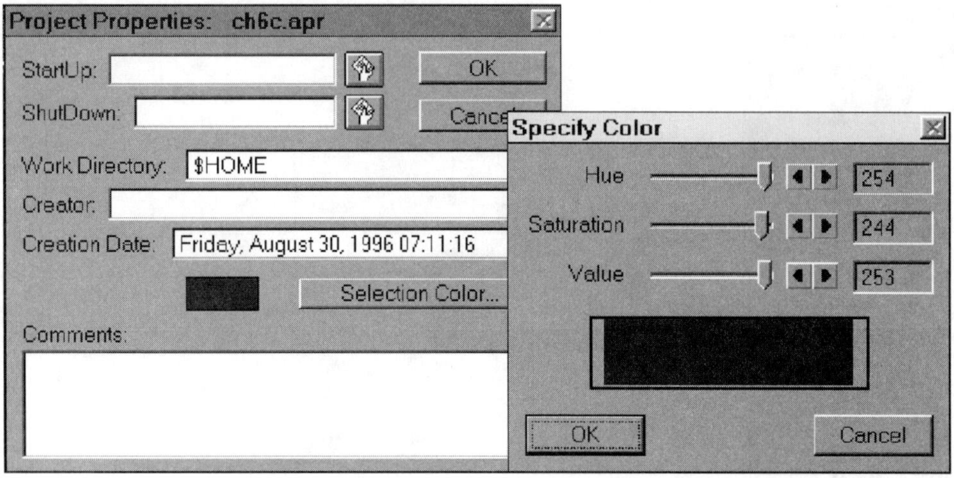

Specify Color sliders from the Project Properties dialog window.

Now that you have identified seven favorable block groups from the customer base, you can learn more about the demographics and lifestyles of some of your best customers. To accomplish this, you will access the MicroVision segmentation data.

MicroVision is a micro-geographic targeting system that uses aggregated consumer demand and census data to classify every household in the United States into one of 50 unique market segments. Each market segment consists of households at similar points in the life cycle that share common interests, purchasing patterns, financial behaviors, and needs for products and services.

To identify the corresponding block groups from the *MicroVision* theme for the seven selected block groups of the *Avg Amount Spent* theme, you can make use of the ability to select features from a second theme based on the location of selected features from the first theme (i.e., Select by Theme).

1. First, make the *MicroVision* theme active.
2. From the Theme pulldown menu, choose Select by Theme.
3. You want to select features that fall completely within the selected features of *Avg Amount Spent.* Do this by first selecting *Avg Amount Spent* from the bottom pulldown list, then selecting Are Completely Within from the top pulldown list. Then click on New Set.



*Select By Theme dialog window.*

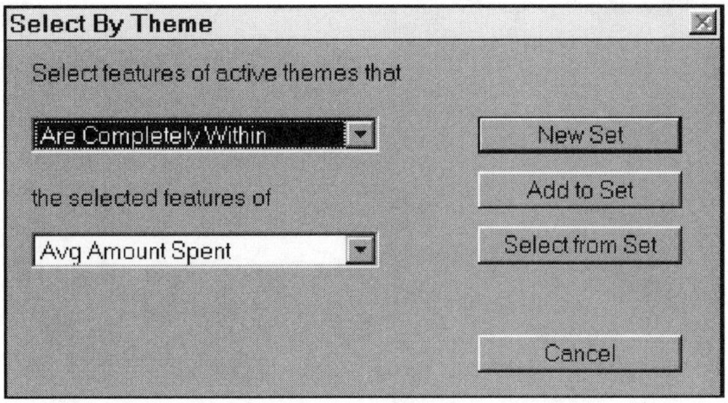

4. Turn off the *Avg Amount Spent* theme, and turn on the *MicroVision* theme to display the results of this selection.

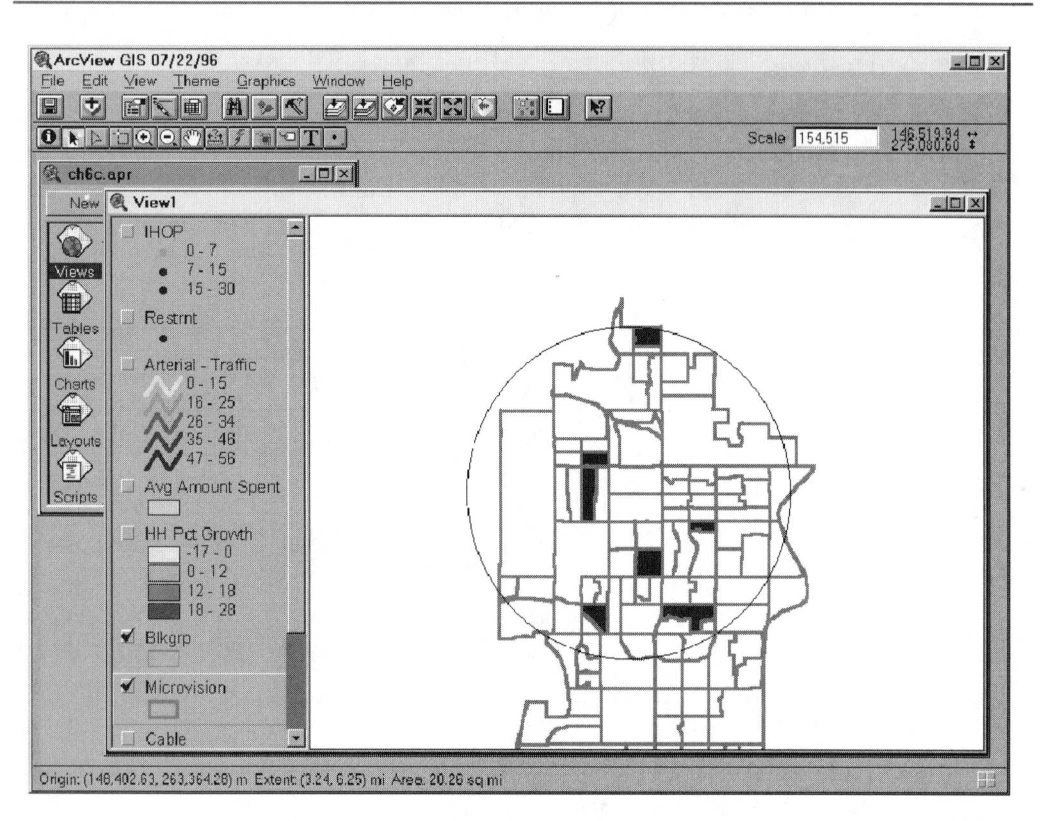


*Selected block groups (from the MicroVision theme) that result from the theme-on-theme selection.*

Although you could use the Identify tool to determine the primary MicroVision segment associated with each block group, instead choose to label each block group with the associated value using the Auto-label tool.

The Auto-label tool labels selected features of the active theme with attributes from the designated field. The field to use for labeling, as well as the placement of the text label, is controlled from the Theme Properties dialog window. To access these properties, and to use the Auto-label tool, take the following steps:

1. Activate the Theme Properties dialog window. Select the Text Labels icon to access the label properties.

2. From the Label Field list, select the *Prim_mv* field and click on OK.

*Text Labels properties from the Theme Properties dialog window.*

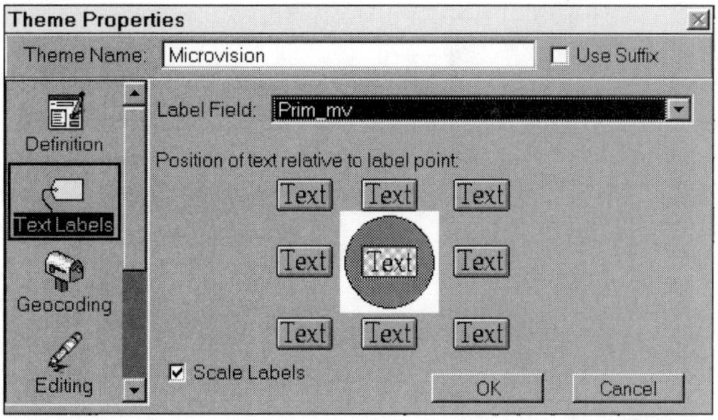

3. From the Theme pulldown menu, select Auto-label. The Auto-label dialog window is displayed. Turn off Find Best Label Placement and turn on Use Theme's Text Label Placement Property, and click on OK. Text labels that correspond to the values for *Prim_mv* are placed in each of the seven selected block groups. Because these text labels are graphic elements, they can be repositioned for greater legibility.

*Auto-label dialog window.*

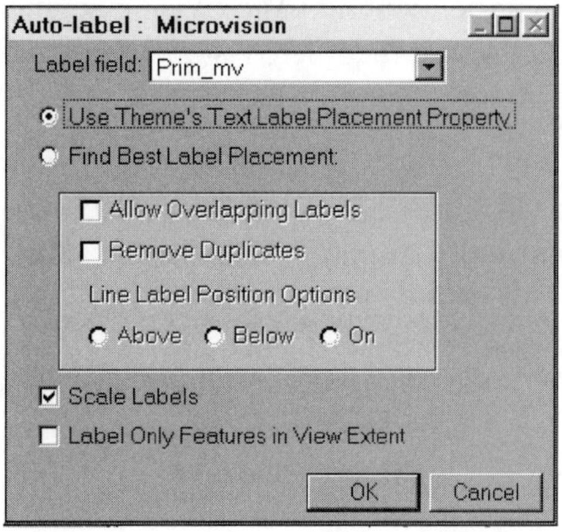

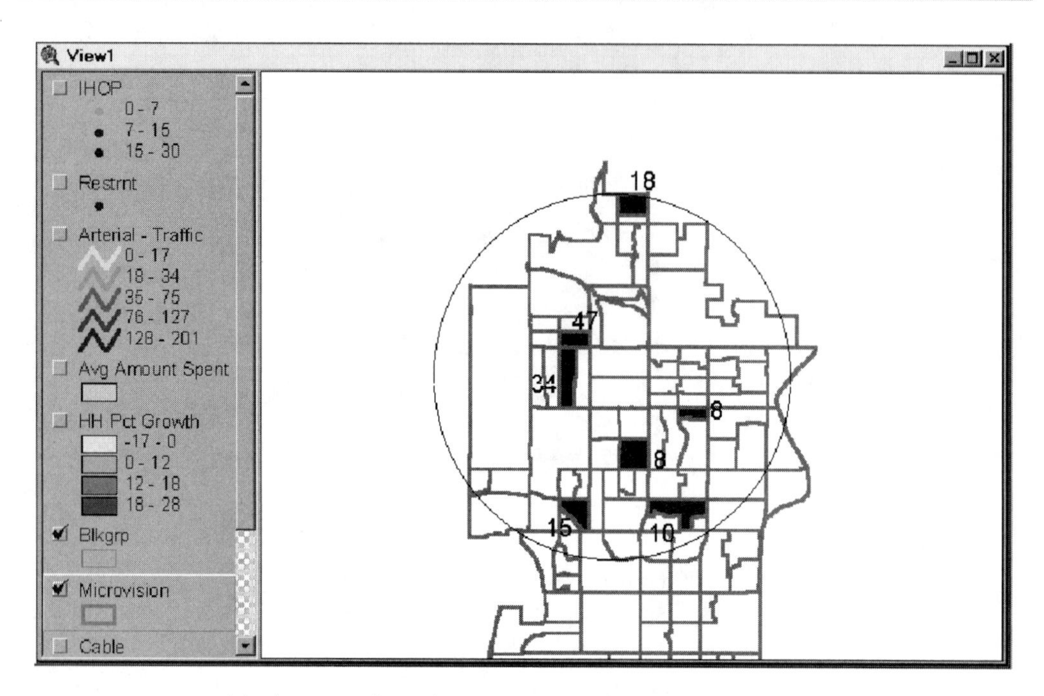


*Seven block groups from the MicroVision theme, labeled by Prim_mv.*

You can now refer to the demographic tables for additional information about each of the identified MicroVision segments. (For details on segments, see Appendix C, "About the MicroVision Segments.")

Now identify heavy traffic arterials adjacent to these block groups as candidate areas for outdoor advertising. Upon examining the initial classification, you determine that a threshold of 40 (40,000 vehicles per day) is a good cut-off point. To enhance the display of the classification, take the following steps:

1. Double-click on the *Arterial - Traffic* theme to access the Legend Editor.
2. Reclassify the theme to two classes, and adjust the class ranges to *0 - 40* and *40 - 56*. To improve the display, use a light gray line for the lower class and a heavy black line for the higher class.

3. To highlight the traffic data, change the symbol for the MicroVision theme so that polygon outlines are not drawn. This is done by selecting a line thickness of None under Outline on the Fill Palette. Thus, only the selected block groups from the MicroVision theme will draw. (The incremental project has been saved as *ch6c.apr*.)

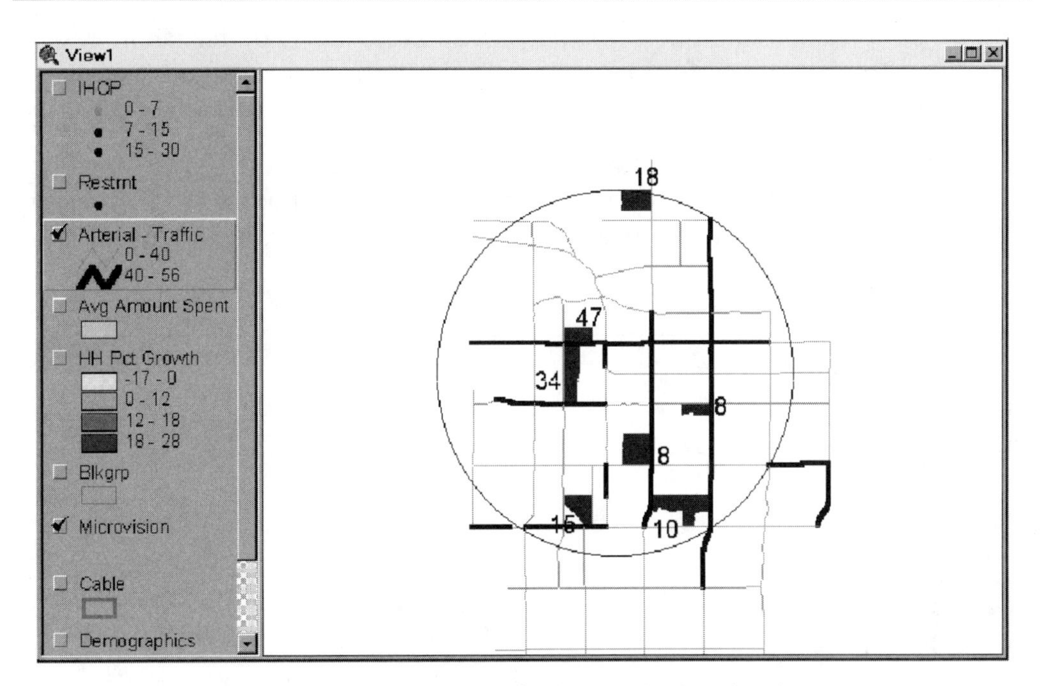


*Classified Arterial - Traffic theme, displayed with the selected block groups from the MicroVision theme.*

4. Your last step is to use the *Restrnt* point theme. Rather than display the location of all restaurants, you can make use of an additional field in this table, one which identifies the type of business. This code, known as the SIC (Standard Industrial Classification), is stored in the Code field of the *Restrnt* attribute table (see endnote). By querying the restaurant theme table, determine that the SIC for IHOP is 581260. It is logical to believe that only those restaurants with an SIC of 581260 would give you a better picture of competitors' locations.

To limit your display to family restaurants, you will apply a logical query on the *Restrnt* theme.

1. Click on the *Restrnt* theme in the Table of Contents to make it active.
2. From the Theme Properties dialog window, click on the Definition icon. Select the Query Builder tool icon.
3. Construct the query *Code = 581260* (the SIC code for family restaurants). Fifteen features out of a total 388 are selected.
4. The *Attributes of Restrnt* table reveals that the selected set of restaurants includes Denny's, JB's, Howard Johnson's, Village Inn, and Shoney's, all of which were previously identified as IHOP competitors. As before, you can adjust the symbol for greater clarity.

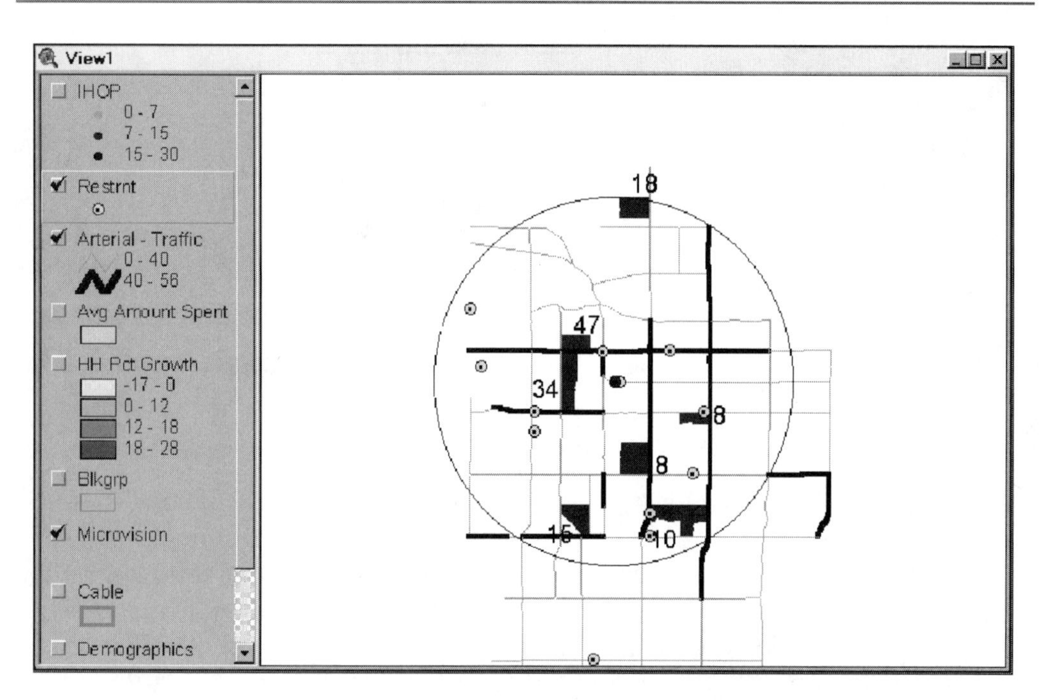


*Previous view, with the location of competing restaurants added.*

The map in the illustration above is usable, but enhancements could improve its visual presentation. For example, you could replace the MicroVision segment number with a brief descriptor, and you could label some of the competing restaurants.

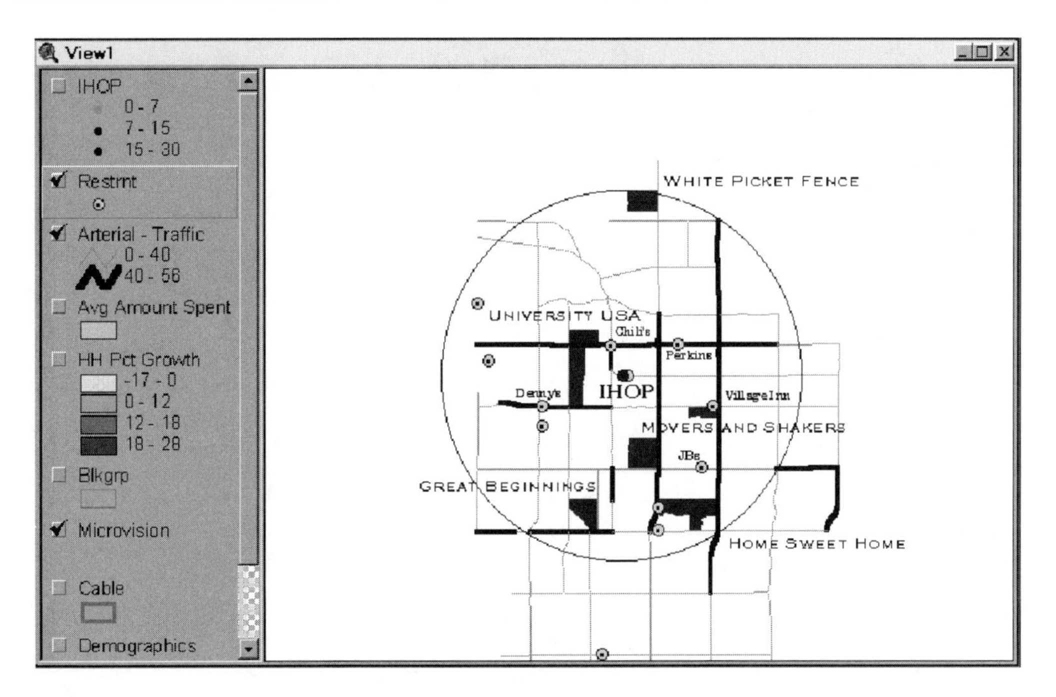


*Same map as above, with enhancements.*

You could do even more here. For instance, you have not yet explored the cable theme, which could yield valuable information about the cable channels most useful for reaching your targeted market through advertising. We think, however, that we have proven our point: spatial analysis and mapping tools can greatly enhance traditional market research.

*Endnote:* This information was derived from Equifax National Decision Systems' Restaurant-Facts™ database. Certain components of the Business-Facts™ database are derived using data obtained from American Business Information, Inc., Omaha, Nebraska, ©1994, and Disclosure Incorporated, Bethesda, Maryland, ©1994. The primary business data source for the Business-Facts™ database is licensed from American Business Information, Inc., Omaha, Nebraska.

## Chapter 7

# Charts

When working with thematic data, the options available for thematic query and classification are often insufficient to fully reveal the relationships inherent in the data. The charting capabilities of ArcView provide additional tools for representing attributes on your map.

One of the strengths of ArcView's charts is their *dynamic* nature. A chart represents the current data in a table. If the table is updated, the chart will reflect the change. A chart can represent a selected set of records in a table as well as the entire table. As the selected set is changed, the chart immediately reflects the change. This dynamic nature allows a chart to function as an especially powerful visualization tool during interactive query.

Six types of charts are available: area, bar, column, line, pie, and XY scatter diagrams. Once a table and records have been selected for charting, you can cycle through the chart types to determine which type most effectively represents the data.

## Creating a Chart

There are two basic approaches to creating a chart. The first approach, which might be considered more direct, is to click on the Chart icon from the project window to make charts the active project component. Click on the New button and then select the table you want from the list of chartable tables.

Alternatively, if you wish to create a chart from a table subset such as the results of a logical or spatial query, you can use another method. In this case, with the table already active and a selected set in place, you simply select Chart from the Table pulldown menu.

Both of these approaches will activate the Chart Properties dialog window, which presents you with additional choices regarding chart design. From the Chart Properties dialog window, the field or fields to chart are selected, as well as the field to be used to label each series or group (optional). When the desired fields have been selected, clicking on OK will create the chart.

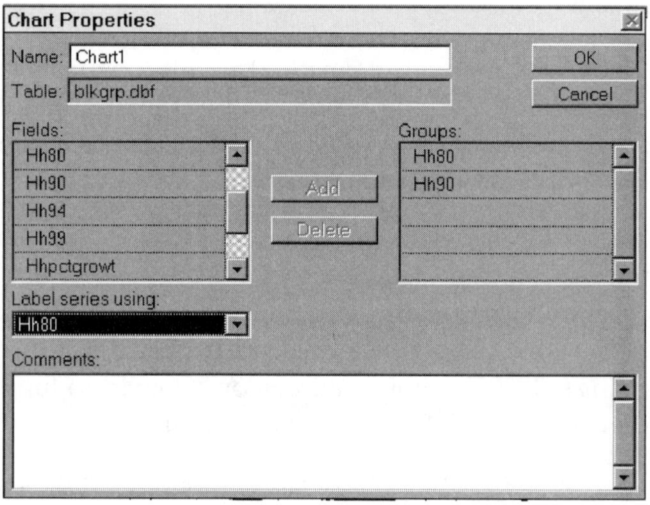

*Chart Properties dialog window.*

By default, a column chart is created with the chart's *data series,* formed from the selected records in the table.

## Clarification of Data Markers, Series, and Groups

When discussing the creation of charts, data markers, data series, and data groups are frequently mentioned. It is easy to become confused, even with the aid of online help and the ArcView manual. While it is possible to create a perfectly usable chart without full comprehension of these terms, the terms appear frequently in the documentation, and it is worth the effort to come to grips with them.

A *data marker* is a feature on a chart—column, bar, area, pie slice, or point symbol—that represents the value of a specific field for a specific record in a table. It is analogous to a cell in a spreadsheet. A data marker can represent the actual value for this cell, or it may express a percentage or logarithm.

To best understand *data series* and *data groups*, let's examine an illustration. In this example, ten *records* are selected from a table. Two *fields* must be charted from these records. This creates 20 *data markers* for the chart. The records represent neighborhoods, and the two fields are 1980 and 1990 population counts.

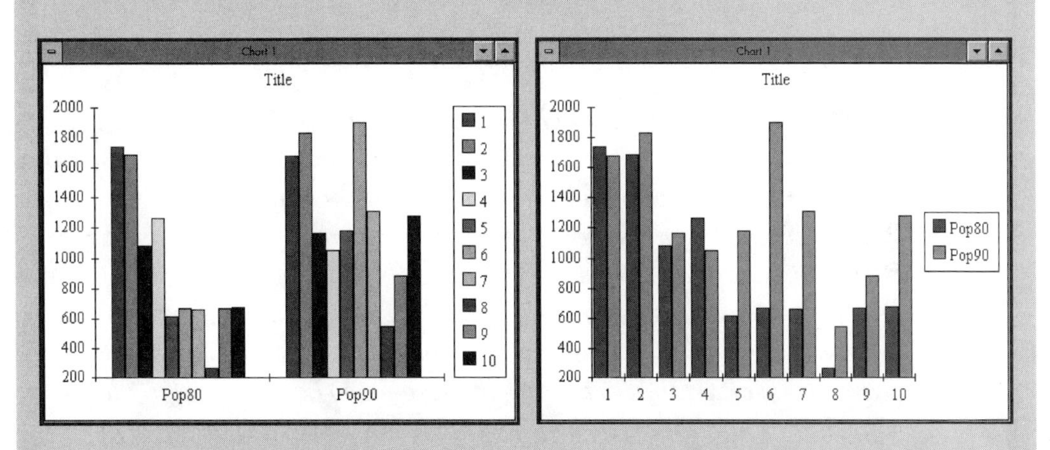


*Two charts created from the same selected records: data series on the left was formed from records; data series on the right was formed from fields.*

In the chart at the left, the data series was formed from the table's records. Each instance, or data marker, is represented by a column. The data group in this example is the field. Two fields are being charted: *Pop80* and *Pop90*. Thus, the chart depicts two data groups, each containing ten instances. In the data series legend, each record in both data groups is assigned a different color.

In the chart at the right, the data series has been formed from the fields of the table. Again, each instance, or data marker, in the series is represented by a column. The data group in this example is the record. Because ten records are being charted, the chart contains ten data groups, each containing two instances that correspond to the two fields, *Pop80* and *Pop90*. In the legend, each field in the ten data groups is assigned a different color.

The chart data series, then, is the comparative unit in the chart. If the data series is formed from records, it contains the full set of selected records for a specific field. If the data series is formed from fields, it contains the full set of selected fields for a specific record. The chart data group is the aggregating unit in the chart.

**NOTE:** *There are several caveats to bear in mind as you begin charting. For example, only numeric fields are available for charting. Next, certain types of charts have limitations on the number of rows they can accommodate. If you receive a message such as "CHART: There is not enough space to plot the chart; check the format parameters and/or resize the chart," switch to another chart format or make your window larger. ArcView does not allow you to scroll around a chart that overflows the screen, so be prepared to alter your format to one that can be adequately displayed all at once.*

# Making Changes to a Chart

There are three categories of changes that can be made to a chart: changing the selected set of records displayed in the chart, changing the chart type, and changing the way the chart elements are displayed.

## Changing the Selected Set

In ArcView, a *data marker* is created on a chart for every selected record in the charted table. Accordingly, changing the number of records selected from the table will accordingly change the resultant chart. You can change a selection interactively by clicking on records in the table, by logical query on the table via the Query Builder, or by spatial selection when charting a theme attribute table. Regardless of the technique you use, the result will be the same: the chart will "forget" references to old selections and depict only those records that match the latest query criteria.

## Changing the Chart Type

The six chart types available in ArcView differ in the way they represent thematic data. Depending on your data, you may find one or two chart types to be superior for illustrating what you wish to portray. For example, bar charts are most useful for comparing relative amount or change, such as population growth, whereas pie charts are most useful for illustrating composition, such as the ethnic makeup of a district.

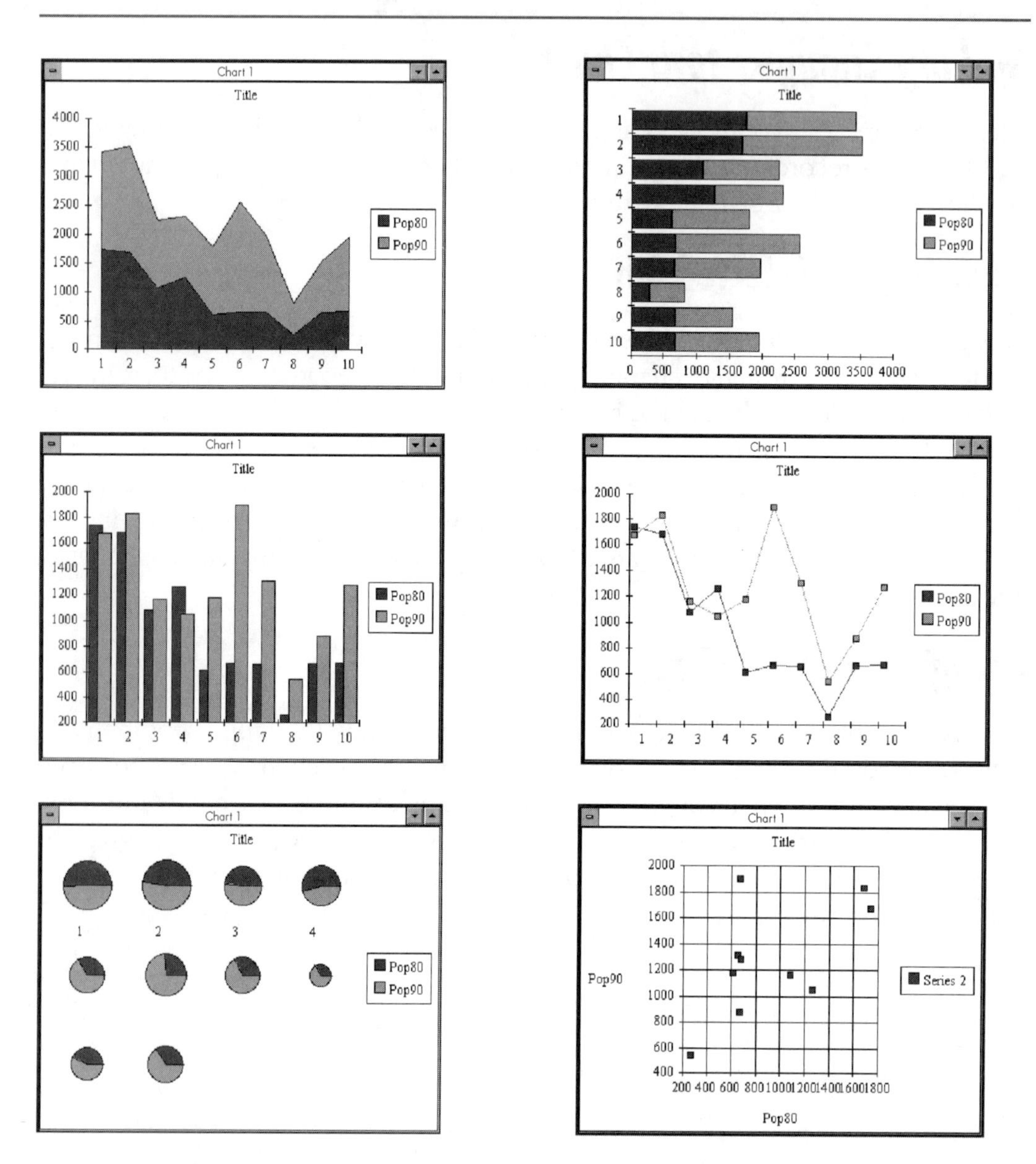


*Same data displayed using six different chart types: area (top left); bar (top right), column (center left), line (center right), pie (bottom left), and XY scatter (bottom right).*

### Changing the Display of Chart Elements

Once the chart type and the desired records from the table have been selected, you can also change how the individual chart elements will appear. Default styles are seen within the chart types in the previous illustration. These choices include how to display individual data markers, whether to display grid lines, and whether to use a linear or logarithmic scale. Chart elements, including the title, legend, and axis properties, can also be edited. Finally, the Color Palette can be used to control the display of chart elements and data markers.

## *Using Charts Interactively*

One of the strengths of ArcView charts is that they are dynamic. As the underlying data changes, or as selected sets change, chart values change. You can use this capability to investigate data, using the charts for immediate display of attributes associated with the currently selected features.

### Locating Theme Elements with the Identify Tool

In a manner that is similar to identifying map elements, the Identify tool allows you to click on a data marker on a chart, locate the element in a table or on a view, and examine the attributes associated with that data marker. With the Identify tool active, clicking on a data marker causes the corresponding record in the table to flash. The Identify window is displayed, showing all the attribute data for the feature. If the chart is built on a theme attribute table that is currently displayed, the feature will blink on the map as well.

### Using Charts with the Select Tool

Because a chart displays the attributes of currently selected features, it can become a powerful tool for visualizing the results of a spatial query, particularly when a chart displays more than one attribute. In the following example, a chart has been constructed based on the attribute table for a theme displaying cable viewership data. From this table, fields have been selected that represent cable viewership relative to the national index of 100 for the CNN, MTV, Nickelodeon, USA, ESPN, and A&E channels.

With the chart displayed alongside the view window, the Select tool is used to select specific census block groups from the cable theme. As the feature is selected, the chart displays the relative viewership for the six cable channels. The first illustration below shows the results for an area near Arizona State University; the second, an older neighborhood in central Tempe. Depending on the type of chart selected, the results of queries on individual features or groups of features can be displayed.

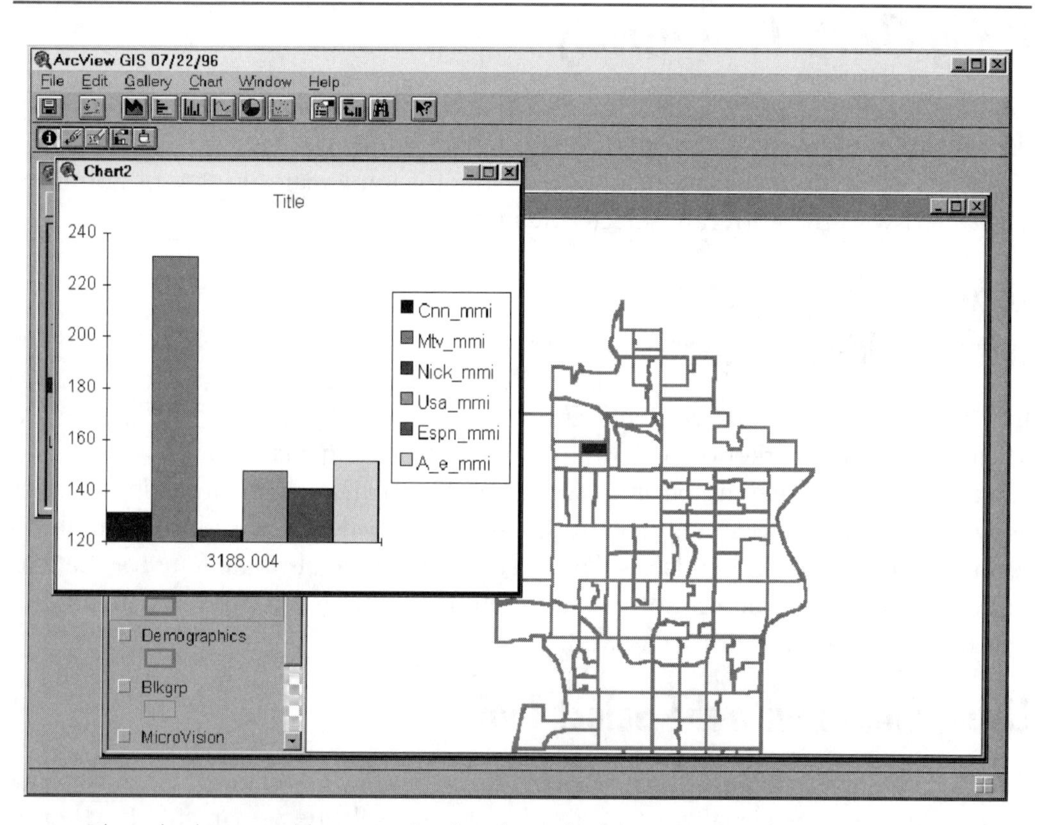


*Chart displaying cable viewership by channel relative to national index of 100 for neighborhood near Arizona State University (Tempe, Arizona).*

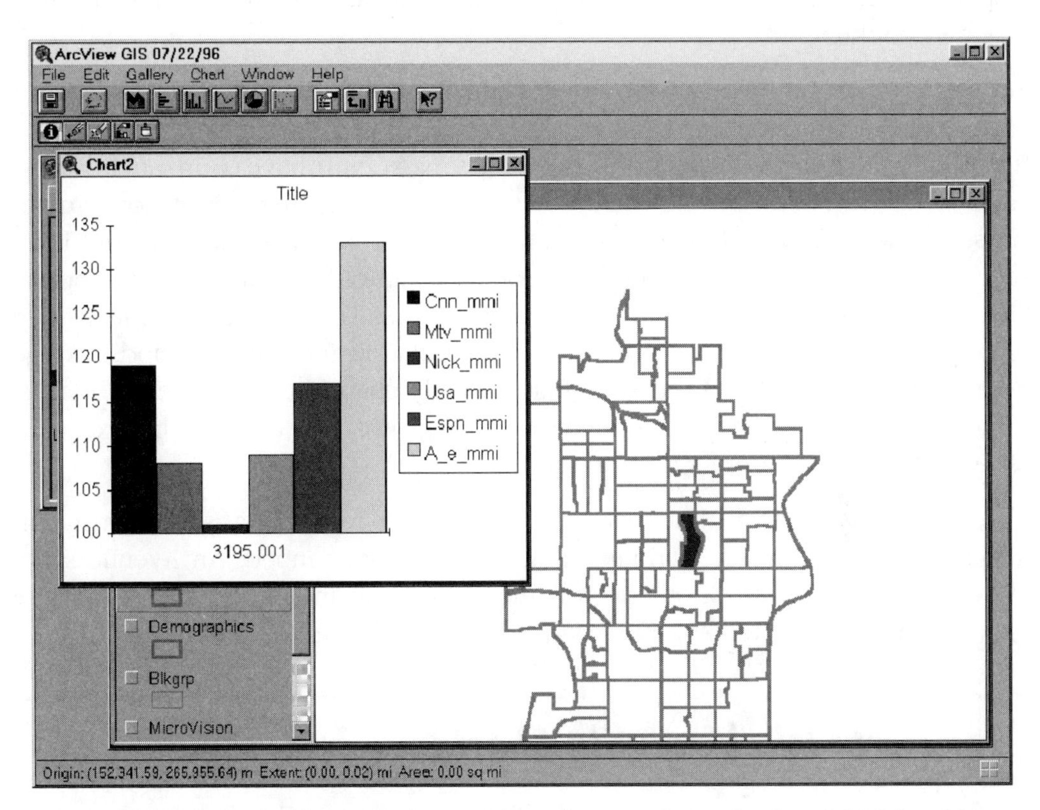


*Bar chart showing cable channel viewership relative to national index of 100 for older, residential neighborhood in Tempe, Arizona.*

# More on Chart Characteristics

You have seen the strengths of charts in ArcView, namely, their dynamic properties and the ability to use them in interactive queries. As mentioned earlier, however, these strengths are not without limitations.

The primary limitation is that charts are tied to specific records in a specific table. While charts are very effective for depicting specific values, they do not handle ranges well. In brief, while you can easily classify themes on attribute values and display the results of these classifications, you will find it difficult to chart attributes grouped on thematic classes.

For example, assume that you want to chart the average household income for each of five classes resulting from a classification on population growth. You must first add an item to the theme's attribute table. Next, you have to manually select each set of features from the theme that corresponds to the population growth classes, code each record in each set with a class identifier, and then use ArcView's statistical capabilities to calculate the average household income for each population growth class, storing the results in a new table. The new table contains five records with fields that correspond to the population growth class and average household income. This is the table that is finally charted to produce the results. Although this operation is not difficult, it is also not one that can be executed repeatedly without careful concentration.

If you need to perform the operation above on a regular basis, customization through ArcView's Avenue scripting language is recommended. An Avenue script can be used to automate an otherwise cumbersome task and integrate it into a larger ArcView project. Selected Avenue features are examined in Chapters 11 and 14.

# Exercise 6: Working with Charts

After the exercise in Chapter 6, you may have thought you had taken this project about as far as it could go. In reality, much territory remains to be explored, some of which is ideally suited for ArcView's charting capabilities.

1. Begin by opening the *ch7.apr* project file in your working directory.
2. Turn off all themes, and then turn on the *MicroVision* theme. Click on the theme to make it active.
3. Copy the *MicroVision* theme using the Copy Themes and Paste selections from the Edit pulldown menu. Click on the copy to make it the active theme.
4. Access Theme Properties and change the name of the new theme to *Mv_spent.*

5. Note that the theme still displays seven block groups, as well as the text labels identifying the MicroVision segments and competitor restaurants. The block groups and text labels were copied along with the theme. To delete these elements, select them using the Graphics Select tool (the arrow), and then choose Delete Graphics from the Edit menu.

6. Click on the Clear Selected Features icon from the button bar to remove all features from the selected set. Turn off the *MicroVision* theme.

7. Although the *Mv_spent* theme is still turned on, the only thing visible is the three-mile radius circle. This is because the color for the *Mv_spent* theme is set to *transparent,* and the outline line width is set to None. The result of these two property settings is that the theme is invisible. Double-click on the theme to access the Legend Editor and change the outline width back to *1.* After you click on Apply, the block group outlines should be redrawn.

To further investigate the restaurant customers, your first objective will be to associate the MicroVision segments with the amount spent. In the previous exercise, you produced a summary table calculating the average amount spent by block group (*spent.dbf*). By joining the summary table to the *Mv_spent* theme, you can examine the average amount spent by each MicroVision segment.

1. Open the attribute table for the *Mv_spent* theme.

2. Click on Tables in the project window. From the list of available tables, select and open *spent.dbf.*

3. Highlight the *Blkgrp* field on both tables. Join the *spent.dbf* table to the *Attributes of Mv_spent* table.

4. Select the Query tool. Using the Query Builder, construct the query *Count > 0,* and click on New Set. This action will select the block groups for which there were survey responses. The status line shows that 20 of 167 features are selected. The selected features are highlighted on the view as well as in the *Mv_spent* attribute table.

Instead of classifying the data as you would for a map, generate a chart for the purpose of examining the relationship between the MicroVision segment and the amount spent.

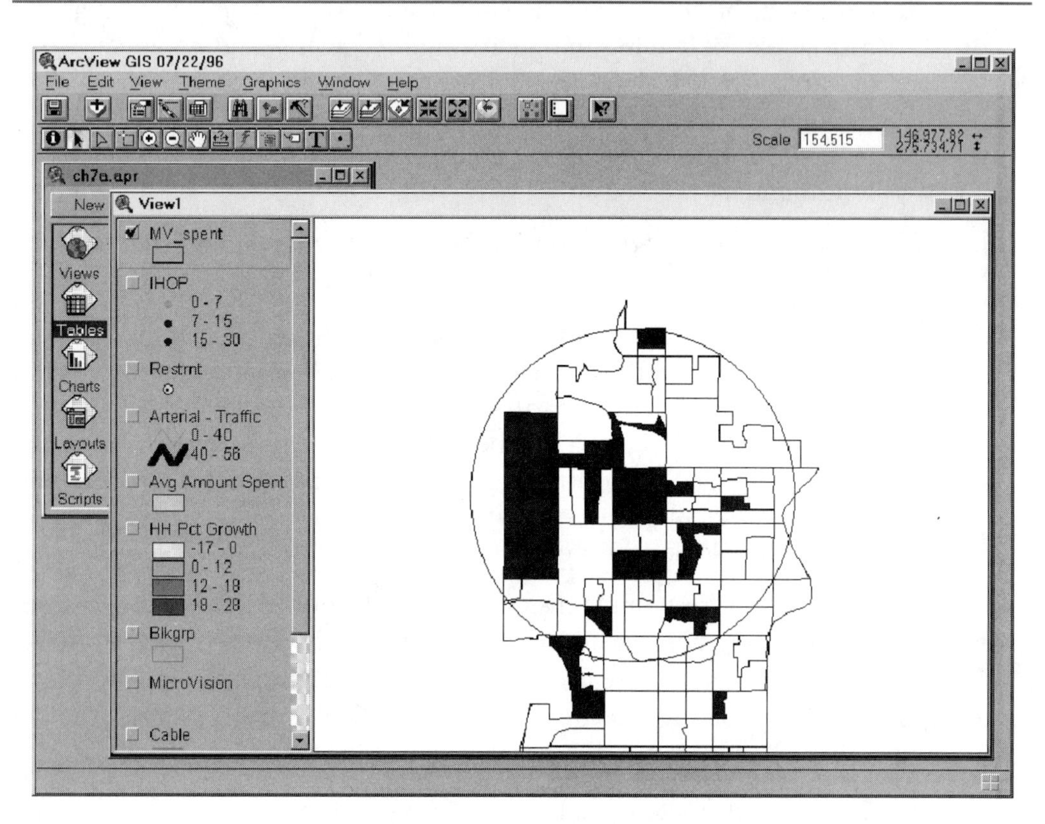


*Selected features from the Mv_spent theme.*

1. From the project window, select the Charts icon. Then click on New.
2. A dialog window prompts you to pick a table. From the pulldown list, select *Attributes of Mv_spent*. Click on OK.
3. You have accessed the Chart Properties dialog window. From the scroll list for Fields, select *Prim_mv* and *Ave_spent*. Click on Add to add each to the list of Groups.
4. From the Label Series Using list, select *Prim_mv*. For the name, enter *MV Spent*. Click on OK to accept these choices.

*Chart Properties dialog window.*

Chart Properties
Name: MV Spent
Table: blkgrp.dbf
Fields:
Blkgrp
Prim_mv
Second_mv
Count
Ave_Spent
Add
Delete
Groups:
Prim_mv
Ave_Spent
Label series using:
Prim_mv
Comments:
OK
Cancel

You may be wondering why you are working with two variables when you only want to plot *Ave_spent.* The reason is that when charting from a single field, ArcView will not allow you to label the bar lines from one field with the values from another. There is a workaround, but it requires beginning with two fields.

By default, you are presented with a bar chart whose data series have been formed from the selected records. Two data groups are present (corresponding to the charted fields), *Prim_mv* and *Ave_spent.*

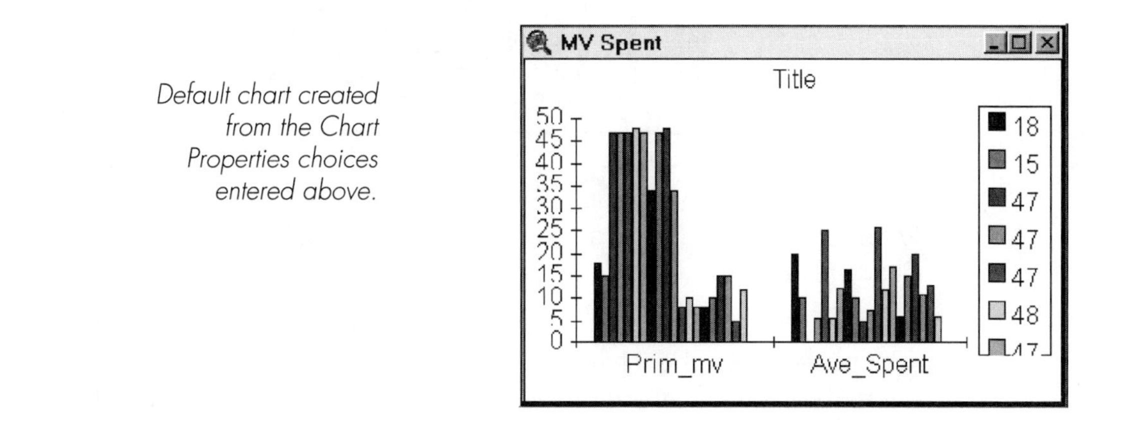


*Default chart created from the Chart Properties choices entered above.*

The chart above is not very effective. You can improve the chart by changing the basis of the series from *records* to *fields.*

1. Click on the Series From Records/Series From Fields icon on the button bar.

*Series From Records/ Series From Fields button icon.*

2. The chart now displays MicroVision expenditures. Resize the window as needed to make the chart more legible.

*New chart with data series formed from fields.*

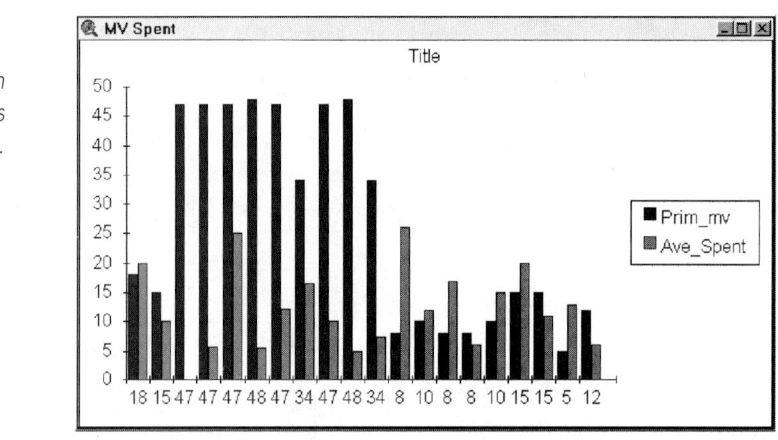


3. Return to the Chart Properties window by selecting Properties from the Chart menu.

4. Click on *Prim_mv* from the list of series. Select Delete and then click on OK to apply this change. The chart series is still formed from fields, but now only one field, *Ave_spent,* is displayed. The data

markers, however, are still labeled with the values for *Prim_mv*, which is just what you want. This is the outcome of the workaround mentioned earlier. (The incremental project has been saved as *ch7a.apr.*)

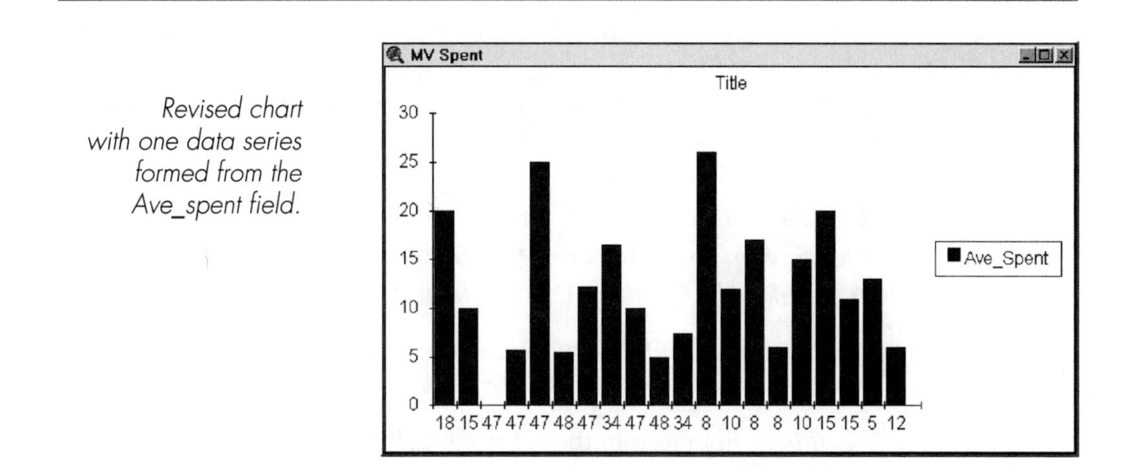


*Revised chart with one data series formed from the Ave_spent field.*

A closer look at the chart above reveals that several records have the same value for *Prim_mv*. What you would *really* like to see, however, is a chart showing average amount spent by *Prim_mv*. To accomplish this, the data must be further manipulated.

1. Close the chart and return to the view.
2. You need a new summary table by block group, but his time you will *sum* rather than *average* the amount spent. To produce this summary table, you first need to recreate the spatial join used in Chapter 6 (Exercise 5).
   - ❒ Open the attribute tables for the *IHOP* and *Blkgrp* themes.
   - ❒ Highlight the Shape field in each table.
   - ❒ Click on the title bar of the *Attributes of IHOP* table to make it the active (or destination) table, and select Join from the Table pulldown menu. As before, the *Blkgrp* attribute is now associated with every response point in the *IHOP* theme.

3. To make *Blkgrp* the summary field, highlight this field in the *Attributes of IHOP* table.
4. Click on the Summary tool to create a summary table.
5. For Field, select Spent; for Summarize By, select Sum.
6. Click on Add. The entry *Sum_Spent* is added to the summary field list. Enter the output file information, naming the file *sumspent.dbf*. Click on OK. (Verify that the *sumspent.dbf* table is added to your working directory.)

**NOTE:** *As before, you need to remove the spatial join before saving the project. Click on the title bar of the Attributes of IHOP table to make it active and select Remove All Joins from the Table pulldown menu.*

7. Close the *Attributes of IHOP* table and then open the attribute table for the *Mv_spent* theme.
8. Use the *Blkgrp* field to join the *sumspent.dbf* table to the *Attributes of Mv_spent* table.

You are now ready to create the new summary table. If the 20 records selected previously from the *Attributes of Mv_spent* table are no longer selected, reselect them using the query mentioned above.

1. Highlight the *Prim_mv* field in the *Attributes of Mv_spent* table, and click on the Summary tool.
2. For Field, select *Sum_spent* and *Count*. For both, select Summarize by Sum. Add both to the summary field list, and save this new table with the name *mvsumsp.dbf*. Click on OK to create the table.

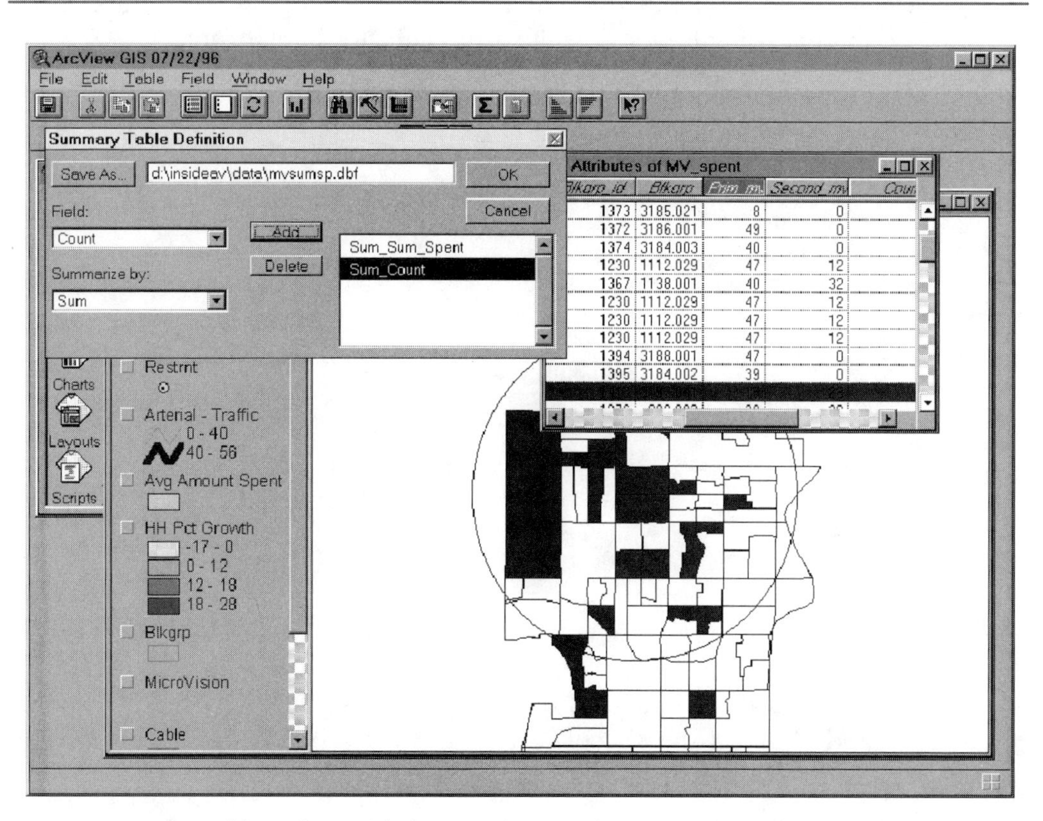


Summary Table Definition dialog window used to create the mvsumsp.dbf table.

The resultant table should contain nine records, each associating a *Prim_mv* class with the total amount spent, as well as reporting the total count of survey responses for the *Prim_mv* class. However, the chart should also contain the per capita amount spent, or the total amount spent divided by the total count. These fields already exist in the table, but you need to add a new field to contain the results of the division.

By default, ArcView prohibits table edits. However, you can add a field to a table by first making the table *editable*.

1. From the Table pulldown menu, select Start Editing. Note that the column heading font style shifts from italic to regular font.

2. From the Edit pulldown menu, select Add Field. The Field Definition dialog window pops up.

3. Enter *Pcap_spent* for the field name, select Number for the type, set the width at *8*, and set decimal places at *2*. Click on OK to add the field to the table.

*Field Definition dialog window.*

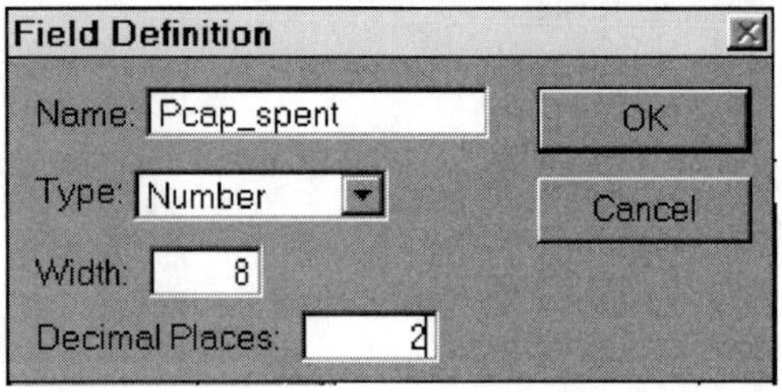

4. To calculate the new values for *Pcap_spent,* highlight the field name and click on the Calculate tool from the button bar (the tool that resembles a calculator).

5. The Field Calculator dialog window looks much like the Query Builder window. Form the expression *Pcap_spent = Sum_Sum_Spent/ Sum_Count* by double-clicking on the appropriate Fields and Requests. Note that the first part of the expression (*Pcap_spent*=) has already been formed within the field calculator. When you have properly formed the expression, click on OK. The results of the calculation are immediately written to the target field.

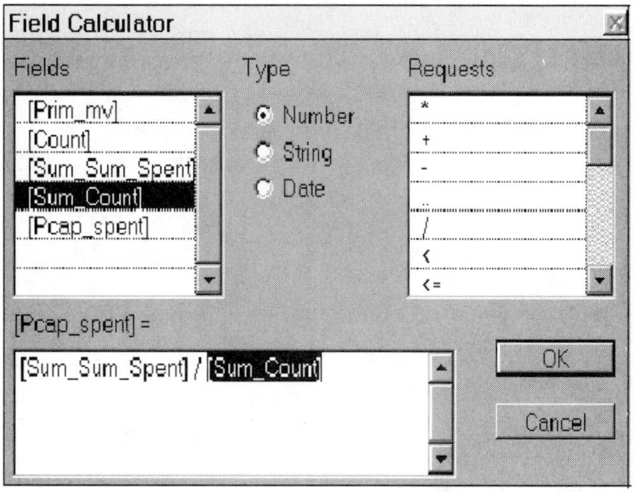

*Field Calculator dialog window.*

6. Examine the table to ensure that your results are as anticipated.

*Results of the Pcap_spent calculation.*

mvsumsp.dbf

| Prim_mv | Count | Sum_Sum_Spent | Sum_Count | Pcap_spent |
|---|---|---|---|---|
| 2 | 1 | 6.0000 | 1.0000 | 6.00 |
| 5 | 1 | 13.0000 | 1.0000 | 13.00 |
| 8 | 3 | 98.0000 | 6.0000 | 16.33 |
| 10 | 2 | 39.0000 | 3.0000 | 13.00 |
| 15 | 3 | 41.0000 | 3.0000 | 13.67 |
| 18 | 1 | 20.0000 | 1.0000 | 20.00 |
| 34 | 2 | 104.0000 | 11.0000 | 9.45 |
| 47 | 5 | 212.0000 | 20.0000 | 10.60 |
| 48 | 2 | 16.0000 | 3.0000 | 5.33 |

7. To lock the table and prevent further editing, select Stop Editing from the Table pulldown menu, answering Yes to the save prompt. When you do so, the field names for the table will return to the italicized font style.

You now have the field you want to chart. To create a new chart, take the following steps:

1. From the Project menu, click on the Charts icon and then the New button. Select the *mvsumsp.dbf* table.
2. For Fields, select *Prim_mv* and *Pcap_spent*. For Label Series Using, select *Prim_mv*.
3. Give the chart a meaningful name, such as *Per Capita Spent by MicroVision Segment*. Click on OK to apply.
4. Switch the series from records to fields by selecting Series From Fields from the Chart menu.
5. Access the Chart Properties window again. Delete the group *Prim_ mv* from the chart, and apply. (The incremental project has been saved as *ch7b.apr*.)

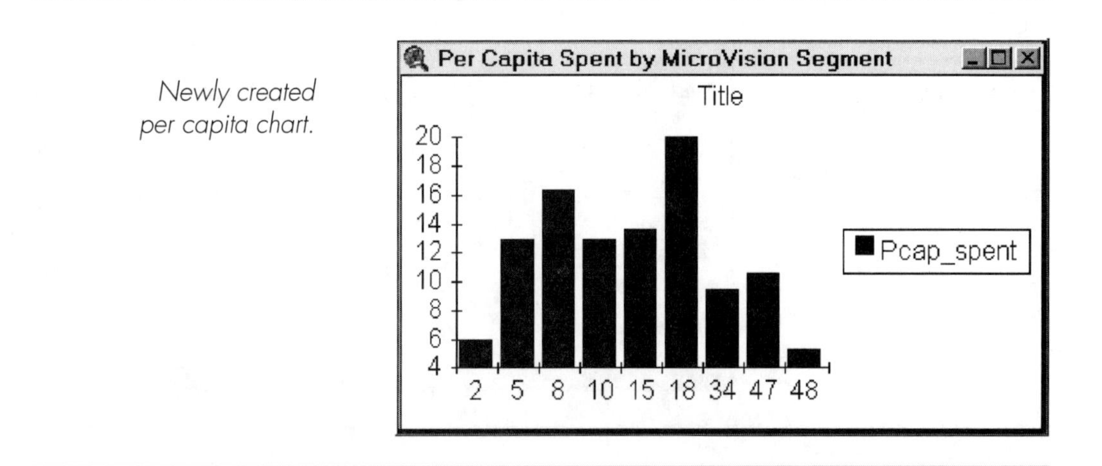


*Newly created per capita chart.*

Two peaks are apparent: one corresponds to MicroVision segment 8, "Movers and Shakers," and one to segment 18, "White Picket Fences." This type of chart will be useful in the future to determine per capita spending in neighborhoods not covered in Tempe. Theoretically, as long as a neighborhood's segment is known, its value can be inferred from the analysis performed here.

Now, to generate some advertising ideas, you can explore the viewing habits of the MicroVision segments. You will create a new table from the attributes of *Mv_spent* that contains only a few selected fields from the 20 records observed in Tempe.

1. Open the attribute table for *Mv_spent.* The 20 records from the earlier query should still be selected.
2. Access the Table Properties window for the attribute table by selecting Properties from the Table pulldown menu. When exporting data to a new table, only those fields checked as visible will be exported. Check the following fields as visible:
   - *Blkgrp*
   - *Prim_mv*
   - *Second_mv*
   - *Ave_spent*
   - *Sum_Spent*
   - *Count*
3. Click on OK to apply the above choices to the table.
4. Select Export from the File pulldown menu. In the Export Table window, select dBase (the default) and click on OK.
5. You will now be prompted to specify a directory and file name. Name the file *mvsel.dbf,* and click on OK to export the table. The table has been created.
6. At this point, you need to add the new table to your project. Click on the project window to make it active. From the Project pulldown menu, select Add Table. Navigate to the directory into which you saved the *mvsel.dbf* file and add the table.

You are now ready to associate the above information with the cable demographics. This process is probably starting to sound familiar.

1. Copy the *Cable* theme. Change the name of the new theme to *Cable_mv.*
2. Open the attribute table for *Cable_mv.* Join the *mvsel.dbf* table to the *Cable_mv* attribute table, highlighting the *Blkgrp* field as the Join field.
3. Using the Query Builder, construct the query *Count > 0* to extract the 20 block groups of the surveyed set.

*Table Properties dialog window showing the field options for the Attributes of Mv_spent table.*

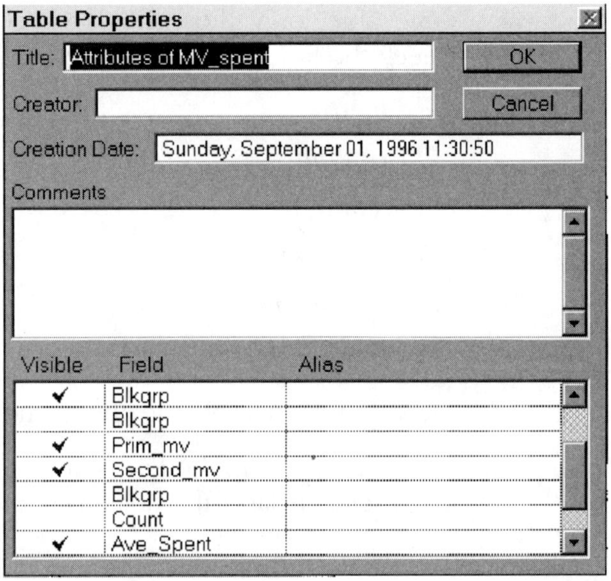

Table Properties
Title: Attributes of MV_spent
OK
Creator:
Cancel
Creation Date: Sunday, September 01, 1996 11:30:50
Comments

| Visible | Field | Alias |
|---|---|---|
| ✓ | Blkgrp | |
| | Blkgrp | |
| ✓ | Prim_mv | |
| ✓ | Second_mv | |
| | Blkgrp | |
| | Count | |
| ✓ | Ave_Spent | |

4. With the MicroVision table set, you now want to create a summary table. Specifically, you want to aggregate the average values for MMI—an index that measures specific market share against a national norm—for several cable channels by MicroVision segment.

5. Highlight the *Prim_mv* field on the *Attributes of Cable_mv* table, making *Prim_mv* the summary field.

6. Click on the Summary tool to access the Summary Table Definition dialog window. For each of the following fields—*Cnn_mmi, Mtv_mmi, Nick_mmi, Usa_mmi, Espn_mmi,* and *A&e_mmi*—select the field, select Average as the Summarize By option, and click on Add to add each to the list.

7. Save the summary table as *cablemv.dbf.*

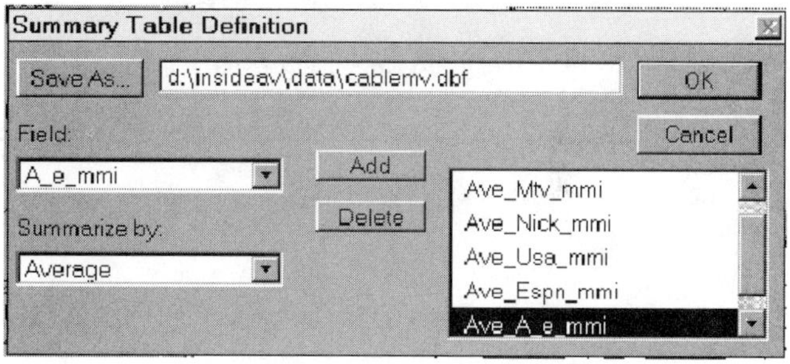

*Summary Table Definition choices for creating the cablemv.dbf table.*

An output table containing nine records, one for each MicroVision segment, is created. You need the per capita amount spent by MicroVision segment, but rather than calculate it, you can simply join the table containing the per capita calculations to the *cablemv.dbf* table. To carry out the join, take the following steps:

1. Click on the Tables icon in the project window and open the *mvsumsp.dbf* table.

2. Highlight the *Prim_mv* field in both tables, and verify that the *cablemv.dbf* table is the destination table. (The active table is the destination table.) Select Join from the Table menu to perform the join.

3. Scroll through the *cablemv.dbf* table to examine the results of the join.

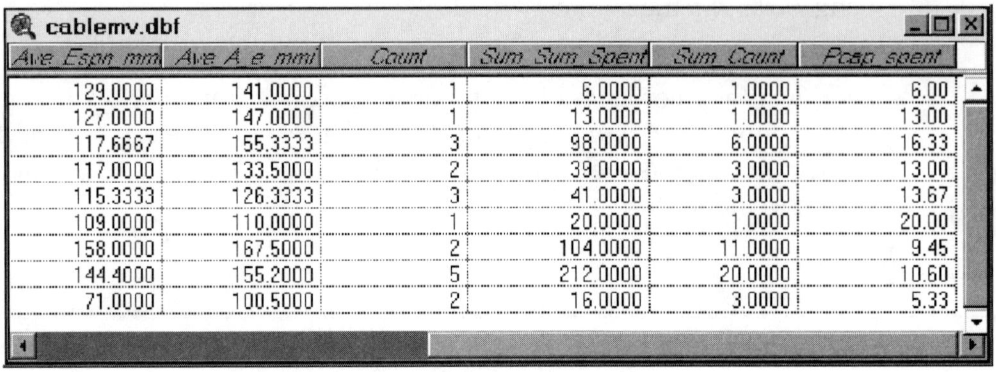

cablemv.dbf

| Ave_Espn_mmi | Ave_A_e_mmi | Count | Sum_Sum_Spent | Sum_Count | Pcap_spent |
|---|---|---|---|---|---|
| 129.0000 | 141.0000 | 1 | 6.0000 | 1.0000 | 6.00 |
| 127.0000 | 147.0000 | 1 | 13.0000 | 1.0000 | 13.00 |
| 117.6667 | 155.3333 | 3 | 98.0000 | 6.0000 | 16.33 |
| 117.0000 | 133.5000 | 2 | 39.0000 | 3.0000 | 13.00 |
| 115.3333 | 126.3333 | 3 | 41.0000 | 3.0000 | 13.67 |
| 109.0000 | 110.0000 | 1 | 20.0000 | 1.0000 | 20.00 |
| 158.0000 | 167.5000 | 2 | 104.0000 | 11.0000 | 9.45 |
| 144.4000 | 155.2000 | 5 | 212.0000 | 20.0000 | 10.60 |
| 71.0000 | 100.5000 | 2 | 16.0000 | 3.0000 | 5.33 |

*Results of joining the mvsumsp table to the cablemv table.*

You are now set to chart the relationship between MicroVision segments and their viewing habits.

1. Click on the Charts icon in the project window, and select New to create a new chart.
2. Select the *cablemv.dbf* table as the one to chart.
3. For Fields, select *Prim_mv, Ave_Cnn_mmi, Ave_Mtv_mmi, Ave_Nick_mmi, Ave_Usa_mmi, Ave_Espn_ mmi, Ave_A&e_mmi,* and *Pcap_spent.*
4. For the Label Series Using option, select *Prim_mv.* Give the chart a meaningful name, such as *Cable - MicroVision Segments.* Click on OK.
5. Toggle to the Series Formed from Fields option to obtain a more useful view of the data.

The bar chart looks promising, but this time try a line chart. Click on the Line Chart Gallery icon on the button bar. Select the second chart style from the six choices available. Click on OK to create the chart. (The incremental project has been saved as *ch7c.apr.*)

*Line chart formed from the cablemv.dbf table.*

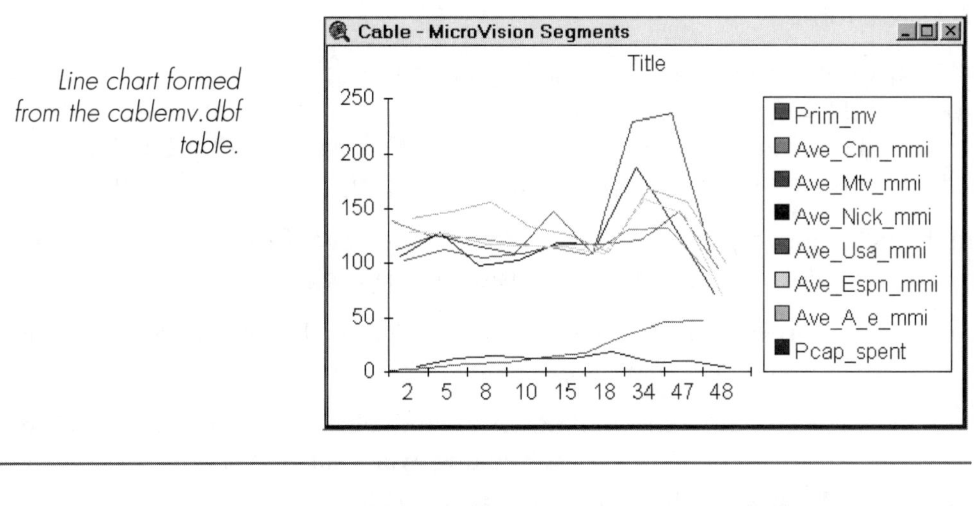


*Same line chart formed from the cablemv.dbf table, but plotted on a logarithmic Y axis.*

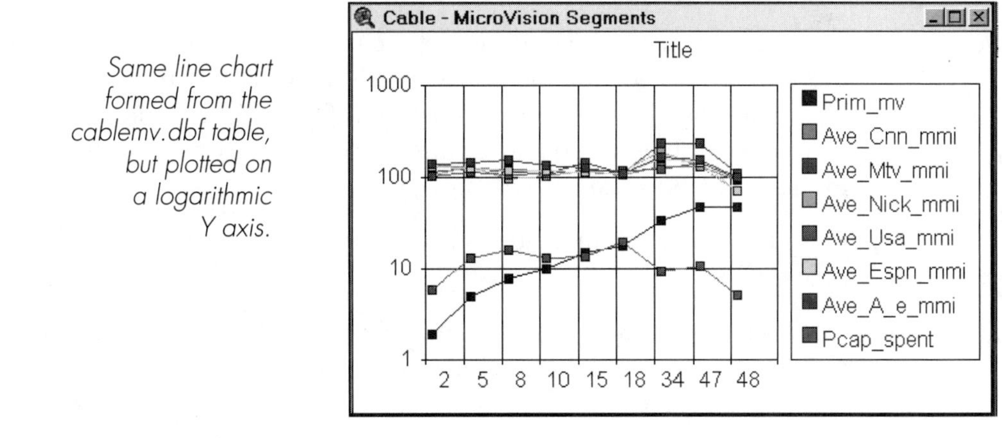


Although you can see clear fluctuations in the cable channel MMI values, the corresponding fluctuation in *Pcap_spent* is difficult to discern. One option is to access the line chart options again and choose the chart style on the bottom in which the Y axis is plotted logarithmically. This option nicely depicts the fluctuation in *Pcap_spent,* but the corresponding fluctuations in the MMI values are now muted.

Take a closer look at the data. The MMI values pertain to local market share relative to the national norm for each channel. The values are percentages, with the national norm expressed as 100. What if you transform the *Pcap_spent* values to the same format?

After performing a simple calculation, you find that the average per capita amount spent is $11.20. To express each *Pcap_spent* value as a percentage of the norm, you need to divide the value by 11.2 and multiply the result by 100. To avoid two calculations for each value, you can multiply *Pcap_spent* by 8.93 (100/11.2 = 8.93).

In the following steps, you will convert the *Pcap_spent* field values to the same format as the MMI values, and then update the chart.

1. Open the *cablemv.dbf* table, or click on the table to make it active if it is already open.
2. From the Table pulldown menu, select Start Editing. As before, the field names that allow editing will change from an italic style to a regular font.

   The field names on the right side of the table are still italicized because the current version of the *cablemv.dbf* table is the result of a join between this table and the *mvsumsp.dbf* table. In ArcView, all edits on tables that have been joined must be performed on the *source* table, in this case, the *mvsumsp.dbf* table.
3. From the Table menu, select Stop Editing, answering No to the save prompt.
4. Open the *mvsumsp.dbf* table, and select Start Editing from the Table pulldown menu. The field names are now shown in the normal font style.
5. From the Edit menu, select Add Field. Name the new field *Pcavg_spent,* making it a number field, with a width of *8*, and *0* decimal places. Click on OK to add this field to the table.
6. Highlight the *Pcavg_spent* field. Select the Calculate tool to call up the Field Calculator dialog window.
7. Form the expression *Pcap_spent * 8.93* by clicking on *Pcap_spent* in the Field list, double-clicking on * in the Requests list, and keying in *8.93*. When you have the expression formed correctly, click on OK.

8. From the Table menu, select Stop Editing, answering Yes to the save prompt. The new field that contains the results of this calculation is written to the *mvsumsp.dbf* table.

9. Close the *mvsumsp.dbf* table. Open the *cablemv.dbf* table.

10. From the Table pulldown menu, select Refresh. This action updates the joined table with changes made in respective source tables. The new *Pcavg_spent* field now appears in the *cablemv.dbf* table, and it contains the newly calculated values.

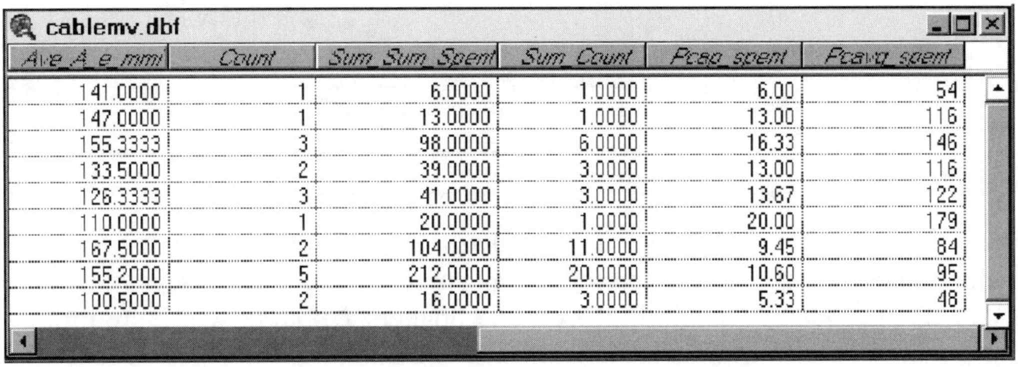

cablemv.dbf

| Ave_A_e_mmi | Count | Sum_Sum_Spent | Sum_Count | Pcsp_spent | Pcavg_spent |
|---|---|---|---|---|---|
| 141.0000 | 1 | 6.0000 | 1.0000 | 6.00 | 54 |
| 147.0000 | 1 | 13.0000 | 1.0000 | 13.00 | 116 |
| 155.3333 | 3 | 98.0000 | 6.0000 | 16.33 | 146 |
| 133.5000 | 2 | 39.0000 | 3.0000 | 13.00 | 116 |
| 126.3333 | 3 | 41.0000 | 3.0000 | 13.67 | 122 |
| 110.0000 | 1 | 20.0000 | 1.0000 | 20.00 | 179 |
| 167.5000 | 2 | 104.0000 | 11.0000 | 9.45 | 84 |
| 155.2000 | 5 | 212.0000 | 20.0000 | 10.60 | 95 |
| 100.5000 | 2 | 16.0000 | 3.0000 | 5.33 | 48 |

*Resultant cablemv.dbf table, showing the edits made to mvsumsp.dbf.*

11. Access the Chart tool again and select *cablemv.dbf* as the table to chart.

12. Select the following fields to chart: *Prim_mv, Ave_Cnn_mmi, Ave_Mtv_mmi, Ave_Nick_mmi, Ave_Usa_mmi, Ave_Espn_mmi, Ave_A&e_mmi,* and *Pcavg_spent.*

13. Under the Label Series Using field, select *Prim_mv.*

14. Give the chart a snazzy name, such as *New Cable - MicroVision Segments.*

15. Choose the line graph option. As before, toggle to Series Formed from Fields. This time, both fluctuations are pronounced.

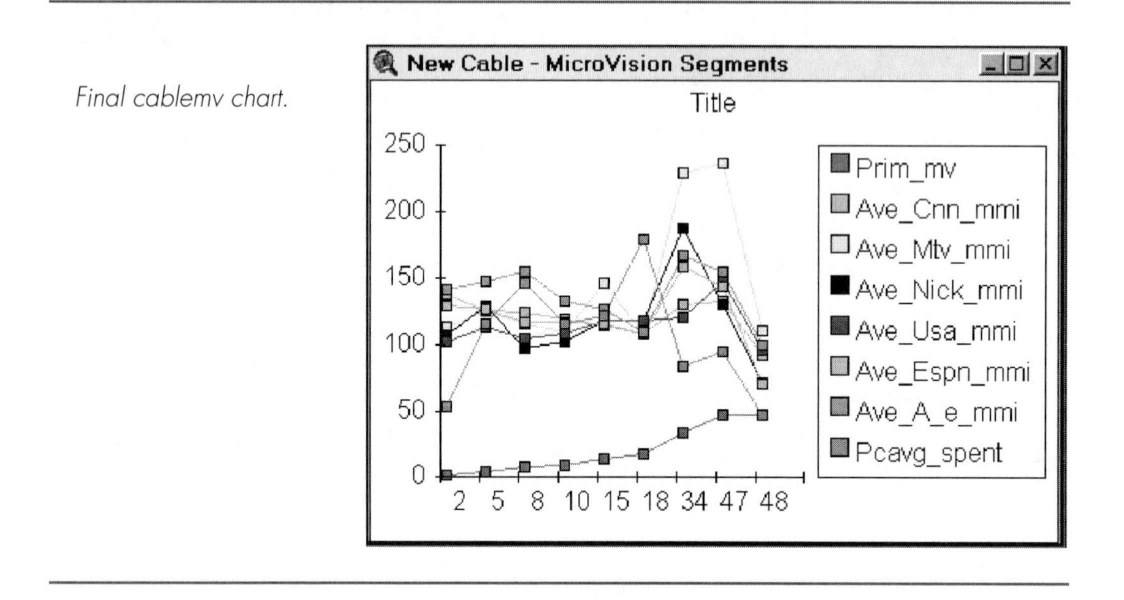


*Final cablemv chart.*

While no cable channel line precisely follows the amount spent line, the A&E line matches the corresponding peaks in amount spent for MicroVision segments 5 and 8 ("Prosperous Metro Mix" and "Movers and Shakers"), both of which are promising groups to target. The CNN Network, while exhibiting lower peaks, follows a similar trend. Both channels warrant consideration as possible candidates for an advertising campaign, coupled, of course, with the outdoor advertising locations identified in the previous exercise.

We urge you to learn and experiment with charting. In a world where "a picture paints a thousand words," charts can be a powerful addition to analyses and presentations.

## Chapter 8

# Layouts

Up to this point, you have worked with views, tables, and charts. In the course of working on a project, you have manipulated themes and linked components together. The next step is to put it all together in a layout.

## What Is a Layout?

A layout is a map that combines ArcView project components—views, charts, and tables—with additional elements, such as legends, scale bars, north arrows, and graphics, to form a single output document.

## Why a Layout?

At this point, you may be wondering why you need to create a new document to contain what you already have displayed on your screen. Doesn't an ArcView project already do a fine job of organizing these components? Why not simply print components by selecting Print from the File menu?

To best answer these questions, we need to discuss not only what a layout is, but how it differs from other ArcView components, and how these differences translate into strengths that you can use.

A layout conceptually prepares your work for hard copy output. It allows you to design your map to fit the format of your specific printer or plotter. You have control over the page size, the page orientation, and the output resolution.

A layout allows you to group ArcView project components and additional map elements through the use of *frames*. A frame is a container for a specific map element. A frame can contain an ArcView project component, such as a view or chart, as well as special map layout elements such as legends or scale bars.

A frame can be dynamically linked to corresponding ArcView project components. For example, if a view is updated in an ArcView project, the view, legend, and scale bar frames will be updated as well. If a chart format is modified, the chart frame will also be updated. Frames also allow you to combine project components that are displayed in different windows, such as views and charts, into a single document.

A layout can also be designed to serve as a *template* for future map production. This is done by creating a layout of empty frames, or frames that have not yet been linked to any ArcView project component. Templates can be stored for future use, at which time the links can be established either interactively or through the use of an Avenue script, thereby facilitating automated map production.

# How to Make a Layout

Begin by clicking on the Layouts icon in the project window. Selecting New will create a default layout.

Next, you need to define the graphics page. The graphics page should correspond to the format you desire for the final output. To access the Page Setup dialog window, select Page Setup from the Layout menu. Page parameters, including page size, page orientation, margins, and output resolution are set in the Page Setup dialog window.

Third, add your map elements (frames and other graphics) and link them to your ArcView project components. For ease of viewing, we recommend that you stretch the layout window to make it as large as possible, and then fit the layout to the page by selecting Zoom to Page from the Layout menu. At this point, map elements are positioned. Frames and primitive graphics alike can be selected and moved by using the Graphics select tool. Save your work and print your map. That is all there is to it.

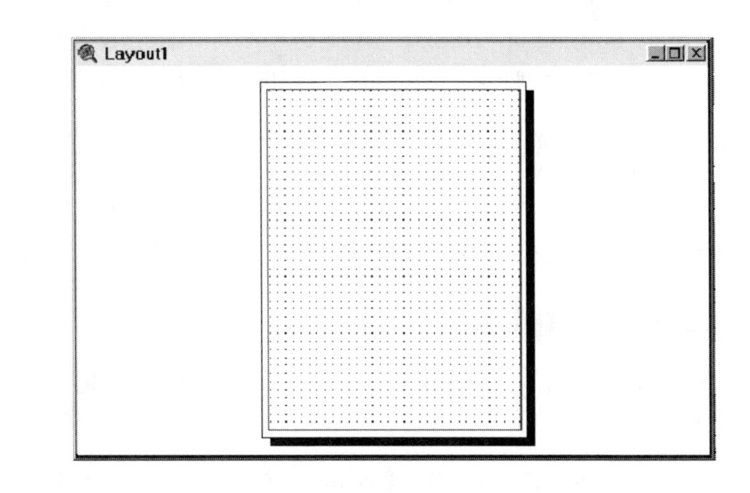


*Default layout format.*

Page Setup
Page Size: Same as Printer
Units: Inches
Width: 8.5 Height: 11
Orientation:
Margins: Use printer border
Top: 0.208333 Left: 0.193333
Bottom: 0.195 Right: 0.166667
Output Resolution: Normal
OK Cancel

*Page Setup dialog window.*

# Layout Frames in More Detail

As described above, a frame is a container for a specific map element. There are seven types available in a layout: view frames, chart frames, table frames, legend frames, scale bar frames, north arrow frames, and picture frames. View frames, chart frames, and table frames contain representations of the corresponding Arc-

View project components. Legend frames, scale bar frames, and north arrow frames contain specialized map elements, usually linked to a corresponding view frame. Picture frames contain graphics created by importing a graphics file.

To create a layout frame, click on the Frame Tool icon at the far right of the tool bar. Hold down the mouse button to reveal several icons for selecting a frame type. Choose the frame icon you want. When you move the cursor to the layout window, it will change shape. Then drag a box on the layout to place the frame.

## Layout Display Options

In addition to the many parameters described below, two display properties can be set for all frame types.

The first property controls when the frame is refreshed. Its two states are When Active and Always. In When Active, the frame is refreshed only when the layout is the active project component. The Always state ensures that the frame is refreshed whenever changes occur in the linked project component. The Always option allows you to make changes in the active view window while watching those changes be applied to the open Layout window.

The second display property controls how the frame is displayed. Its two states are Presentation and Draft. When set to Presentation, the associated project component or map element will be displayed. When set to Draft, a shaded box is displayed for each frame. These show the location of each frame on the layout, along with the name of the frame. The Draft state can speed redrawing of the screen, thereby accelerating map composition.

## View Frames

*View Frame tool icon.*

A view frame contains a representation of a view from your ArcView project. The frame can be either linked to the view at the time of creation, or left empty to be linked at a later date. View frame links can either be *live* or *static*. A live link results in dynamic changes between the view frame and the corresponding view: changes in the view will be reflected in the view frame in the layout. If the live link is not selected, the view is static, representing a snapshot of the view at the time the view frame was created.

You can also set the scale parameters of a view frame. By default, the entire view is scaled to fit in the corresponding view frame in the layout. By selecting Preserve View Scale from the Scale pulldown list, the view will be displayed at the same scale in both the view and the layout. Therefore, depending on the size of the view frame in the layout, only a portion of the view may be visible.

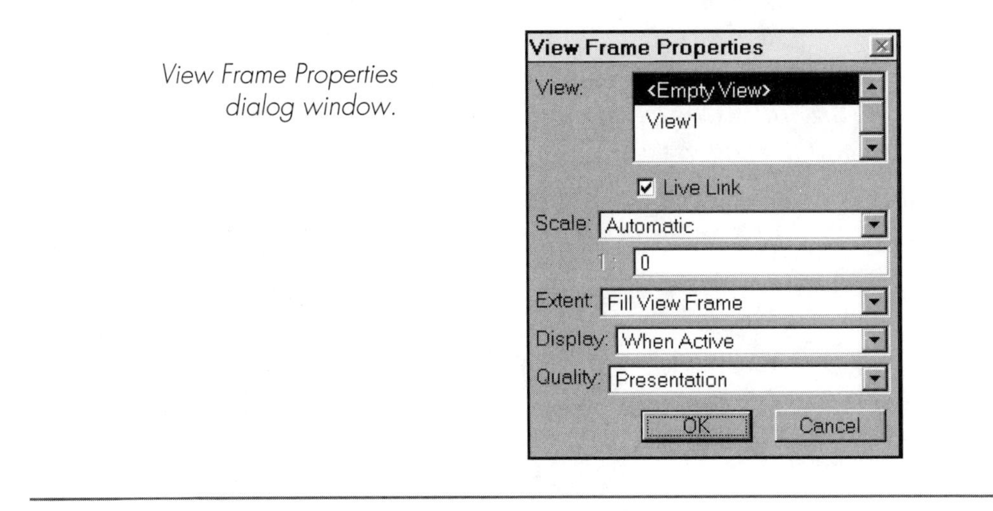


*View Frame Properties dialog window.*

# Legend Frames

*Legend Frame tool icon.*

A legend frame allows you to replicate a view's Table of Contents on a layout. Only those themes currently drawn in the view will be included in the view legend. Legend frames will generally correspond to a view frame. They can either be linked to a view frame at the time of creation, or left empty to be linked at a later date. If the legend frame is linked to a view frame that is "live-linked" to its view, the legend will be updated any time the view and associated view frame are updated. If the view frame is static, the legend will stand as a snapshot of the view at the time of creation.

> **NOTE:** *The Simplify selection from the Graphics menu allows frames, such as the legend frame, to be decomposed into separate graphic elements. In this manner, the legend frame can be broken into separate graphic elements that can then be edited, and legend entities can be moved or deleted as desired. However, when simplifying the legend, the live link to the view frame is broken and cannot be re-established, even if the edited legend elements are subsequently regrouped into a single legend graphic. Layout frames that can be simplified include charts, tables, scale bars, north arrows, legends, and views.*

*Legend Frame Properties dialog window.*

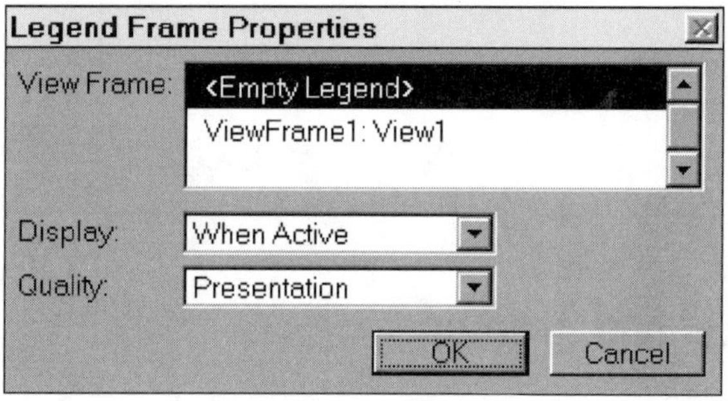

# Scale Bar Frames

*Scale Bar Frame tool icon.*

A scale bar frame creates a scale bar on your layout. Several styles are available for creating scale bars. Like legends, scale bars usually correspond to a view frame. They can either be linked to a view at the time of creation, or left empty to be linked at a later date. As with a view frame, the link can be either live or static. If the link is live, the size of the scale bar will change if the extent of the view is changed by zooming in or out. The size of the scale bar will also be changed if the size of the corresponding view frame is changed on the layout.

*Scale Bar Properties dialog window.*

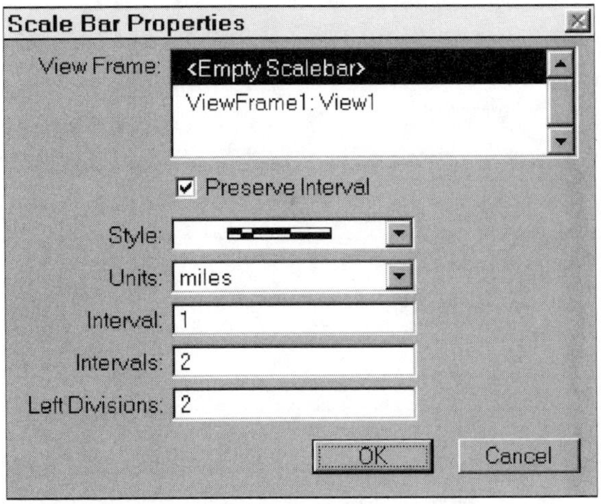

Scale Bar Properties

View Frame: <Empty Scalebar>
ViewFrame1: View1

Preserve Interval

Style:
Units: miles
Interval: 1
Intervals: 2
Left Divisions: 2

OK Cancel

# North Arrow Frames

North Arrow Frame tool icon.

A north arrow frame creates a north arrow on your layout. A scrolling list displays the styles of north arrows available on your system. If you desire, you can specify a rotation angle when the north arrow is created.

➾ **NOTE:** *The Store North Arrows selection from the Layout menu allows the currently selected graphic element to be stored as a north arrow. The stored graphic is then available to add to any layout by adding a north arrow frame to that layout. In this manner, graphics that are used frequently, such as locator maps, corporate logos, or standard legends, can be added quickly to new maps, or incorporated into canned map templates.*

# Chart Frames

Chart Frame tool icon.

Like other frames, a chart frame contains a representation of a chart from your ArcView project. A chart frame can either be linked to a chart at the time of creation, or left empty to be linked at a later date.

A chart frame is always live-linked to the corresponding chart in your project. If the chart in your project is closed, the chart frame will represent the chart as a solid rectangle that contains the name of the chart. When the chart in your project is open, the chart frame will display the current state of the chart. When the chart format is changed, the format of the chart in the chart frame will change as well.

*Chart Frame Properties dialog window.*

Chart Frame Properties
Charts: <Empty Chart>
Cable - MicroVision Seg
MV Spent
Display: When Active
Quality: Presentation
OK Cancel

# Table Frames

*Table Frame tool icon.*

A table frame contains a representation of a table from your ArcView project. This frame can either be linked to a table at the time of creation, or left empty to be linked at a later date.

A table frame is always live-linked to the corresponding table in your project. If the table in your project is closed, the table frame will represent the table as a solid rectangle that contains the name of the table. If the table is open, note that the table frame displays the same data that is visible in the table. The table will be reproduced as it appears in the project, including the shading of selected records. In addition, if the visible fields in the table exceed 80 characters in width, only the field at the far left will be displayed up to a width of 80 characters. If the displayed fields or selected records are changed in the table, the display in the table frame will be changed as well.

*Table Frame Properties dialog window.*

Table Frame Properties

Tables: <Empty Table>
Attributes of Arterial - Tr
Attributes of Avg Amoun

Display: When Active

Quality: Presentation

OK Cancel

# Picture Frames

*Picture Frame tool icon.*

A picture frame allows you to import graphics into a layout. The graphic could be a photo, a document image, or work from other application software, such as a spreadsheet or a database form. Imported graphics cannot be edited, although the picture frame containing the graphic can be resized.

Supported graphic file formats include PostScript (including EPS), GIF, Windows Bitmap, SunRaster, TIFF, MacPaint, ERDAS GIS, ERDAS LAN, RLC, BIL, BIP, BSQ, IMPELL Bitmap, Nexpert Object Image, and Windows Metafile (on Windows platforms), and Macintosh PICT (on Macintosh platforms).

*Picture Frame Properties dialog window.*

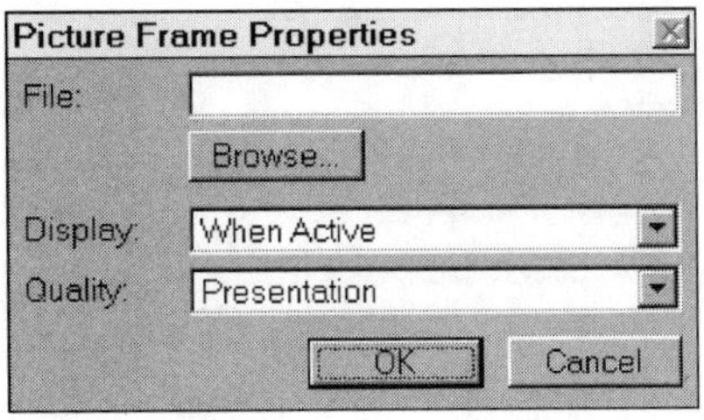

# Map Composition

In ArcView, map composition involves creating the required frames or map elements, adding other graphics such as titles or neatlines, and arranging everything on the page until you achieve the format you desire.

## Graphics

The same graphics tools that are available when working in a view (covered in Chapter 5) are available in a layout. Within layouts, the two most frequently used graphics tools are the Box tool for creating neatlines and the Text tool for adding text or annotations.

Graphics manipulation tools are also available in layouts. The Group option allows you to group selected graphics and subsequently manipulate them as a unit. Selected graphics can include frames as well as graphics primitives. This feature can greatly facilitate map composition by allowing a frame and its associated graphics, such as a title and neatline, to be repositioned as a unit.

The ability to move selected graphics to the background is another useful function. Using this feature, you can create a box, shade it with a solid fill, and then place it behind a view frame to provide added emphasis.

Graphic Size and Position

Graphic

1.25 in from top
2.75 in height
7 in from bottom

1.75 in from left
4.25 in width
2.5 in from right

Maintain Aspect Ratio

OK
Cancel

One last reassuring note regarding map composition. The Undo button can always be used to reverse the last action you performed. This includes adding or deleting frames or graphics, grouping or simplifying elements, and repositioning or resizing elements. Unlike the Undo feature used in editing shapefiles, however, the Layout Undo only works on the last action performed.

*Undo button icon.*

## Decomposing Map Legends

By default, a legend frame displays legend entries for all themes currently drawn in the view. As the complexity of your view increases, the odds that the default legend does not entirely meet your needs also increase. Help is at hand—it is possible to edit the legend to better fit your specific needs.

Begin by selecting the legend frame with the graphic Select tool. With the legend frame selected, choose Simplify from the Graphics menu.

The single graphics frame will be decomposed into a collection of separate graphic entities, one for each element in the legend. Initially, all graphic elements are selected, and the selection handles for all graphic elements are visible.

At this point, individual legend elements can be repositioned, resized, or deleted. One word of caution before you begin moving things around. If you move elements separately, it can be very difficult to re-establish the original relationship between symbols and class labels, particularly for a theme displayed with multiple classes. Before you start repositioning or resizing elements, you may wish to select all the elements associated with a specific theme and group them as a single graphic.

Another word of caution. Because the Undo feature in Layout mode is only one level deep, when editing decomposed legends, you may wish to park legend elements in an unused area of the layout rather than delete them immediately, in case you need them later on.

Lastly, once a legend is decomposed, it becomes a collection of graphic primitives, and the link back to the view is severed. If the live link from the view frame to the view is maintained, the legend will not be updated. It will be necessary to create a new linked legend—and to decompose this new legend—for updates to the Table of Contents to be reflected in the legend. Accordingly, we recommend that simplifying the legend be the last step you perform in your map composition, ideally after the live link has been broken between the view frame and the view.

## Positioning Map Elements

Several tools exist to help you place map elements in your layout. The primary tool is the *map grid.* The map grid serves as a guide for alignment of map elements. It can serve as a visual guide only, or as a *snapping grid* to which all graphic elements will be snapped when placed.

Grid properties are controlled from the Layout Properties dialog window, which you access from the Layout pulldown menu. The horizontal and vertical spacing of the grid can be specified, as well as whether map elements will be snapped to the grid when positioned. Another feature available from the Layout menu is the ability to show or hide the grid. If grid snapping is turned on, graphic elements will be snapped to the grid, even if the grid is not visible.

The Align tool, which you may have used for aligning graphics in a view window, is also available from the layout window. By using this tool, you can align selected graphics vertically or horizontally, even if grid snapping is turned off. In addition, you can align selected graphics, including frames, with layout margins.

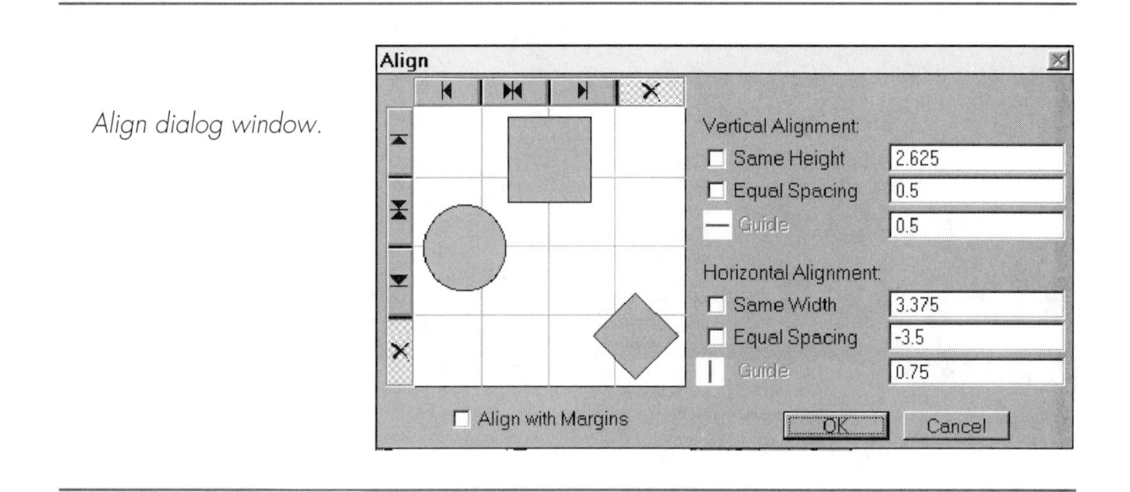


*Align dialog window.*

Two methods are available for moving selected graphics: positioning them with the mouse, and specifying the position relative to the layout page using the Graphic Size and Position dialog window. The Graphic Size and Position dialog window allows you to specify a graphic element's height and width, and to position it relative to the page edges.

## On Cartographic Design

Cartography is defined by the International Cartographic Association as the "art, science, and technology of making maps." In GIS, it is safe to say that the science and technology of making maps consumes the majority of our project development time, and the creative process of design—the art of making maps—is often ignored or poorly planned.

The old adage of "A picture tells a thousand words," was never more true than in the art of cartography. People are visual creatures, and given that maps are a natural choice when visually presenting data relationships to others, the importance of carefully designing maps to effectively communicate your intent and purpose is evident.

### The Five W's of Communication

Maps are a highly specialized form of communication. One useful way to begin the cartographic design process is to answer the five basic questions known in communication planning as the five *W*'s:

- ❒ *Who* is your target audience?
- ❒ *What* message do you have for them?
- ❒ *Why* are you presenting this message to the target audience?
- ❒ *Where* would this message best be given to the audience?
- ❒ *When* is the best time to present this message?

### Who Is Your Audience?

Marketing professionals, who spend millions of dollars trying to sell a product or service, know the strategic importance of pinpointing their audience. Even if you are not a marketing professional, you are still making the effort to communicate to a certain group of people. Who are they? What do you know about them as a group? What are their demographics? Understanding who your audience is will lend insight about how to reach them at their level.

## What Message Do You Have for the Target Audience?

Perhaps you are presenting the results of an analysis project, or the conclusions of important research. Whatever your message may be, it must be clearly focused in your own mind before you can expect to effectively present it to others.

## Why Are You Presenting This Message to This Audience?

What do you wish to gain from this communication? There is a purpose behind every map. It may be used for navigating from A to B, or it may be used for influencing others and gaining support for a new project. For the former, a map design that is easy to read and pleasing to behold is paramount, while the latter will benefit from a careful consideration of the psychological impact associated with the color of each map element.

## Where Will You Present This Message to the Audience?

Knowing, or not knowing, the conditions under which your map will be presented should influence the design of your map. If your map is part of a larger presentation, you will want to consider the conditions under which the presentation is taking place. What type of lighting? Is the presentation room large or small? How many people will be there? What type of projection systems are available?

The media to be used influences map design. When using slides, the level of detail should be kept minimal. The short viewing time—typically less than 20 seconds—favors dramatic use of color and layout. The size and resolution of the media will affect the amount of detail that can be presented. If the map is part of a publication, it may be necessary to reduce the use of color.

## When Is the Best Time to Present This Message?

You will rarely have full control over the time available to create the map or the time available for presentation. It is crucial to use the control you do have over the process to your best advantage.

## General Points to Consider when Designing

When you are ready to design your map product, consider the following general guidelines:

- ❒ Be observant of other maps and graphic design styles. What colors work best? What text styles are easiest to read? Take note of what "works" and what you like and dislike. This will help you develop your own style.
- ❒ Develop an organizational style, or standards that you consistently follow. If your design is strong, others will eventually come to recognize it and to associate your organization with the maps you make. This can be a very simple and effective marketing tool.
- ❒ Once you have developed a style, create ArcView templates that can be used when you have to produce maps in a hurry. Keep the templates simple, and allow for flexibility within the design to keep that creative spark alive.

Some specific points to consider when designing maps and pages:

- ❒ Take advantage of the power of white or negative space; that is, areas that are not part of your map. White space is a very powerful design tool.
- ❒ Build contrast between the geographic figure (the focus of your map's purpose) and the ground (the rest of the map or the page background).
- ❒ Prioritize the geographic layers of your map based on the map's purpose. What elements are most important? These elements should visually stand out more than other elements of less importance. Create a worksheet that establishes this "cartographic order."
- ❒ Use one or more visual variables, such as color, shape, texture, and size, to create levels of contrast between the map

elements. The point here is to create the visual illusion of the cartographic order.

- At the same time, ensure that your map consistently presents related data. Verify that features of the same type of element are presented similarly. For example, stick to one style per feature type for annotation. Avoid labeling similar objects with different fonts, sizes, or colors.
- Consider the entire page. Where does the map fit best, considering its shape? Where will the title and legend fit best? Think in terms of the map's flow. What does your eye focus on first, and where does it naturally go from there? Use this natural flow to guide the viewer through the logical sequence of your map's "story."
- As you near completion, work through a mental checklist of the basic map elements. Have you placed a prominent title? Included an adequate legend with your name, date, and north arrow? Have you listed both map and data sources? Have you considered a "locator" map? Each of these items has the potential to contribute to and clarify your presentation.

# Printing

The final step to creating a layout is sending the finished product to the printer. The Page Setup dialog window gives you initial control over what will be sent to the printer. From the Page Setup dialog window, you can set page size (standard page sizes from A through E, along with the option of entering a custom page size), page orientation (landscape or portrait), page margins, and output resolution.

✓ **TIP:** *Depending on your output device, graphics printed with a line weight of 1 may not correspond to the finest line possible from your printer or plotter. The remedy is to set the line weight to a fraction of 1.*

*For graphics in a view frame, this necessitates changing the theme symbology by accessing the Legend Editor. In the box for Pen Size of Fill Outline Size, type in a new smaller value, such as* 0.01. *While this value will not be reflected in the line weight as drawn on your display, it will be reflected when the view or layout is printed.*

Additional options are available from the Print and Print Setup dialog windows, which are accessed from the File menu. These options are platform-specific, and include the ability to print to a file rather than directly to an output device. Entering a file name in the To File input box will override the selected printer option. The output format can be either CGM or PostScript, and is specified by adding the appropriate file extension to the file name (e.g., *.cgm* or *.eps*).

An alternative to printing a file is to select Export from the File menu. Both views and layouts can be exported. When you export a view, however, the Table of Contents will not be exported. Supported export formats include PostScript (.eps); CGM Binary, CGM Character, and CGM Clear Text (.cgm); Adobe Illustrator (.ai); and, on the Windows platform, Windows Metafile (.wmf) and Windows Bitmap (.bmp).

---

## Exercise 7: Working with Layouts

For those who have patiently followed along through the previous six exercises while we created views, manipulated themes, queried tables, and prepared charts, the time has come to pull it all together. In this exercise we examine layouts, the final component of an ArcView project.

In previous exercises you used survey responses from restaurant customers to provide information about the demographics of the customer base and how the customers were distributed. In this exercise you will pull some of these elements together into a final map. Again, you will find that we have made some modifications, such as a new classification on the *Cable_mv* theme.

1. Open the *ch8.apr* project file.
2. Click on the Layouts icon in the project window, and select New to create a new layout.

   A layout window containing a default layout will be opened.

3. Resize the layout window as desired to ensure that you have room to work.

4. From the Layout menu, select Page Setup and click on the icon that represents landscape page orientation.

5. From the Layout menu, select Zoom to Page.

   The layout will expand to fill the layout window.

6. From the Layout menu, select Properties. In the Layout Properties dialog window, set the horizontal and vertical grid spacing to 0.2 in. At this point, the layout is ready for input.

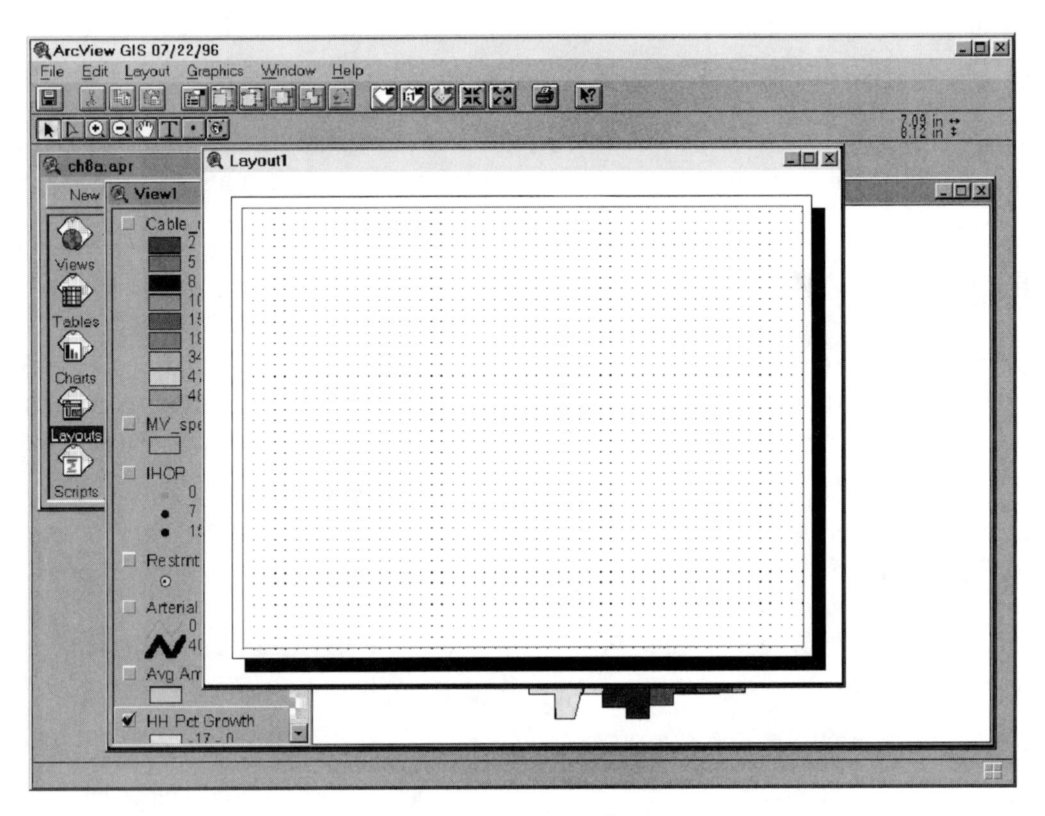


*Prepared layout ready for input.*

You are now ready to begin adding frames, the boxes that will contain your finished map elements, such as views and legends.

1. With the Layout1 window active, select the View Frame tool from the tool bar. Drag a box in the lower left portion of the layout. Do not worry about the exact size and proportions because you can always resize the box later.
2. Accept all defaults in the View Frame Properties dialog window, including the option to create this frame as an empty view. A shaded rectangle labeled *Empty View* will be created on the layout.
3. Add a second view frame to the right of the first in the lower center of the layout. Create this frame as an empty view frame as well.
4. From the Frame tool's pulldown list, select the Chart Frame tool, and drag a box in the upper left portion of the layout. Accept all defaults from the Chart Frame Properties dialog window.
5. Select the Legend Frame tool from the Frame tool pulldown list. Drag a legend frame in the lower right portion of the layout. Accept all defaults. Your layout should now look approximately like the layout below. If necessary, use the Graphics Selection tool to resize or reposition. (The incremental project has been saved as *ch8a.apr.*)

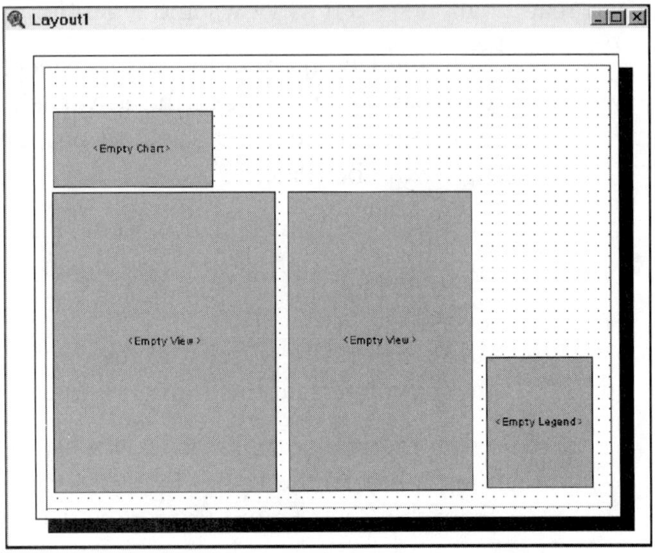

Layout with all frames in place.

You will not use the prepared layout just yet. Instead, you will store it as a *template* for future use.

1. From the Layout menu, select Store as Template.
2. Key in *Exercise - CH 8* for the layout name. Click on OK to accept these choices.

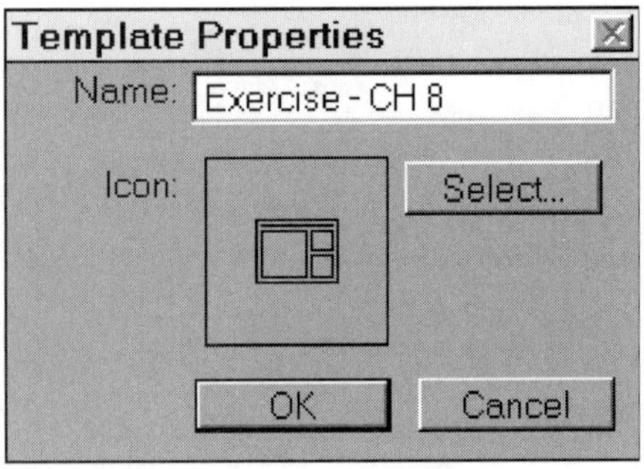

Template Properties dialog window.

In order to use the layout template, you have some additional work to do with your views.

1. Close the layout window.
2. Create a new view window by selecting the Views icon in the project window and clicking on the New button.

   A new view window, titled *View2*, will be opened.
3. Move and resize this window until it is approximately the same size as the View1 window.
4. Place your cursor in the View2 window. Note that the coordinates returned in the tool bar are not the same as those returned in View1. This is because a new view has been added and, as such, it is unprojected and uses geographic coordinates by default. Before proceeding, open the View Properties dialog window. Change the projection to Arizona State Plane coordinates (the same as View1) by selecting State Plane - 1927 from the Category list and Arizona, Central from the Type list.
5. Return to the View1 window and click on the *HH Pct Growth* theme to make it active. From the Edit pulldown menu, select Copy Themes.
6. Return to the View2 window and select Paste from the Edit menu. A copy of the *HH Pct Growth* theme is inserted into View2. Turn this theme on to display it in the view.
7. To differentiate this theme from that in View1, you will apply a different classification. Double-click on the theme in the Table of Contents to initialize the Legend Editor. From the Classification Field pulldown list, select Hhincmedia. Click on the Classify button and select *5* for the number of classes. Select the Blues to Reds dichromatic color ramp, and click on Apply.
8. Select Properties from the Theme menu, and change the theme name to *HH Median Income*.

You have completed work in the views, and you are ready to create the new layout.

1. Click on the Layouts icon in the project window, and select New to create a new layout. A new layout window, titled Layout2, is opened. Move and resize this window as needed.

2. It is is time to use the layout template you stored earlier. From the Layouts pulldown menu, select Use Template. The Template Manager dialog window is opened. Click on the template titled *Exercise - CH 8,* and select OK.

3. The stored template is opened in the layout window. Note that View1 and View2 have been linked to the corresponding view frames. ArcView takes the initiative to link currently open view windows to corresponding view frames in a layout template. Because the legend frame could not be uniquely associated to one of the two open view windows, it was left empty. (The incremental project has been saved as *ch8b.apr*.)

---

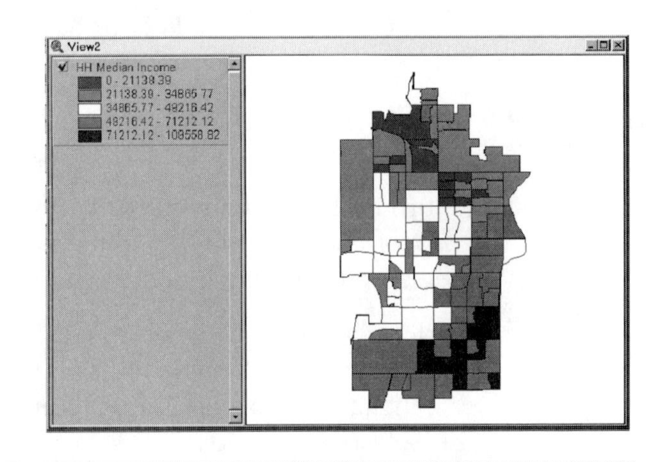


*View2 window containing the copied and newly classified HH Median Income theme.*

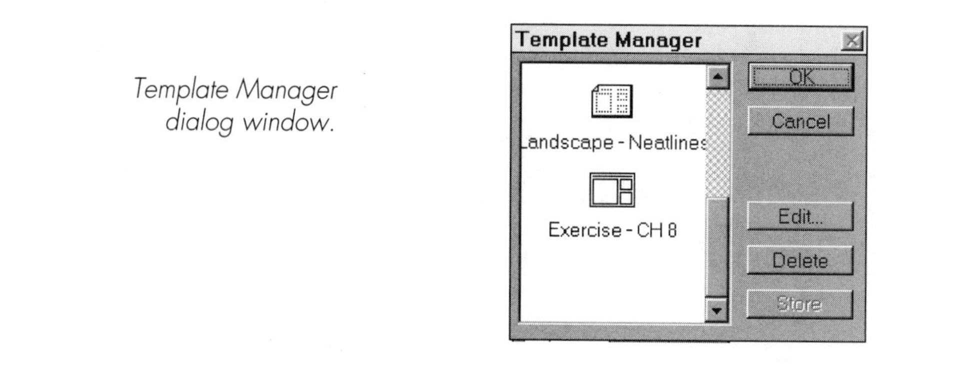


Template Manager dialog window.

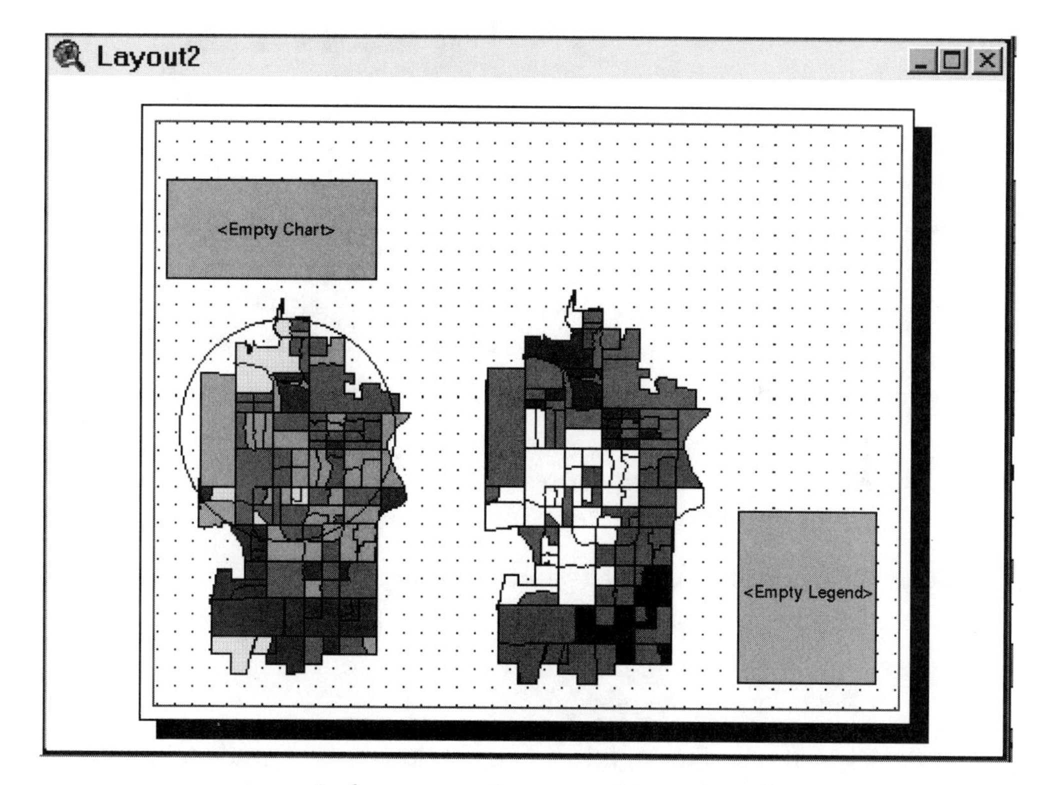


Layout2 after restoring the previously saved template.

4. With the Select tool, double-click on each view frame to access the View Frame Properties dialog window. Change the display property to Draft. Note that the view frames have changed shape, reflecting the drawing area of the corresponding view windows. Reposition and resize the frame boxes according to what you feel looks best.

5. Continue by double-clicking on the legend frame, linking it to *ViewFrame2: View2*. Double-click on the chart frame, and link it to the chart titled *Per Capita Spent by MicroVision Segment.*

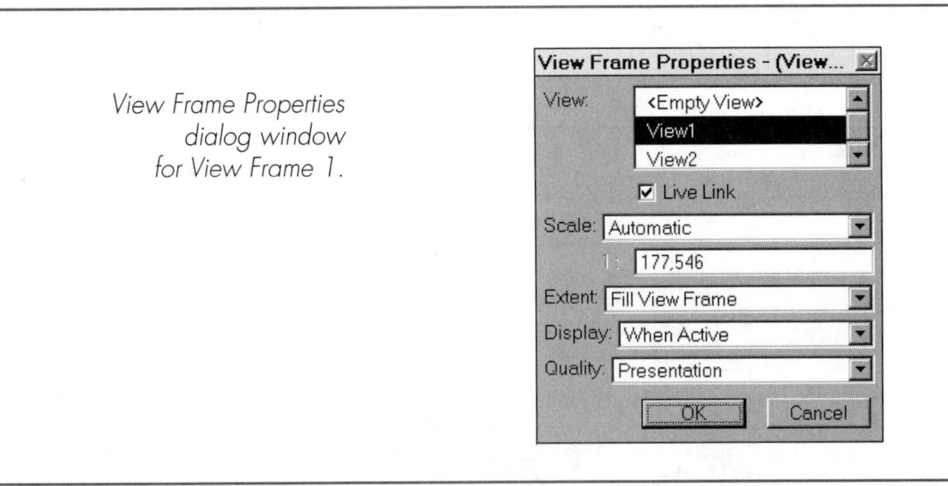


*View Frame Properties dialog window for View Frame 1.*

**NOTE:** *In order for a chart or table to display on a layout, the chart or table must be open in the corresponding view.*

6. When linking is complete, the name of each project component will be displayed in each frame box.

7. Select the Box tool from the Graphics Tool pulldown list and drag a neatline around the layout.

8. Using the Text tool, add a title by clicking in the upper right area of the layout. Enter the following text: *Demographics of Market Survey Responses*. Use two lines for the text entry. Indicate that Horizontal

Alignment is centered. Resize and reposition the text as needed. When complete, your layout in draft form should (or at least *could*) resemble the layout in the illustration below.

**9.** Finally, double-click on each frame element, and switch the display from Draft to Presentation. If you have difficulty selecting the individual frame elements, select the neatline graphic and, from the Graphics pulldown menu, select Send to Back. (The incremental project has been saved as *ch8fin.apr*.)

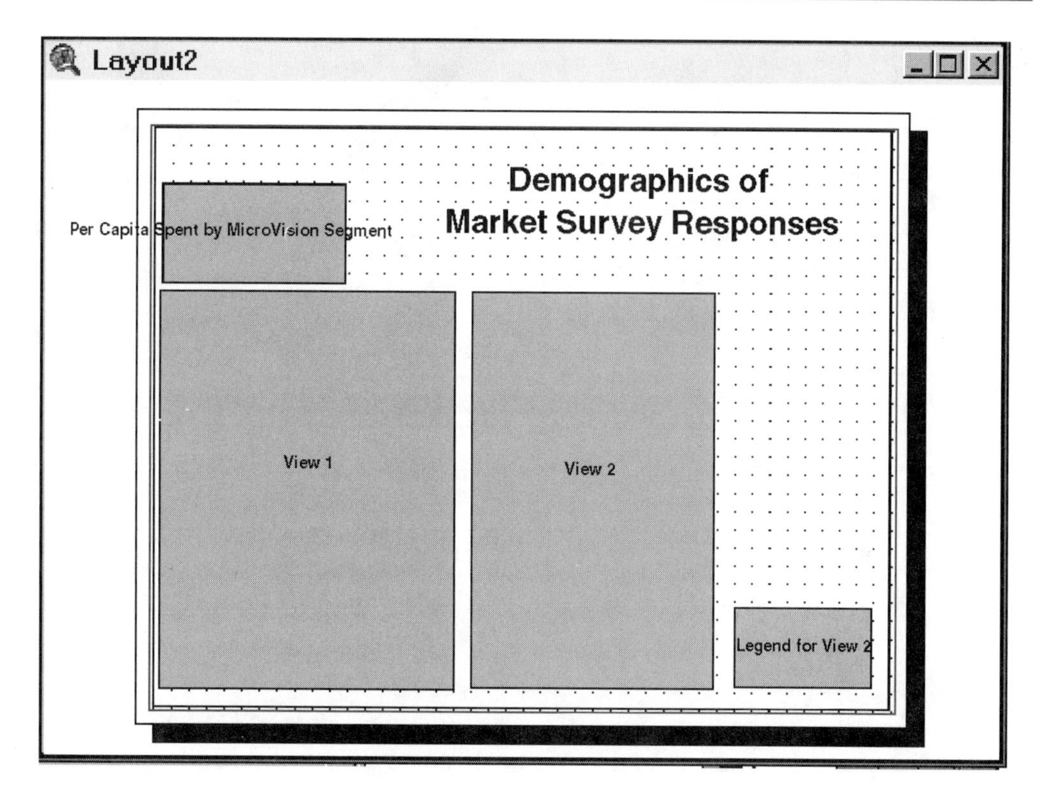


*Completed layout in draft form.*

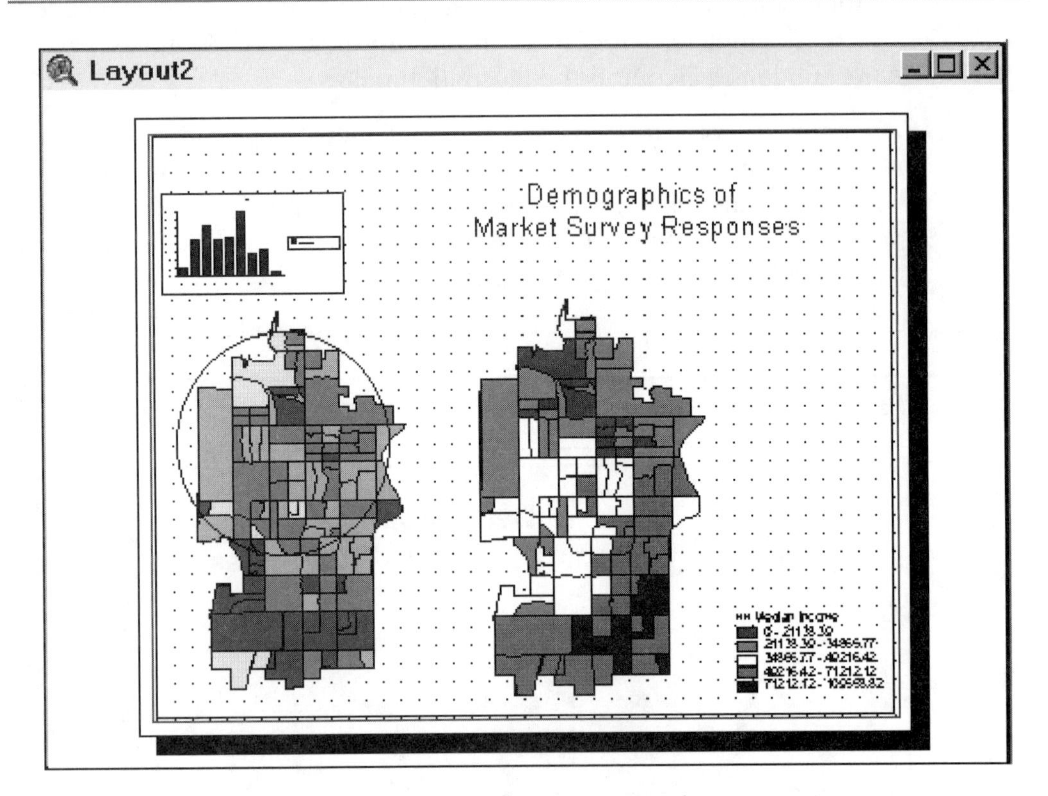


*Completed layout in presentation form.*

10. Your presentation is complete and ready to print. If you have a printer on your system, print the layout by selecting Print from the File menu.

You have now explored the four primary functional areas of ArcView: views, tables, charts, and layouts. With the use of these components alone, it is possible to design projects that accomplish a wide variety of tasks.

In the remaining chapters, you will explore additional functionality in ArcView, but we encourage you to visit the earlier chapters as often as you feel is necessary until you are comfortable with the basics. Although "basic," everything you need in order to work in the real world appears in Chapter 3 through Chapter 8.

## Chapter 9

# CAD and Image Themes

Through the first eight chapters of this book, we introduced you to ArcView and presented the basic functionality in a progressive fashion. Beginning with creating a new ArcView project, we have covered the basics of loading data, developing themes, and working with the ArcView document types—views, tables, charts, and layouts—to create a finished project. At this point you should have an understanding of the basic functionality in ArcView, and how the pieces of this functionality fit together.

In this chapter, and the next three, you will move into the area of advanced functionality, to experience more of the power available within ArcView for exploring and modeling your spatial data. You will look at such areas as CAD and image themes, editing and digitizing shapefiles, editing tables, advanced classification, overlay operations, and hot links. You will also discover the additional functionality available through Network Analyst, Spatial Analyst, Spatial Database Engine, and the Data Automation Kit.

## Extensions

Everything you have learned thus far is part of ArcView's core; in other words, it is all available as soon as you start the software and open a project. ArcView also makes additional functionality available through the use of extensions. Extensions allow you to add specific features to the software as you need them, and then unload this functionality when it is no longer needed. ArcView only delivers specialized tools and menu choices as needed, thereby simplifying the basic user interface. It also allows for the more efficient use of system resources, reducing the memory that would be required to load software functionality you might never use.

Some ArcView extensions are provided with the core product. These include the CadReader, Digitizer, Database Themes, and the IMAGINE and JPEG Image Support extensions. Other extensions, such as the Network Analyst and Spatial Analyst, are optional and must be purchased separately.

To load an extension, make the project window active and select Extensions from the File menu. A dialog window presents you with the available extensions; click on the appropriate boxes to load the extensions you want, then click on OK. A check in the box indicates that the extension is already loaded; remove the check and click on OK to unload the extension.

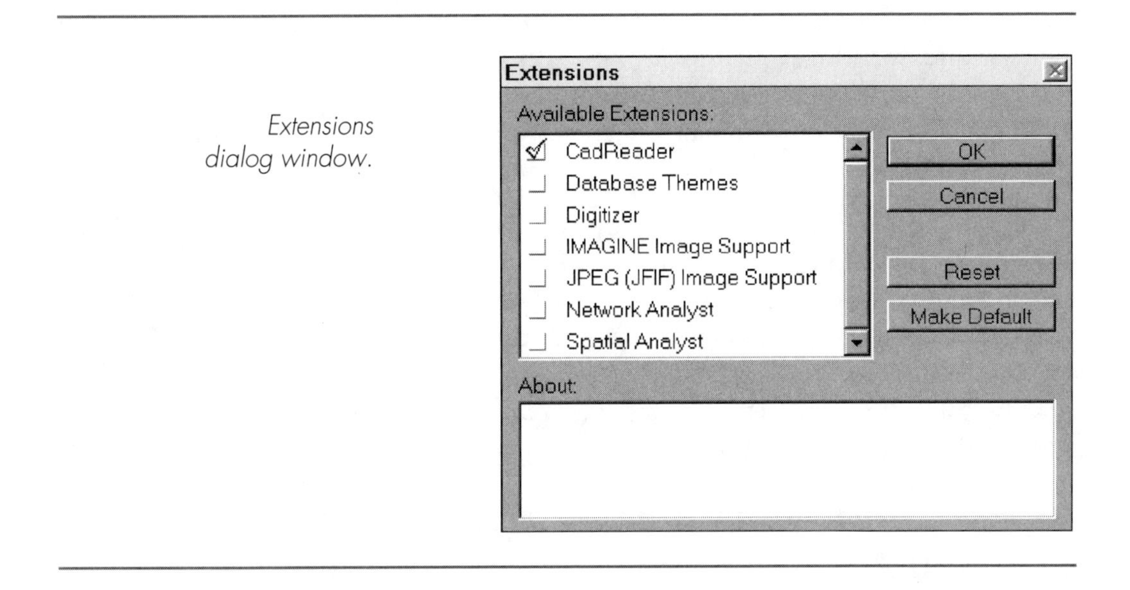


*Extensions dialog window.*

# CadReader

The CadReader extension allows themes to be created directly from CAD drawing files. Supported formats include MicroStation 5.0/5.5 .dgn files, AutoCAD Release 12/13 .dwg drawings, and ASCII and binary DXF files. (AutoCAD .dwg files are not supported on UNIX platforms.) The CadReader is not available for the Macintosh.

When a CAD theme is added to an ArcView project, that theme is created directly from the CAD file—there is no conversion to an intermediate format such as an ARC/INFO coverage or ArcView shapefile. Accordingly, it is desirable to understand the basics of how features are created and organized in CAD drawings.

All features in a CAD drawing are constructed from basic elements, which are referred to as entities. Available AutoCAD entities include points, lines, polylines, arcs, circles, shapes, solids, and text, among others. Entities in a CAD drawing are stored in layers. Typically, many layers are created in a CAD drawing, each layer serving to organize entities according to the features represented. In such a manner, one layer may be created to store parcel boundaries, another to store building footprints, another for sewer lines, and so on.

Unlike in an ArcView theme or shapefile, there are no restrictions on the types of entities that can be contained in a single layer. Thus, a single sewer layer may contain polyline entities representing sewer lines, point entities representing valve locations, and text entities labeling each line with its capacity and a maintenance date.

When an ArcView theme is created from a CAD drawing, each entity type is represented as one of four ArcView feature types: points, lines, polygons, or annotations. Because an ArcView theme can contain only one feature type, entities from a single CAD layer may be distributed among four ArcView themes when brought in to the view.

When adding a theme from a CAD drawing, the Add Theme dialog window will present a folder for each available drawing. When the folder is opened, up to four feature classes will be presented (one each for line, point, polygon, and annotation), depending on what entities are present in the drawing. For each feature class, the new theme will contain features from all applicable entities from all layers present in the drawing.

*Add Theme dialog window, showing the four feature types within the folder for the CAD drawing titled Sw16n4.dgn.*

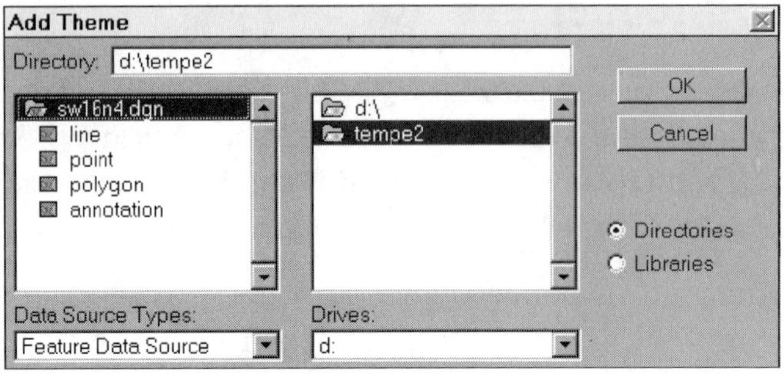

As with themes from ARC/INFO coverages or ArcView shapefiles, each theme from a CAD drawing contains an associated feature attribute table. This table includes Entity, Layer, Elevation, and Color fields. In this manner, the features present in the theme can be related back to their source in the original drawing; these fields can be used subsequently as the basis for a thematic classification or logical selection.

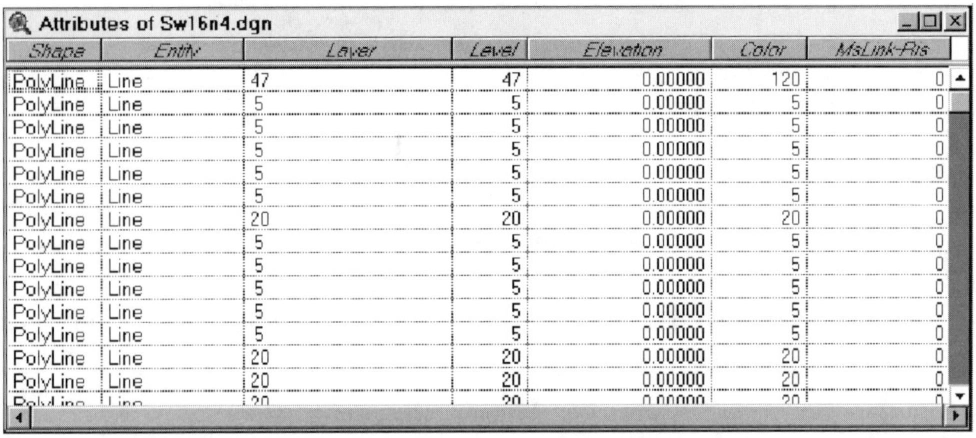

Attributes of Sw16n4.dgn

| Shape | Entity | Layer | Level | Elevation | Color | MsLink-Ris |
|---|---|---|---|---|---|---|
| PolyLine | Line | 47 | 47 | 0.00000 | 120 | 0 |
| PolyLine | Line | 5 | 5 | 0.00000 | 5 | 0 |
| PolyLine | Line | 5 | 5 | 0.00000 | 5 | 0 |
| PolyLine | Line | 5 | 5 | 0.00000 | 5 | 0 |
| PolyLine | Line | 5 | 5 | 0.00000 | 5 | 0 |
| PolyLine | Line | 5 | 5 | 0.00000 | 5 | 0 |
| PolyLine | Line | 20 | 20 | 0.00000 | 20 | 0 |
| PolyLine | Line | 5 | 5 | 0.00000 | 5 | 0 |
| PolyLine | Line | 5 | 5 | 0.00000 | 5 | 0 |
| PolyLine | Line | 5 | 5 | 0.00000 | 5 | 0 |
| PolyLine | Line | 5 | 5 | 0.00000 | 5 | 0 |
| PolyLine | Line | 5 | 5 | 0.00000 | 5 | 0 |
| PolyLine | Line | 20 | 20 | 0.00000 | 20 | 0 |
| PolyLine | Line | 20 | 20 | 0.00000 | 20 | 0 |
| PolyLine | Line | 20 | 20 | 0.00000 | 20 | 0 |

*Attribute table for the line theme derived from the CAD drawing titled Sw16n4.dgn.*

To add a theme from a CAD drawing file, first load the CadReader extension. Do so by making the project window active and selecting Extensions from the File menu. Check the box for the CadReader extension in the dialog window.

Once the CadReader is loaded, clicking on the Add Themes button will cause available CAD drawings to be listed (when the Data Source Type is set to Feature). Open the folder for the CAD drawing, select the desired feature classes, and click on OK when done. The corresponding themes will be added to the active view. By default, a Unique Values classification is applied, one that is based on the value in the Color field.

> ✓ **TIP:** *The default classification applied by ArcView, based on the Color field, does not map the color number to the corresponding color from the CAD software. If you wish to apply a standard legend, with each color represented by its CAD layer color, create the legend for a representative theme, then save the legend by clicking on the Save button in the Legend Editor. This writes the legend out as a system file that can be subsequently loaded and applied to new CAD themes.*

# Properties Unique to CAD Themes

Putting aside the differences that are apparent when translating CAD entities to ArcView features, there are some other properties of CAD themes that differentiate them from other themes.

## Blocks

A block is a group of entities in an AutoCAD drawing that have been aggregated into a single complex object. Blocks are commonly used to represent standard elements. These elements could be standard cartographic elements, such as map symbols for section corners or highway markers, or they could be design elements, such as water distribution valves.

Blocks can be composed of dissimilar element types. A block that represents a section corner symbol might be made up of polyline, point, and text entities. Blocks typically contain attributes pertinent to the feature.

Depending on the application, it may be appropriate to explode a block so that its components become part of the theme, or it may be appropriate to represent the block as only a point feature containing the attributes associated with the block. ArcView provides the option of exploding blocks as necessary to access their component geometry.

## Feature and Attribute Editing

Themes based on CAD files do not support either feature or attribute editing. Editing of features and associated attributes must be performed within the originating CAD software. The Refresh option from the Table menu can be used to update the CAD themes with any changes performed on the drawing subsequent to opening the ArcView project. (It is necessary to open the attribute table for the target theme before selecting Refresh.)

## Transformation

In survey applications, it is common to reference locations using an arbitrary coordinate system. For example, a subdivision plat may be surveyed relative to a known starting point, or benchmark, which is assigned an arbitrary X,Y coordinate of *100,100*. ArcView supports the ability to apply a transformation—using one or two points—to reference coordinates in drawing units to real-world coordinates, such as UTM. This transformation is accomplished through the creation of a world file. A world file is a text file that is given the same name as the drawing file, but with a *.wld* extension. The file contains one or two pairs of X,Y coordinates. Inside this file, the coordinates are listed in drawing units, followed by the corresponding world units, in the following format:

```
<x,y drawing coordinates> <x,y world coordinates>
```

For example, `120.4,20.3 488120.6,303490.1`.

> **NOTE:** *Make sure both coordinate pairs support the same level of accuracy. Drawing units accurate to 0.1 meter will not transform accurately to world coordinates in geographic units that are only accurate to 0.01 decimal degree.*

# CAD Theme Properties

In addition to those theme properties that are shared with coverage and shapefile themes, such as Definition and Display, CAD themes contain an additional set of drawing properties. To access CAD theme Drawing Properties, select Theme | Properties. Scroll down the list of properties and select Drawing.

The Drawing page of the Theme Properties dialog window controls three theme properties: layers, transformation, and block handling.

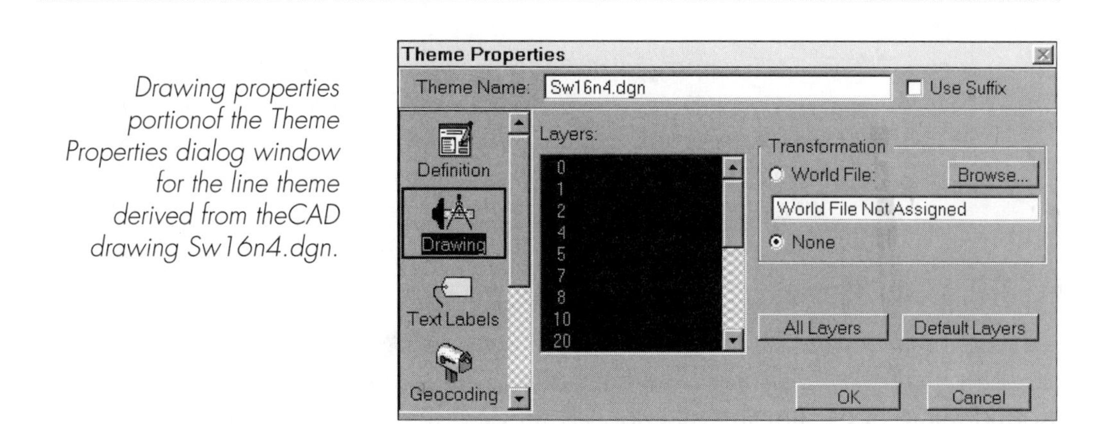


*Drawing properties portionof the Theme Properties dialog window for the line theme derived from theCAD drawing Sw16n4.dgn.*

## Layers

By default, when a CAD theme is added to a view, only those layers currently active in the drawing are visible in the theme. The Drawing page of the Theme Properties dialog window contains a scrollable list of available layers, and those layers that are currently displayed are highlighted. Additionally, a button allows you to select all layers. The Default Layers button will reselect only those layers that are currently turned on in the drawing.

Selecting the layers to display has the same effect as applying a logical selection on the theme via the Query Builder portion of the Theme Properties dialog window. Only those layers that are currently displayed are available for query, display, and classification. Correspondingly, the number of records in the associated theme attribute table will vary as layers are turned on or off.

By default, when a theme is added to a view, blocks are exploded and block entities are rendered as theme features.

## Transformation

This gives you a way to associate a theme with a world file on the system, and to turn the transformation on or off as desired. A browse button lets you assign a different world file, or a file located in a different directory, for association.

> **NOTE:** *CAD themes are not refreshed automatically. To refresh a CAD theme after changes have been made to the drawing, open the Attribute Table for the CAD theme, then select Refresh from the Table menu. Subsequently, you may need to update the legend classification and/or add new layers via the Theme Properties dialog window.*

## Exporting CAD Themes

Although it is based on a CAD file, a CAD theme is like any other theme in that it can be converted to an ArcView shapefile. This provides a means for working around many of the limitations imposed on CAD themes, specifically the inability to edit features or attributes directly in ArcView. Additionally, if only a subset of features is currently selected—whether by logical selection, turning selected layers on or off, or selecting features using the Select tool—only those features will be written out to the resultant shapefile. Thus, converting to a shapefile provides a means for extracting a subset of features from a large and complex drawing file. To convert a CAD theme to a shapefile, make the theme active, then select Convert to Shapefile from the Theme menu.

## Optimizing CAD Files for Use Within ArcView

Due to the dissimilar nature of CAD and GIS with regard to feature types and software function, adding a CAD theme from a CAD drawing file does not always produce the desired results. Commonly encountered problems derive from the manner in which layers are created and used in the CAD application, and the manner in which CAD entities are translated into line and polygon features.

GIS software is inherently *feature-based.* Data capture performed within the GIS software is performed within the constraints of the software and the geospatial environment. Digitizing within ARC/INFO necessitates organizing data into coverages; inputting data using specific feature types such as points, lines, and polygons; and maintaining a link between graphic elements and their attributes. Additionally, the GIS tools are robust and well documented for maintaining the proper connectivity between features that is necessary for subsequent modeling in a GIS environment.

CAD software does not constrain the user in the same manner when capturing data. The fact that a single layer can contain many entity types leads to a more function-based than feature-based approach to data organization. Layers may be used to separate data by projects or maintenance as easily as by feature type.

In addition, the whole concept of feature connectivity is less critical in the CAD environment than in the GIS environment. A CAD drawing, whose primary purpose is to provide a map or picture of the data rather than model the data, is much less constrained with regard to feature connectivity. This is particularly evident when examining network or polygon themes derived from CAD drawings. In order to ensure proper modeling in ArcView and to maintain feature connectivity, all CAD entities used to construct the database, such as arcs and polylines, must be snapped to adjacent features during input. If CAD entities are not snapped, gaps may be present and yet not visible, precluding proper translation into ArcView feature types.

If you have the luxury of designing your CAD applications from the outset with an eye to their incorporation into ArcView, keep in mind the following to help ensure successful translation: assignment of layers to correspond with thematic feature types, and feature snapping when inputting polylines and polygons.

If you have to use an existing CAD drawing that was not created with GIS incorporation in mind, you may need to perform some post-editing on the drawing so that features model properly. Additionally, translation into an ARC/INFO coverage, and the tools available therein for building and maintaining topology, can help mitigate connectivity problems that are difficult to resolve by other means.

# Image Themes

ArcView provides the ability to import image and GRID data into a project, as well as the basic tools for manipulating this data. The optional Spatial Analyst extension greatly extends your ability to model and manipulate data stored in ESRI's GRID format.

## Image Data

An image is a form of raster, or grid-cell, data. Raster data is stored as cells in a uniformly spaced matrix of rows and columns in which each grid cell contains a value that is representative of the feature being depicted. The resolution of raster data is dependent on the size of the grid cell.

As used in ArcView, image data may include scanned data, such as images of photographs or maps, or remotely sensed data, such as satellite imagery. ArcView supports the following image formats:

- TIFF
- TIFF/LZW compressed
- GeoTIFF
- ERDAS .gis and .lan
- ERDAS IMAGINE
- BSQ, BIL, BMP
- Sun raster files
- Run-length compressed (RLC)
- JPEG
- GRID

## Image Types

There are four general image types:

1. Monochrome

2. Grayscale
3. Pseudocolor
4. Multiband/true color

Monochrome, grayscale, and pseudocolor images are all single-band images. In a monochrome image, each grid cell, or pixel, is either black or white. Grayscale images represent features with a range of gray shades, typically 256 shades. Pseudocolor images make use of a color map to map each pixel value to a specific value. This color map may be in the header of the image file, as with Sun raster files, or in a separate file, as with ERDAS .gis images.

Multiband (or true color) images are actually comprised of several images, each representing a single band in the color spectrum. For example, a satellite image may include separate bands in the blue, green, red, near infrared, and far infrared portions of the spectrum. To display a multiband image, each band is associated with one of the three primary colors in the video display: red, green, or blue. Accordingly, only three bands of a multiband image may be displayed at one time.

To add an image to a view, click on the Add Theme button and select Image Data Source from the Data Source Type dropdown list. Navigate to the directory containing the image, highlight the image in the list of available files, and click on OK.

## Controlling the Display of Images

Images do not contain feature attributes. Consequently, most of the operators available to feature-based themes are not available for images. A legend editor is available for images, however, providing you with the ability to manipulate the linear stretch, the interval lookup, the color map for single and multiband images, and the display colors for multiband images.

To access the Image Legend Editor, double-click on the image name in the view's Table of Contents. The Image Legend Editor dialog window will be displayed. The exercise at the end of this chapter explores the options available from the legend editor.

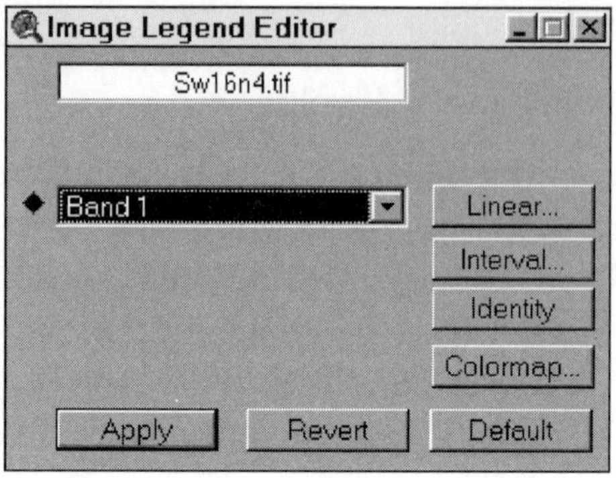

*Image Legend Editor for the single-band image Sw16n4.tif.*

## Limiting the Image Extent

You may encounter images and GRIDs that cover a larger area than the other themes in a view. To reduce the drawing time that results from the amount of data contained in the image, the display extent can be set to a value less than the full extent of the image. The Theme Properties dialog window for the image, accessed from the Theme menu, allows you to set the image extent in one of four ways:

1. To the extent of all themes in the view
2. To the extent of a specific theme in the view
3. To the current extent of the view display
4. To user-specified coordinates for the lower-left and upper-right corners of the display extent

Additionally, a minimum and maximum scale can be set for the display of the image.

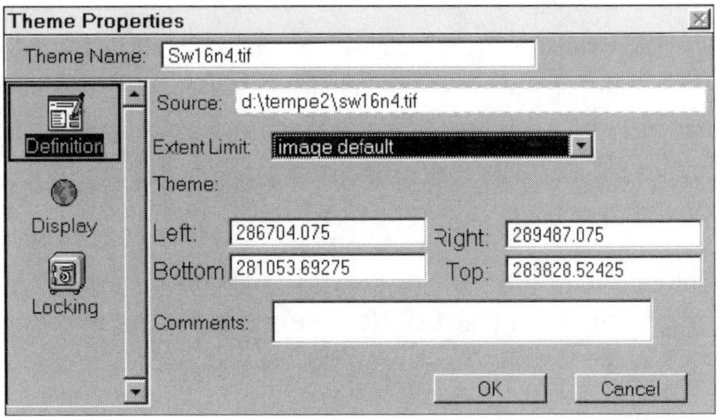

*Theme Properties dialog window for an image theme, showing the current settings for image extent.*

## Image Catalogs

An image catalog is an ARC/INFO-supported table that allows multiple images, each covering a portion of the total area, to be displayed as a single image. While an image catalog is typically created within ARC/INFO, it can be created with ArcView as well. Any image whose format is supported in ArcView can be placed into an image catalog.

An image catalog is an INFO or dBASE table that contains five fields: IMAGE, XMIN, YMIN, XMAX, and YMAX. If you are defining the table yourself, the IMAGE field is a string field with a length of 50, and the XMIN, YMIN, XMAX, and YMAX fields are numbers, each with a length of eight digits plus four decimal places.

The IMAGE field contains the full pathname to the image. The XMIN, YMIN, XMAX, and YMAX fields contain the coordinate values for the image's minimum and maximum extension along the X and Y axis.

To add an image catalog to a view, click on the Add Theme button and select Image Data Source from the Data Source dropdown list. Navigate to the directory that contains the image catalog table, highlight the file name in the list of available files, and click on OK. The image catalog is added to the view's Table of Contents.

## Registering an Image

Internally, each grid cell in an image is referenced by its row and column number. To register an image so that it displays with the other themes in a view, each row/column location must be translated into real-world coordinates. This transformation may be stored within the header of the image file, or in a separate file known as a *world* file.

Also, images can be registered within ARC/INFO using the REGISTER command. Alternatively, a third party Avenue script is also available to register images within ArcView.

## GRID Data

GRID is ESRI's proprietary raster-based grid cell data format. The GRID data format can be used to represent discrete or categorical data (such as land use), or continuous data (such as reflectance or elevation).

ArcView's Spatial Analyst contains a robust set of tools for manipulating GRID data. Even without the Spatial Analyst extension, GRIDs can still be displayed in ArcView as a single-band image theme.

If a GRID is an integer GRID, the theme's attribute table, referred to as a Value Attribute Table (or VAT), is available for query and modeling within ArcView. Accordingly, individual grid cells in an integer GRID can be queried, and the VAT is available for thematic classification, or for joining to additional tables within ArcView.

In the following two exercises you will work with a MicroStation drawing and a scanned aerial photo that covers a quarter section—160 acres—in northwest Tempe. This data is part of an actual data set maintained by the city of Tempe as part of their city-wide GIS.

# Exercise 8: Working with CAD Themes

1. To begin, start ArcView, and open a new project.
2. With the project window active, select Extensions from the File menu. Click on the box for CadReader in the list of available extensions, and click on OK. The CadReader extension is now loaded and available.
3. Make the Views icon in the project window active. Click on New to create a view window. Resize this view window to fit your available display area.
4. Click on the Add Themes button, and then navigate to the *$IAPATH/data* directory in the Add Theme dialog window. With the data source type set to Feature Data Source, click on the *folder* next to the entry for *Sw16n4.dgn.* Four feature types will be listed: line, point, polygon, and annotation. Select all four and click on OK.
5. Four themes, one for each feature type, will be added to the view. A default classification is applied to each theme, applying the Unique Values legend based on the value for Color for each layer in the drawing.

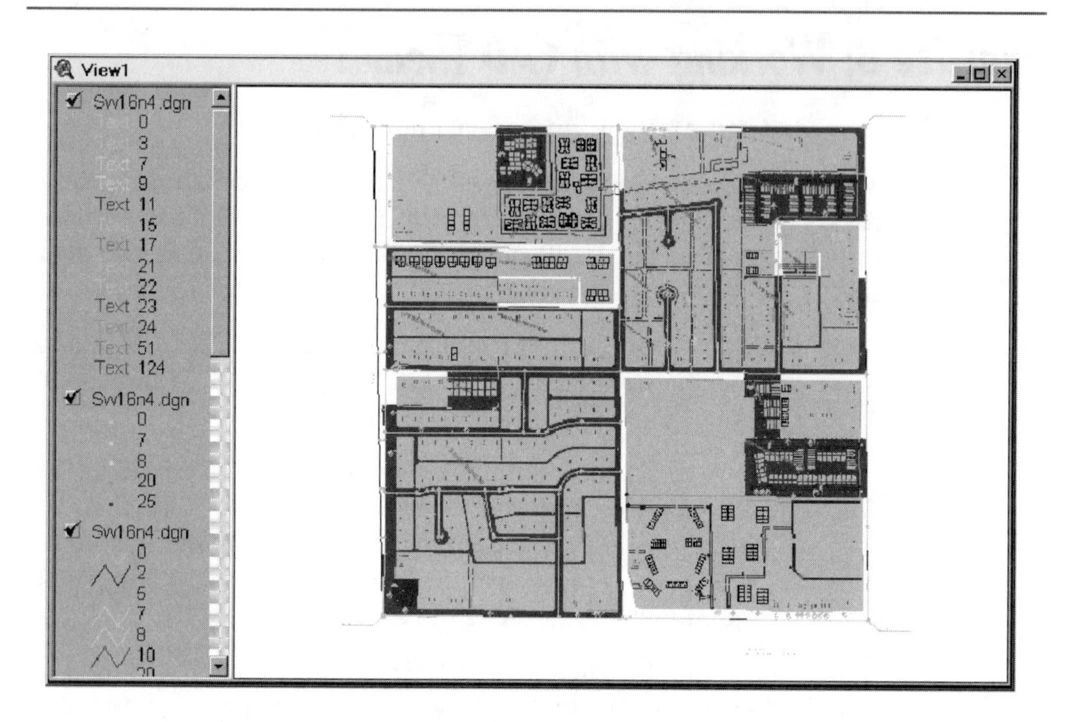


*Four themes added from CAD drawing Sw16n4.dgn, with default classifications. The polygon theme has been dragged to the bottom of the Table of Contents so that the other themes display properly.*

For your initial operations, you will work with the polygon theme from this drawing.

1. The legend for the polygon theme initially presents five classes, with the values 0, 8, 25, 56, and 60. Make the polygon theme active, and double-click on the legend to access the legend editor. Leave the Legend Type set to Unique Value, and select Layer from the Values Field pulldown list. Click the Random colors icon until a distinctive palette is obtained, and apply the classification.

2. Turn on the polygon theme to display it in the view. The theme contains an assortment of parcel lots and building outlines. As the theme draws, notice that the polygon shapes appear to overlap. Also notice

that although Layer 56 is present in the legend, no corresponding features are visible in the view.

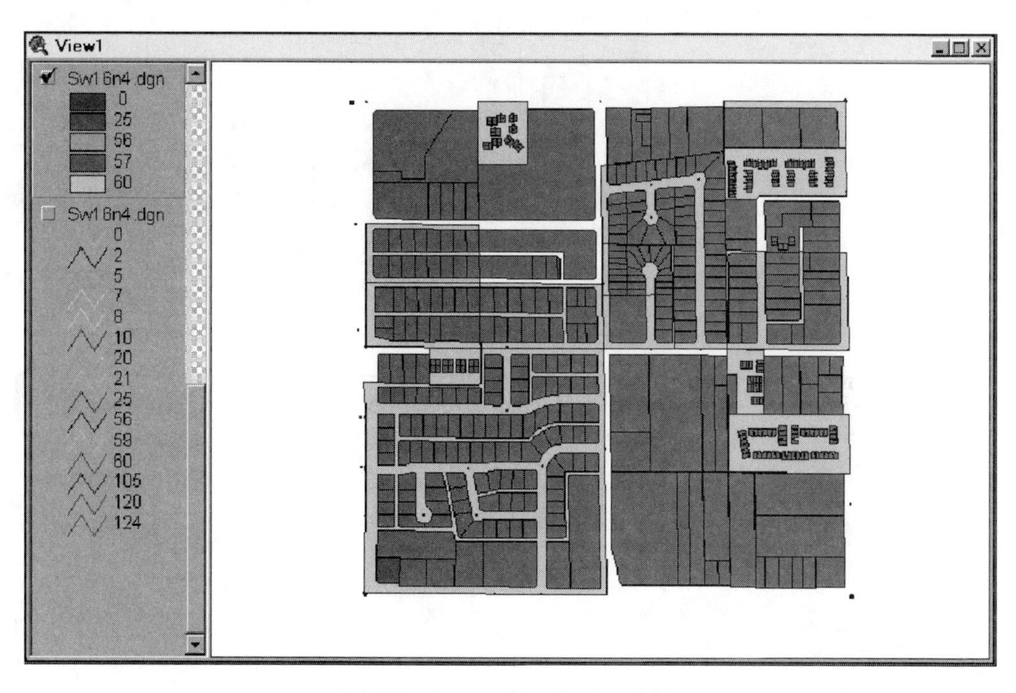


*Polygon theme classified on layer.*

3. To examine this more closely, copy the theme by making it active and selecting Edit | Copy Themes followed by Edit | Paste. A copy of the polygon theme is placed at the top of the Table of Contents.
4. Make the newly copied theme active, then open the Theme Properties dialog window by selecting Properties from the Theme menu. First, to avoid confusion, change the theme's name to Layer 56. Next, scroll down the list of icons at the left and click on the Drawing icon. Note that all layers are currently highlighted. From the Layers list, select 56, and click on OK to apply.

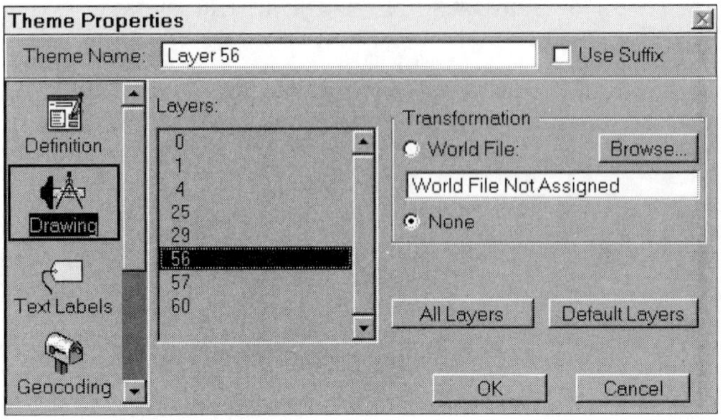

*Theme Properties dialog window for the Layer 56 theme with 56 highlighted in the list of layers.*

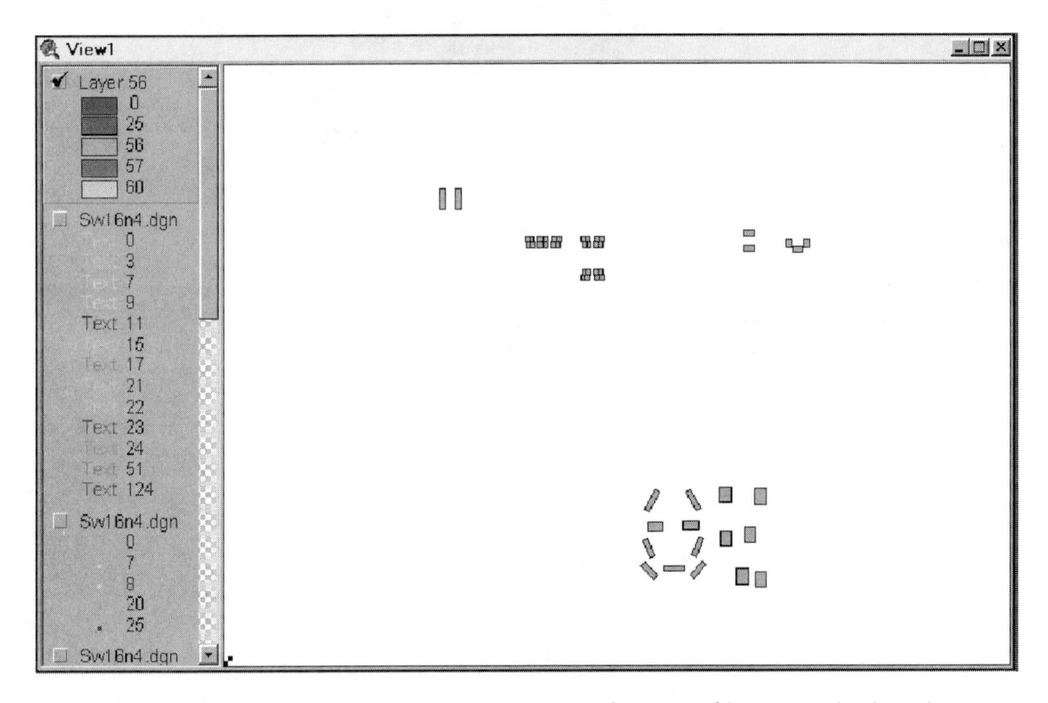


*Rsultant Layer 56 theme with only the polygon features of layer 56 displayed.*

5. Turn off the *Sw16n4.dgn* theme, and turn on the *Layer 56* theme to display the results. The building outlines placed on *Layer 56* are now visible. Turn on the *Sw16n4.dgn* theme. With the *Layer 56* theme uppermost in the Table of Contents, the building outlines now draw over the features drawn from the *Sw16n4.dgn* theme. Note, however, that the legend for the initial classification is still visible in the Table of Contents.

6. With the *Layer 56* theme active, click on the Zoom to Active Themes icon from the button bar. Note that the redrawn view is centered on the extent of only those features contained in *Layer 56*. Next, click on the Open Tables of the Active Themes icon on the button bar to open the table for the *Layer 56* theme. Fifty records are displayed. As you scroll through the table, notice that only the records for *Layer 56* are present.

   Setting the active layers from the Drawing section of the Theme Properties window, then, operates in a manner similar to applying a logical selection via the Query Builder from the Definition portion of the Theme Properties dialog window. The result is that only selected features are available for query and display.

7. Finally, double-click on the *Layer 56* entry in the Table of Contents to access the Legend Editor. From the Values Field pulldown, click again on Layers. When you do so, the legend changes to only one entry, for *Layer 56*. Click on Apply to apply this change to the view.

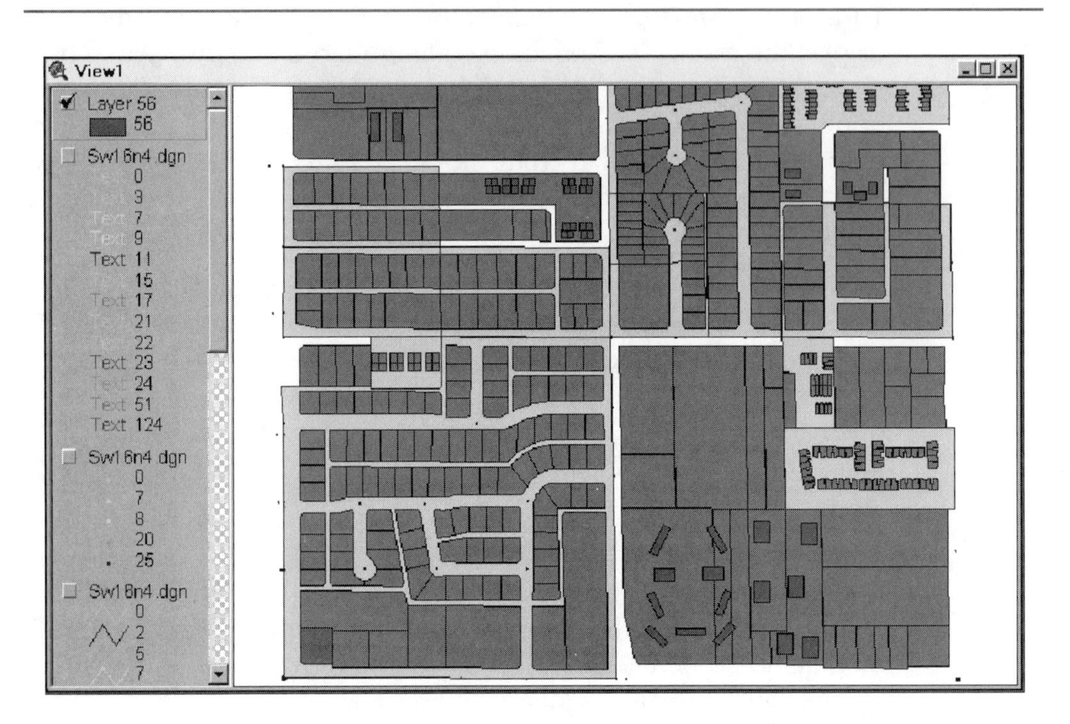


*Reclassified Layer 56 theme displayed over the original Sw16n4.dgn polygon theme.*

8. Repeat the steps above to create a new theme that displays only polygon features of *Layer 25*. Only one feature should be present in the *Layer 25* theme. (The incremental project has been saved as *ch9a.apr*.)

You will now work with the line theme from this drawing.

1. Turn off all other themes, and click once on the line theme in the Table of Contents to make it active, and then double-click on the theme in the Table of Contents to access the Legend Editor. Using the Unique Values legend type, classify on the Layer field. Click the Random Colors icon until a suitable palette is obtained, and apply.

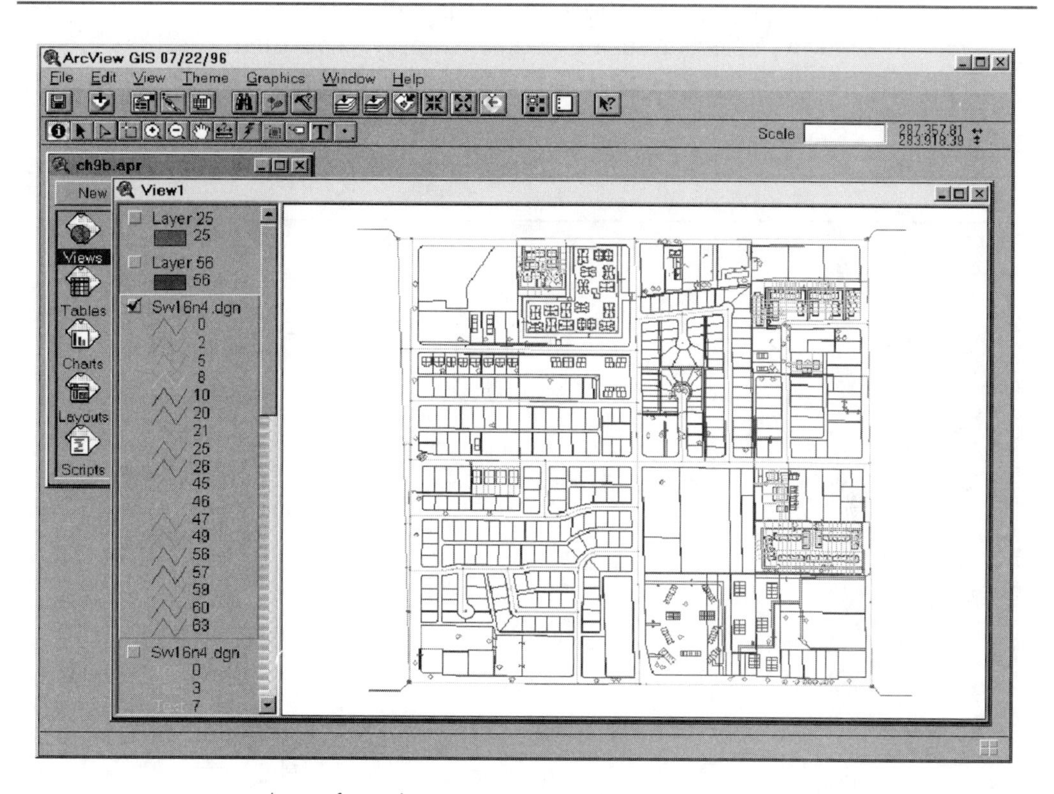


*Line theme from drawing Sw16n4.dgn, classified on layer.*

**2.** With 18 different layers present, note that the resultant legend is somewhat unwieldy. One way to examine features present on specific layers is to set all layers to display with the same color, then change the target layer to a contrasting color. Still in the Legend Editor, double-click on the box for the first symbol to access the Color Palette window.

**3.** Click on the box for the first symbol so that it is highlighted. Next, scroll to the bottom of the class list and, with the <Shift> key held down, click on the box for the last symbol. All 18 classes should now be highlighted. From the Color Palette window, select a medium green color. Return to the Legend Editor and apply.

4. Next, highlight only the symbol for *Layer 25*. From the Color Palette window, select a strong red and apply. The theme is now drawn with all lines from *Layer 25* drawn in red, and all other lines drawn in green.

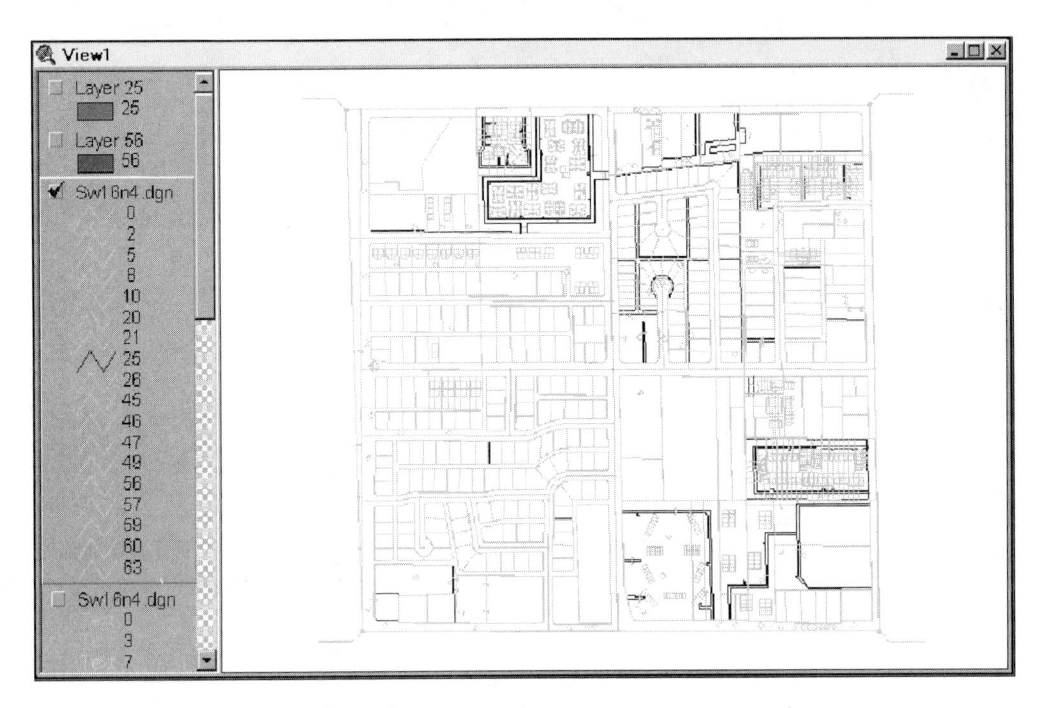


*Line theme with Layer 25 drawn in a contrasting color.*

5. Display the line theme, along with the polygon theme for *Layer 25*. Note how many features draw on *Layer 25* from the line theme as opposed to the polygon theme. Open the attribute table for the line theme, and use the Query Builder to select those features for which *Layer = 25*. Promote these records to the top—you should have 333 of 4,781 records selected—and examine the value for Entity. Most of the features on *Layer 25* are Line Entities. A small number are Complex Chain Entities, and one feature is a Shape Entity.

6. Next, make *Layer 25's* polygon theme active and open the attribute table for this theme. One record is displayed. The entity type for this feature is Shape.

7. To examine this further, copy the line theme. On the copy, set the Theme Properties via the Drawing section so that only *Layer 56* is displayed. Via the Legend Editor, reset the theme's legend so that only *Layer 56* is displayed in the legend, and apply. Display the resultant theme in conjunction with the polygon theme for *Layer 56* created earlier.

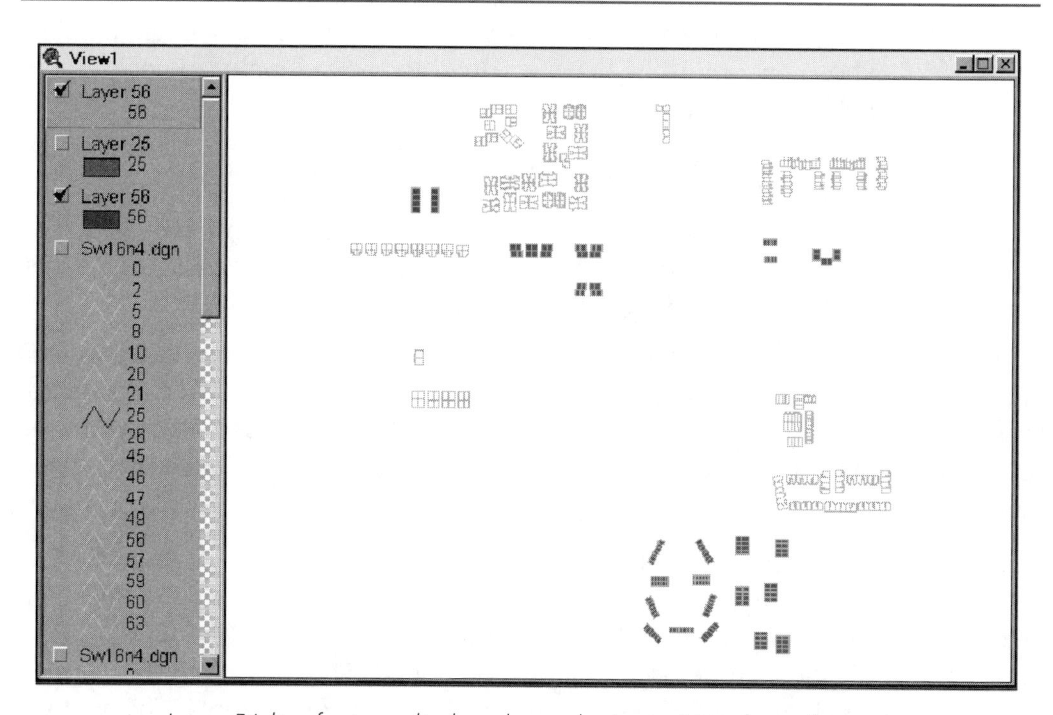


*Layer 56 line features displayed over the Layer 56 polygon features.*

8. Note that the additional line features on *Layer 56* all describe polygons, unlike the additional line features present on the line theme for *Layer 25*. Open the attribute table for the *Layer 56* theme, and exam-

ine the values for Entity. The majority of the entities are Line entities, with a much smaller number of entity type Shape. Next, access the Query Builder, and select the set for which *Entity = Shape.* Note that the selected features correspond with the features present in the *Layer 56* polygon theme. (This incremental project has been saved as *ch9b.apr.*)

Two observations can be made from this test:

1. Only specific features will translate as polygon features in an ArcView theme. In order to translate as a polygon feature, the corresponding CAD element must be comprised of a *single* closed feature. This can be a geometric shape, such as a circle or square, or a single polyline whose endpoint has been snapped to the beginning point. An element comprised of multiple entities, such as a polygon defined by several polylines, will be translated as line features, even if all polylines are snapped and the resultant element is closed.
2. All entities translated as polygon features are also translated as line features. If it is not acceptable to have duplicate features among themes, some post-processing of the theme must be performed.

In conclusion, if you want to create ArcView polygon themes based on a CAD drawing, special attention must be given to how these features have been captured in the CAD environment. The constraints on how features are drawn in the CAD drawing may be markedly different from the methods ordinarily used when drafting or digitizing.

Alternatively, intermediate translation from a CAD drawing to an ARC/INFO coverage can provide increased flexibility when translating CAD elements as polygon features. Translation to an ARC/INFO coverage, followed by ARC/INFO's CLEAN or BUILD command, can be used to allow the connectivity described by snapped CAD entities (such as polylines) to be resolved as polygon features. This is done through the building of the topological data structure for the coverage. The modules to convert a DXF file to an ARC/INFO coverage, and to build topology for the coverage, are also contained in the Data Automation Kit.

# Exercise 9: Working with Image Themes

If you saved and closed the project you were working on in Exercise 8, open it again, or open the incremental project titled *ch9b.apr*.

1. You are now ready to add an image to the view. Click on the Add Themes icon in the button bar; select Image Data Source from the Data Source Types pulldown list, navigate to the *$IAPATH/data* directory, and select *sw16n4.tif*. The resultant theme is added to the top of the Table of Contents.

2. Drag it to the bottom of the Table of Contents and turn on this theme to display the TIFF image. If you still have the *Layer 56* themes turned on, notice how the building outlines align with the buildings in the image.

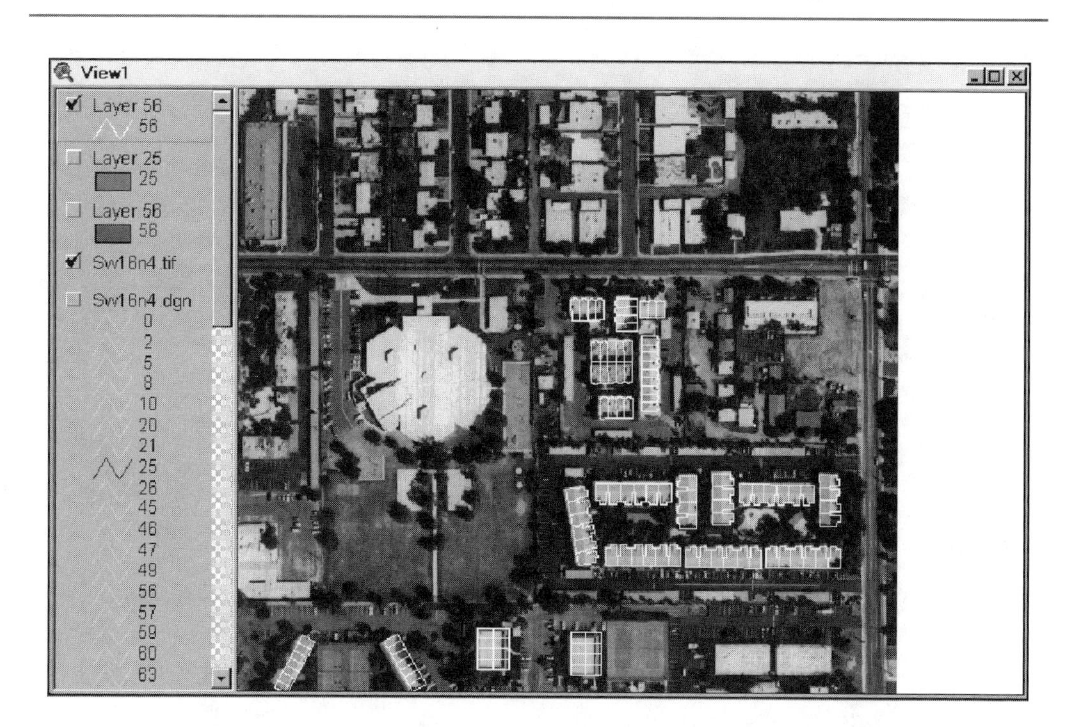


*Portion of the sw16n4.tif image with the Layer 56 building outlines displayed on top.*

The *sw16n4.tif* image is a grayscale image that was scanned from an aerial photograph. This format provides high resolution (one-foot pixel size), which is desirable for use in conjunction with engineering data such as CAD drawings. Although grayscale images are limited in terms of display manipulation when compared to multiband color images, there are still some adjustments that can be made to enhance their display in the view.

1. The Legend Editor for image themes is accessed in the same manner as that for feature themes, by double-clicking on the theme entry in the Table of Contents. So, double-click now on the image theme to access the Image Legend Editor.

2. The primary adjustment made to grayscale images is to adjust the brightness or contrast of the image. To perform brightness or contrast adjustments, click on the Linear button in the Image Legend Editor dialog window. This brings up a separate Linear Lookup window.

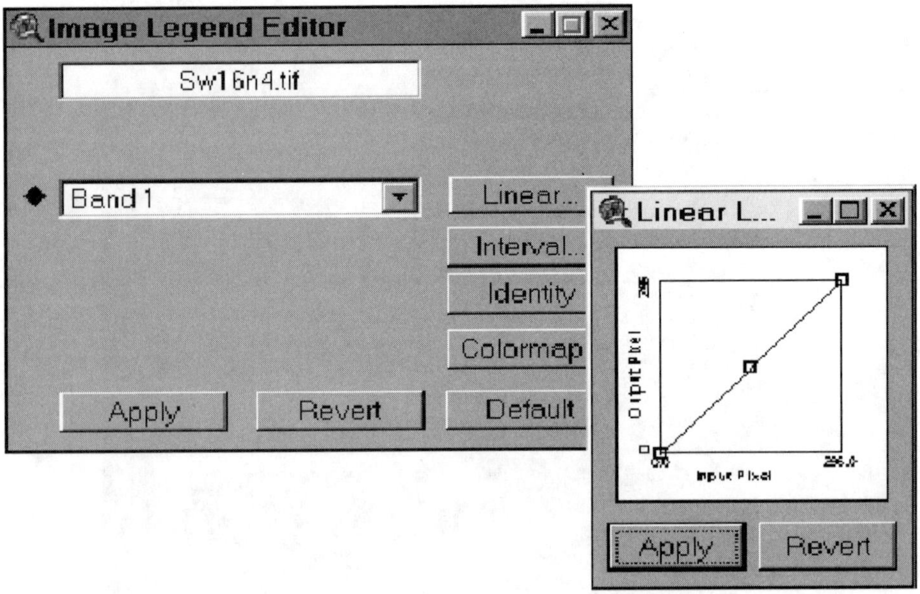

*Image Legend Editor with the Linear Lookup window displayed on top.*

3. In the Linear Lookup window, the diagonal line displays the input pixel value relative to the output pixel value. A diagonal line connecting the lower left and upper right corners indicates that the output pixel value is the same as the input pixel value, or that there is currently no transformation.

   By controlling the position and slope of the line, the output pixel value can be transformed. Three handles are provided along the line to effect this transformation. The middle handle is used to drag the line to the left or right, without changing the slope of the line. This has the effect of increasing or decreasing the overall brightness of the image without affecting image contrast.

4. Grab the middle handle and move the entire line to the left. Click on Apply to transform the image. You will notice that the image becomes lighter. Conversely, moving the line to the right causes the image to become darker. Try it now. Move the line back to the left, to lighten the overall image, and apply again.

5. By grabbing the top or bottom handles on the line, the slope of the line can be increased or decreased. Decreasing the slope of the line decreases the contrast of the image, while increasing the slope of the line increases the contrast of the image. Grab the top handle, and drag it to the right so that the top of the line ends at or just to the right of the upper right corner. Apply this change and notice that the overall contrast of the image is improved.

   This particular enhancement is often useful because the darkest values, which are all mapped to the same value of black, are essentially discarded. In this manner the remaining values in the image—say, from 30 to 255—are displayed using the full range of 256 shades of gray, effectively increasing the contrast range in the useful part of the image.

The Image Legend Editor dialog window contains the Colormap button, which is used to define the colors that display each of the 256 values in an image. By default, these are assigned to a scale of grays between black and white; however, this can be changed.

1. Click on the Colormap button.

2. From the Image Colormap dialog window, double-click on the black cell that is assigned to the 0 value in the image. This brings up the Symbol Palette.

3. Access the Color Palette, and select the pure blue value to assign to 0 in the image. Next, click on the Ramp button and apply. The image is now drawn in bluescale rather than grayscale. This can be useful for reducing the overall visual impact of the image, and for increasing the contrast between the image and any features displayed against the image.

To produce specific effects, the color for any image value can be arbitrarily changed. Next, examine a rudimentary schema that assigns all values in the image to one of three colors.

1. First, scroll down the values until you locate the cell for the 100 value. Highlight this cell and, holding down the <Shift> key, scroll back toward 0 until all cells holding values between 0 and 100 are highlighted.

2. Select the strong blue color from the Color Palette to assign this color to all selected cells.

3. Next, select values 101 through 18 and assign the magenta color to them; select values 190 through 255 and assign the yellow color to them. After applying this schema, examine the view. The result is a posterized image in which buildings appear in yellow, sidewalks and driveways and similar features appear in magenta, and the remaining features are drawn in blue.

Depending on the imagery and the features involved, this process, referred to as image slicing, can produce useful results.

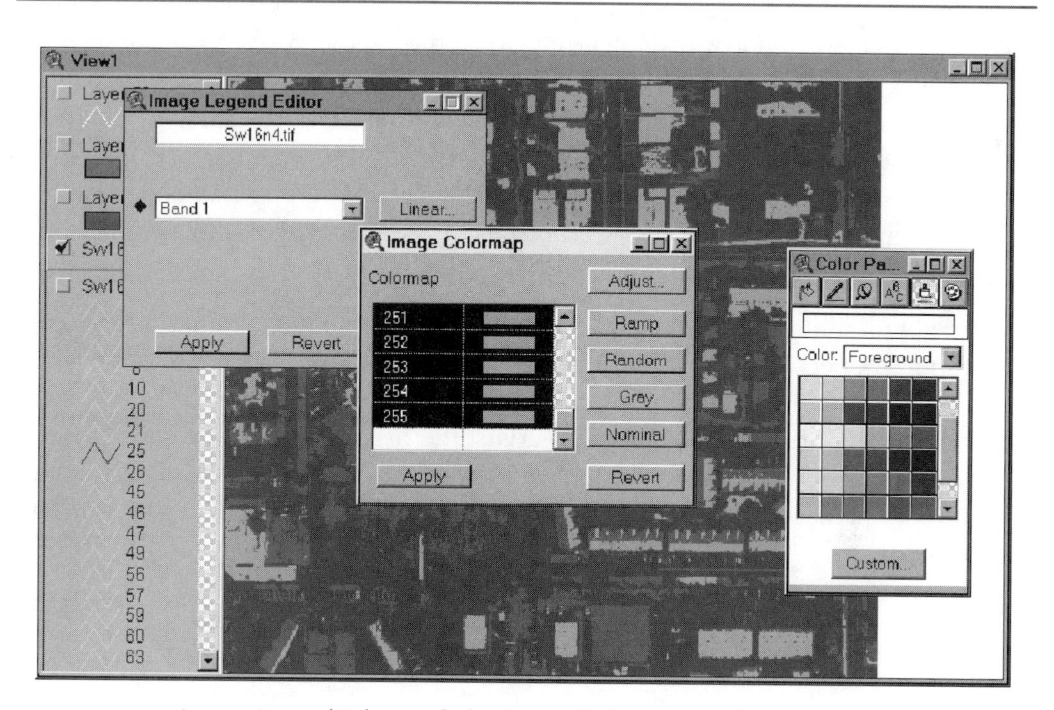


*Image Legend Editor with the Image Colormap window applied. The three-color classification has been applied to the image.*

The Random button on the Image Colormap dialog window can be used to trash your image in a hurry, but don't worry—clicking on the Gray button in the Image Colormap dialog window will return the image display to the default grayscale.

The principles for working with multiband images are basically the same as the ones we have covered for grayscale images. You still use the Linear Lookup control for adjusting image contrast and brightness—only now there are three bands instead of one.

Additionally, the Image Legend Editor for a multiband image displays three band pulldowns, one each for the red, green, and blue guns in the display. In this way, you can control which color in the display is used to portray each band in the image. In satellite imagery, typically more than three bands are provided; accordingly, only three bands of this image can be displayed at any one time.

If you have access to the sample data that shipped with ArcView 2.*x*, you can experiment with a multiband satellite image of the Atlanta area. Navigate to the /*atlanta/images* subdirectory and load the *Eosatimg.bil* image. Experiment with assigning bands to colors on the display, and with altering the image brightness and contrast. Assigning bands 1, 3, and 7 to the red, green, and blue guns of the display will produce a realistic color image, whereas assigning bands 5, 3, and 2 to the colors red, green, and blue will produce a useful false color image, in which vegetation appears in reddish colors.

The remaining tools in the Image Legend Editor—Interval and Identity—are used to manipulate specialized types of imagery. The Interval tool is used to control the class breaks for the display of continuous range imagery, as seen in digital elevation images. The Identity tool is used to adjust the display of discrete class images, as seen in images classified to display vegetative types or land use. The Identity tool can also be used to manipulate the display of discrete class data stored in the GRID format.

## Chapter 10

# Editing Shapefiles

Despite the plethora of data available in ArcView-compatible formats, at times you have no choice but to edit existing data or create new spatial data from scratch. As of Version 3.0, ArcView provides a robust toolkit of functions to create or edit data maintained in the shapefile format. Editing can be performed in a "heads up" manner by using the mouse to manipulate features on the display. You can also edit with a digitizer by loading the Digitizer extension. In our discussion of editing shapefiles, we will refer to editing with a mouse; the same functionality is available when editing with a digitizer.

## Heads-Up Editing and Digitizing Compared

Those coming to ArcView from the CAD or workstation GIS world need no introduction to the use of digitizing tablets. For others, whose exposure has been focused in the desktop environment, a brief overview might be in order.

A digitizing tablet is, fundamentally, another input device. It consists of a tablet and a pointing device, or puck. The pointing device can be configured to function identically to a mouse. In this "relative" mode, moving the puck moves the cursor on the screen. As with a mouse, it is not important where the puck is located initially, because moving the puck places the cursor at a new location *relative* to the starting point.

In *absolute* mode, each point on the surface of the digitizing tablet is mapped to a corresponding point on the screen. The puck must be moved to that specific point on the tablet each time the point on the screen is addressed.

When digitizing from a hardcopy map, the map is taped to the tablet and *registered* so that each absolute point on the tablet can be associated with a specific point on the map. This is done by establishing a *transformation* between the tablet coordinates and the map coordinates. The specifics of how this is accomplished will be covered in the section titled "Using a Digitizer" later in this chapter.

The advantage of editing with a digitizer is that features can be precisely located as they are input. In this manner, the accuracy present in the original map can be maintained. Digitizing tablets range in size from 12" × 12" to 48" × 60". Even a small tablet can be used to digitize large maps if the map is re-registered section by section. When using a tablet, coordinate input can be switched between absolute and relative mode at any point, allowing the puck to function as a mouse when needed.

# Getting Started

Before you begin editing, it helps to think a bit about the task you are about to perform. Is there a chance you may wish to return to the original state of the shapefile after you have completed your editing? If so, you may want to make a copy of your data before you begin.

> **NOTE:** *Copying the theme in the view does not copy the underlying data—both the old theme and the new theme point to the same shapefile on disk.*

One easy way to copy your data is to make the target theme active, then select Convert to Shapefile from the Theme menu. Give the shapefile a new name, add it as a theme to the view, and begin editing the new shapefile.

To begin editing, make a theme active, then select Start Editing from the Theme menu. A dashed line appears around the check box for this theme in the Table of Contents, indicating that the theme is currently being edited.

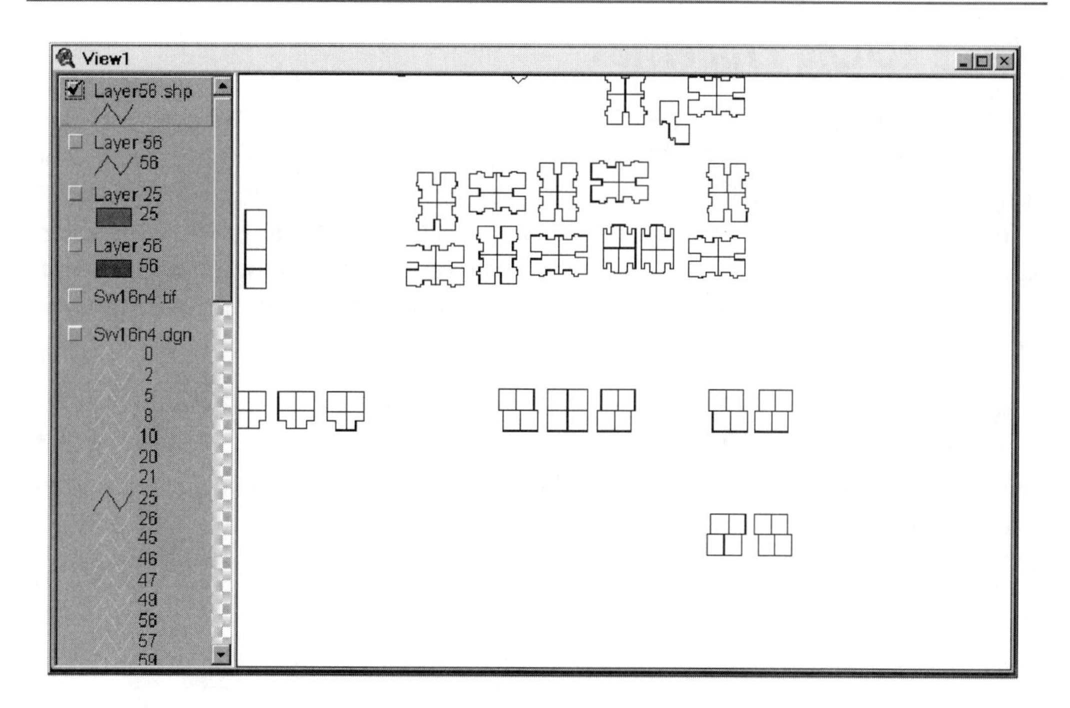


*Layer56.shp theme—the dashed line around the check box indicates which theme is currently being edited.*

Which tools are available for editing depend on the feature type for the theme being edited. Editing operations that are applicable to polygons may not be applicable to points or lines. All feature types share several basic editing properties and tools; we will cover these first, then move to the additional functionality specific to the feature type being edited.

When you are done editing, select Stop Editing from the Theme menu. You will be prompted to save your edits. Alternatively, you may save your edits incrementally—without stopping the editing process—by selecting Save Edits from the Theme menu. If you wish to save the edited theme as a new shapefile and leave the original unedited, select Save Edits As from the Theme menu.

# Theme Editing Properties

The Theme Properties dialog window is accessed by making the theme active and selecting Properties from the Theme menu. Clicking once on the Editing icon brings up the Editing section of the theme properties. In this dialog window, properties can be set that govern feature snapping while editing, and how attributes are handled when line or polygon features are split or merged. Snapping and attribute handling will be discussed in the following sections, along with the editing properties that are specific to each feature class.

## Editing Tools

Two editing tools, the Draw tool and the Pointer tool, are common to editing all feature types. A third, the Vertex Edit tool, is common to editing line and polygon features. Together, these comprise the basic tools available for feature editing.

The Basic Draw tools:

*From left to right, Point Draw, Straight Line Draw, Polyline Draw, Rectangle Draw, Circle Draw, and Polygon Draw tool icons.*

The Editing Draw tools:

*From left to right, Split Lines with Line Draw, Split Polygon with Line Draw, Append Adjacent Polygon Draw tool icons.*

The Selection tools:

*From left to right, Vertex Edit and Pointer tool icons.*

The Pointer tool is used to select features in a theme. After selection, a feature's selection handles are revealed. Once selected, a feature can be moved, resized, or deleted as desired.

To delete a selected feature, simply press the <Delete> key on the keyboard.

To move a selected feature, move your cursor inside the area encompassed by the selection handles. The cursor's appearance will change to a pair of crossed arrows. Click and drag the feature to a new location. Release the mouse button to place the feature at the new location.

To resize point and line features, place the cursor on top of one of the selection handles. The cursor will change to a single arrow. Click and drag the handle in or out to increase or decrease the size of the feature. In the case of polygon features, dragging on a corner handle will resize the feature proportionally, whereas dragging on a side handle will stretch the feature.

The Vertex Edit tool is used to reshape a selected feature by moving its individual shape points, or vertices. Vertices can be added, deleted, and moved as desired. The specifics of vertex editing will be covered in the following sections.

The Draw tools are used to add new features to the theme. Those drawing tools applicable to that feature type will be displayed in black; unavailable drawing tools will be grayed-out. The draw tools specific to each feature type will be discussed in the sections to follow.

Lastly, at any time during editing, selecting Undo Feature Edit from the Edit menu will allow you to undo your last edit. Selecting Undo Feature Edit more than once allows you to step back through the changes you made since you started the editing session. A Redo Feature Edit allows you to restore an editing change you inadvertently "undid."

## Editing Point Features

Editing point features is very straightforward. Because they lack dimension, they cannot be resized or split, only moved or deleted.

Use the Pointer tool to select point features. When a point is selected, its selection handles will be visible. Place the cursor inside the selection handles and click and drag the point to a new location. Pressing the <Delete> key when you have a point selected will cause that feature to be deleted.

To add point features, click on the Draw tool's pulldown list and select the Point tool. Using the mouse button, click in the view to add new points. Points can be snapped to existing points in the theme by enabling General Snapping through the Theme Properties menu. Click on Editing to access the edit properties for the theme. After enabling General Snapping and turning on the accompanying check box, a type-in field will become available. This field allows you to specify the snapping tolerance, in map units. Alternatively, with the Snap tool active, a circle can be drawn on the display to describe the snapping tolerance.

At any time during editing, pressing the right mouse button brings up a popup menu of editing options. From this menu, you can choose Undo Edit, Redo Edit, Clear Selection, General Snapping On, General Snapping Off, Zoom In, Zoom Out, Zoom to Selected, and Pan.

The attribute table for the point theme can be edited at the same time the point features are edited. Fields can be added or deleted, and new values can be entered as needed. When you stop editing a theme, the changes to the table also stop, and vice versa.

## Editing Line and Polygon Features

ArcView provides a robust set of tools for creating and editing line and polygon features. Lines and polygons can be added and deleted. Individual vertices can be added, deleted, moved, or snapped to other features. Vertices common to two lines can be moved or deleted either separately or as a unit. Lines can be split or merged and intersections resolved as desired. Polygons can be split or merged, or they can be digitized by defining only the new portion of the boundary and using the existing boundaries of adjacent polygons to complete the polygon.

Before editing line or polygon features, you may want to enable feature snapping. With snapping enabled, any vertex entered within a specified snap tolerance will be snapped to an existing vertex or line.

Two forms of snapping can be enabled: *general* snapping and *interactive* snapping. General snapping controls how vertices are snapped after a feature has been added or moved. Interactive snapping controls how vertices are snapped as a feature is added.

To set a theme's snapping properties, select Properties from the Theme menu. Click on the Editing icon to access the theme editing properties. Two check boxes are present, one for General Snapping and one for Interactive Snapping. Click on each box as desired to enable snapping; if either type of snapping is enabled, a type-in field will allow you to set the snapping tolerance. Alternatively, with the Snap tool active, a circle can be drawn on the display to describe the snapping tolerance. General Snapping and Interactive Snapping can be enabled separately, or together, as desired.

*Setting the editing properties from the Theme Properties dialog window.*

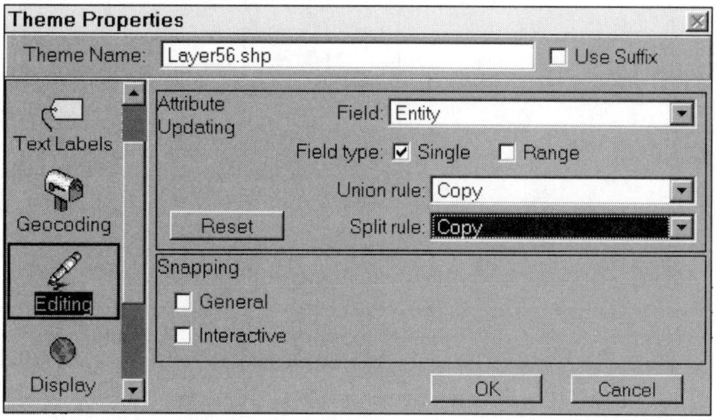

Use the Pointer tool to select line or polygon features. When a feature is selected, its selection handles will be visible. To move a feature to a new location, position the cursor in an area inside the selection handles. The cursor's shape will change to a pair of crossed arrows. Click and drag the feature to its new location.

*In this polygon theme, one feature is selected, and its selection handles displayed.*

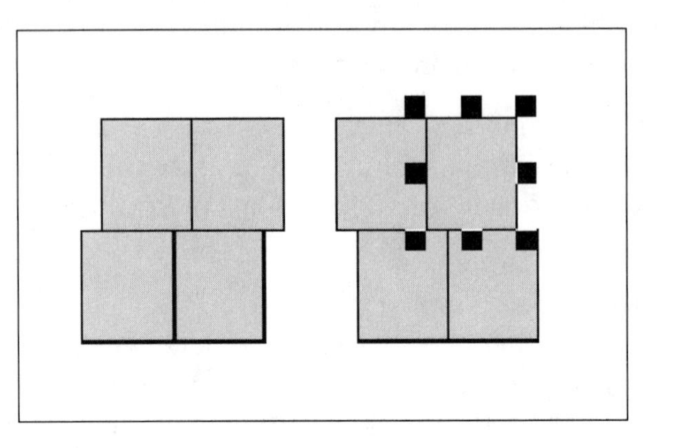

*Same polygon, dragged to a new location with the mouse.*

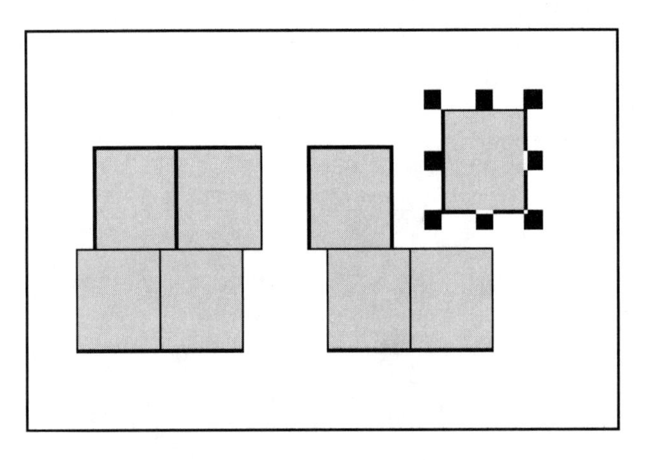

To resize a feature, move the cursor over one of the selection handles. The cursor will change to a single arrow. Click and drag to resize the feature. Resizing with a corner handle will resize the feature proportionally; resizing with a side handle will stretch the feature in that direction.

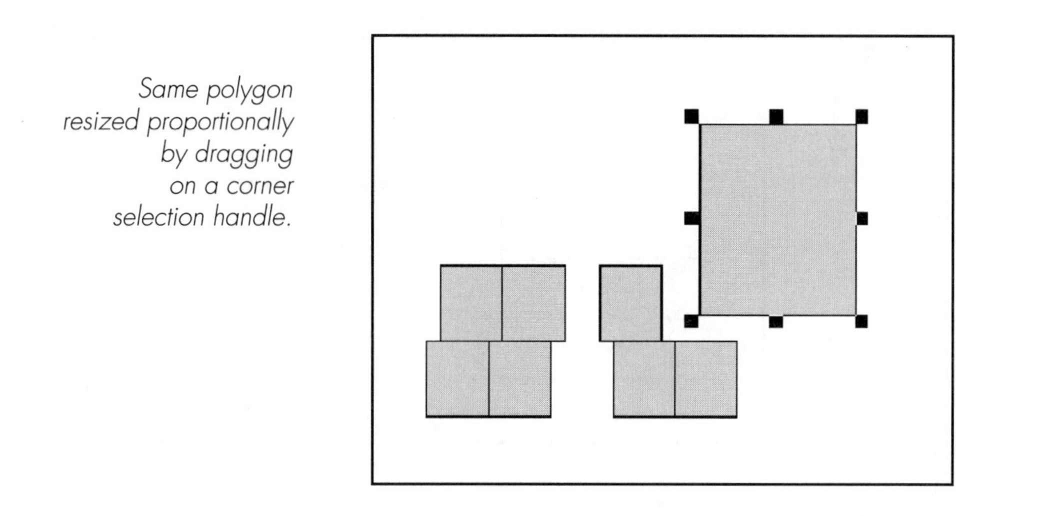

*Same polygon resized proportionally by dragging on a corner selection handle.*

The Vertex tool is used to edit individual vertices on a line or polygon feature. Clicking on the Vertex tool causes the vertices of the first of the currently selected features to be revealed; the Vertex tool can be used like the Pointer tool to select new features as well.

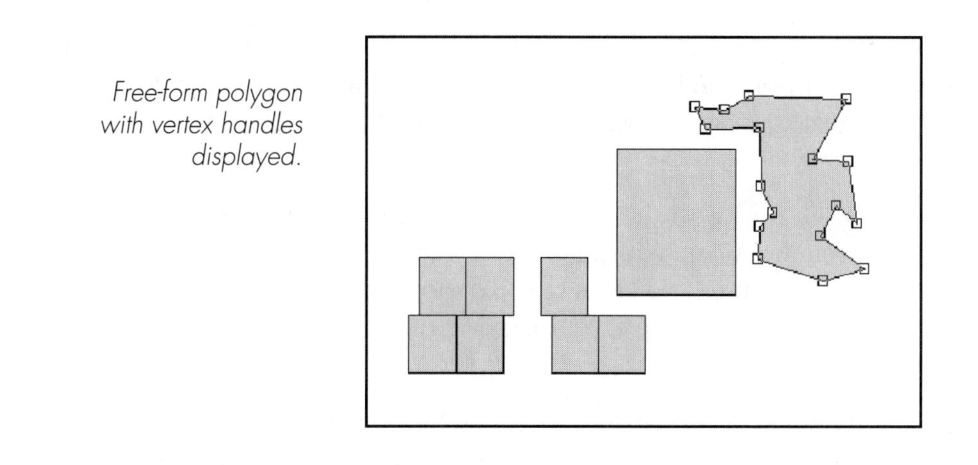

*Free-form polygon with vertex handles displayed.*

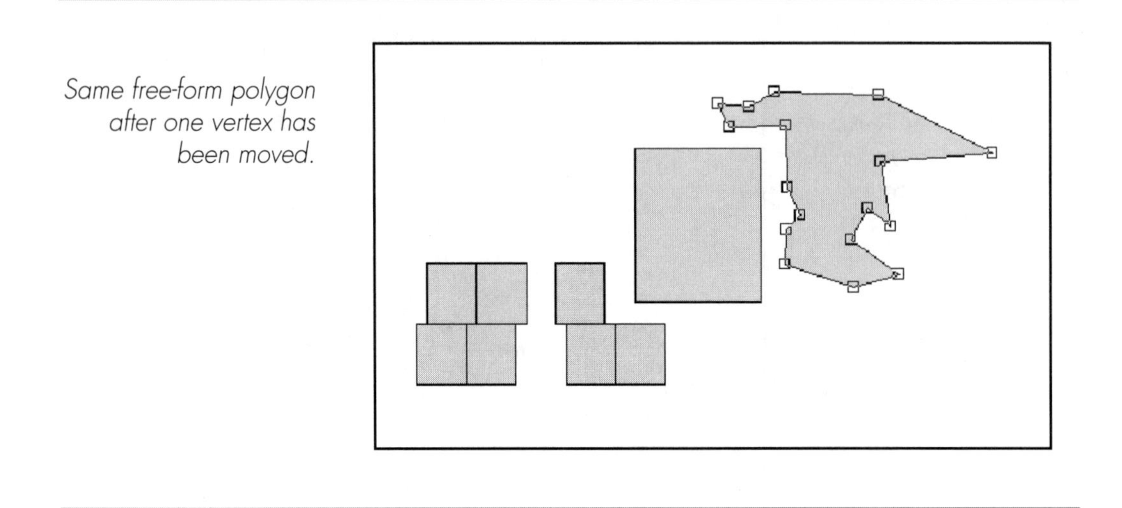

Same free-form polygon after one vertex has been moved.

***NOTE:** Holding down the <Shift> key while selecting features with the Vertex tool will cause the latest selection to replace the previous one as the currently selected feature. Conversely, holding down the <Shift> key while selecting features with the Pointer tool will cause any new features to be added to the set of currently selected features.*

With two exceptions, vertices can be edited for only one feature at a time; the exceptions are line or polygon features that share common nodes or common segments. For this initial discussion, we will describe editing vertices with no common nodes.

When a feature is selected with the Vertex tool, a hollow square is drawn at the location of each vertex. When you place your cursor over a vertex, the cursor shape changes from a pointer to a cross hair. To move a vertex, click and drag the vertex to a new location. To delete a vertex, place the cursor over it and press the <Delete> key.

It is also possible to add new vertices to a line or polygon. Place the cursor over a segment of the selected feature. The cursor shape will change from a pointer to a circle containing a cross hair. With the cursor thus displayed, click to add a vertex at that point. The new vertex can then be repositioned to reshape the feature as desired.

The Vertex tool also supports editing of line or polygon features that share common nodes. To move a node that is held in common by more than one feature, click directly on the node. The node will be appear as a hollow square. Additionally, the next vertices on all features sharing this node will be drawn as circles. When the vertex is dragged to a new location, all features sharing that vertex will be reshaped.

*Editing a node that is shared by two polygons. The common node is a hollow square; the adjacent nodes are circles.*

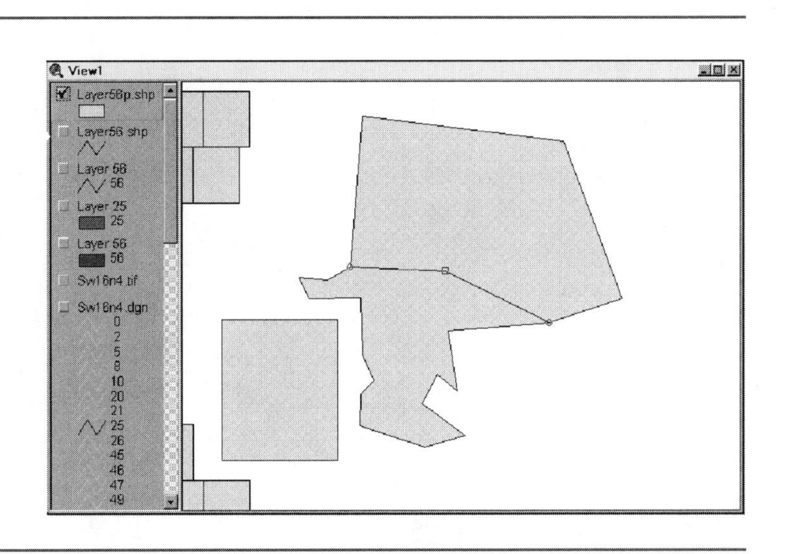


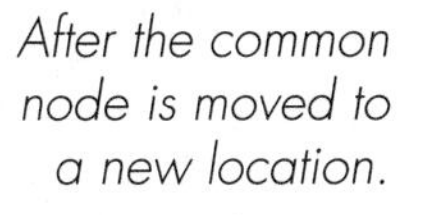

*After the common node is moved to a new location.*

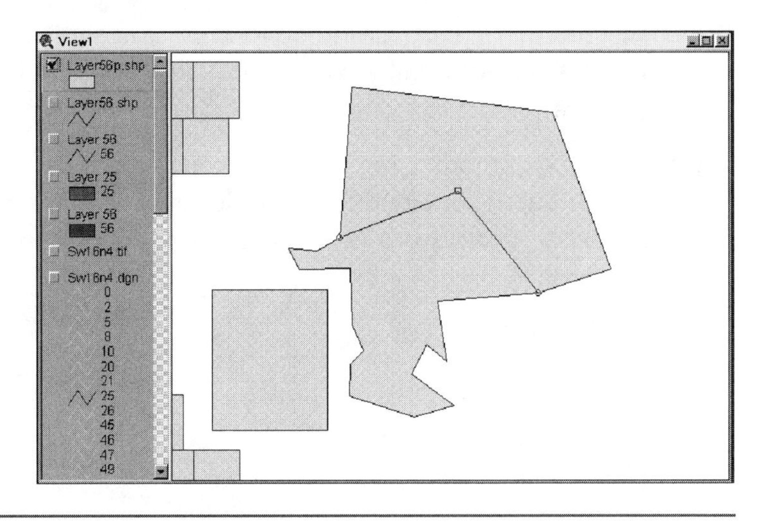

Additionally, clicking on a segment that is common to more than one feature will cause all edits to be performed on all common segments. Click on the common segment away from the vertices; the vertices in common to the line or polygon features will be drawn with squares, and the vertices at the ends of the common segments will be drawn with circles. Any vertices added, deleted, or moved will be edited simultaneously for all common features.

*Menu for polygon feature editing, which pops up with a click of the right mouse button.*

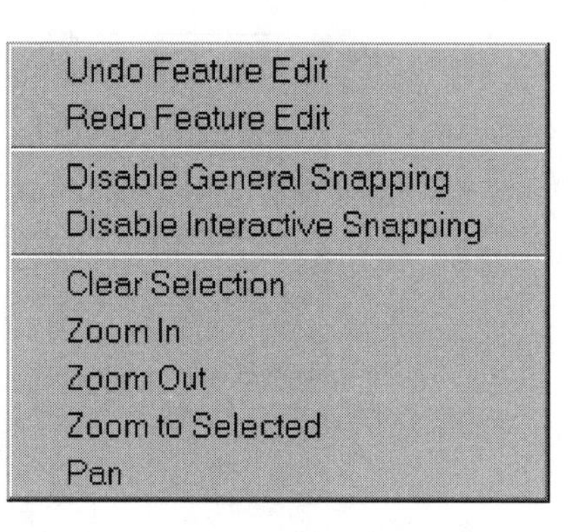


## Adding Line Features

The Draw tool can be used to add new line features to a line theme. Two tools are active on the Draw tool pulldown list: the Line tool and the Line Split tool.

When adding lines with the Line and Line Split tools, you begin each line with a single click of the mouse. Intermediate vertices are designated through a single click, and the line is terminated with a double-click. When adding lines, clicking the right mouse button brings up a popup menu of the following digitizing options:

- Delete Last Point
- Undo Edit
- Redo Edit
- Clear Selection

- General Snapping On
- General Snapping Off
- Interactive Snapping On
- Interactive Snapping Off
- Zoom In
- Zoom Out
- Zoom to Selected
- Pan

Additionally, if Interactive Snapping has been enabled under Theme Properties, the popup menu will also contain the following snapping options: Snap to Vertex, Snap to Boundary, Snap to Intersection, and Snap to Endpoint. These options govern how the next added vertex will be snapped.

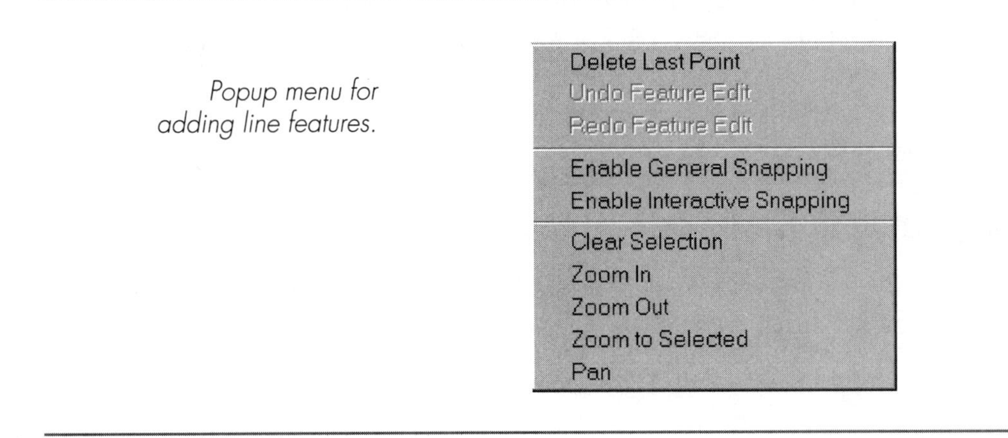


*Popup menu for adding line features.*

If a line created with the Line tool crosses another line, the two lines do not topologically intersect; they remain as discrete two lines with no common nodes. The Line Split tool can be used to create lines where all intersections are resolved. When a new line created with the Line Split tool crosses an existing line, both the new line and the existing line are split at the point of intersection, resulting in four line features.

Conversely, two or more line features can be joined into a single shape. Using the Pointer tool, select the lines to join, then select Union Features from the Edit menu to join the features into a single shape.

## Line Editing and Attribute Handling

When line features are either split or "unioned," rules can be set that will govern how the attributes are handled in the resultant feature. By default, when a line feature is split, the attributes are copied to each new line. When line features are unioned, the attributes from the first record in the table are copied to the resultant line feature. The rules for handling attributes can be changed so that the values in the new fields are left blank. Additionally, rules can be set for numeric fields such that the new value for the field will be proportional to the length of the new lines. For unioned lines, the values for numeric fields can be added or averaged as well. The specific rule that governs attribute handling can be set for each field in the feature attribute table.

Additionally, special rules can be set for fields that need to be considered together when determining attribute values for new features. These fields, such as address fields, are referred to as *range fields.* In range fields, two fields are paired for attribute handling. In the example of address fields, paired fields could include the Left Address From field and the Left Address To field, as well as the Right Address From field and the Right Address To field. Two options are available for handling range field attributes: Address, which maintains the even or odd values necessary for address geocoding; and Continuous, for continuous values.

## Adding Polygon Features

The Draw tool can be used to add new polygon features to a polygon theme. The following tools are available from the Draw tool pulldown list:

- Circle
- Rectangle
- Polygon
- Polygon Split
- AutoComplete

When adding polygons with the Circle tool, position the cursor at what will be the center of the circle. Click and drag in any direction to define the radius. Release the mouse button when the desired radius is obtained.

To add polygons with the Rectangle tool, place the cursor at what will be one corner of the rectangle. Click and drag in any direction to form the shape of the rectangle. Release the mouse button when the desired rectangle is obtained.

When adding polygons with the Polygon tool, begin outlining the polygon with a single click of the mouse. Indicate any intermediate vertices by further single clicks. Close the polygon with a double-click. Clicking the right mouse button brings up a popup menu of digitizing options that can be seen below: Delete Last Point, Undo Edit, Redo Edit, Clear Selection, General Snapping On, General Snapping Off, Interactive Snapping On, Interactive Snapping Off, Zoom In, Zoom Out, Zoom to Selected, and Pan. Additionally, if Interactive Snapping has been enabled under Theme Properties, the popup menu will also contain snapping options: Snap to Vertex, Snap to Boundary, and Snap to Intersection. These options govern how the next added vertex will be snapped.

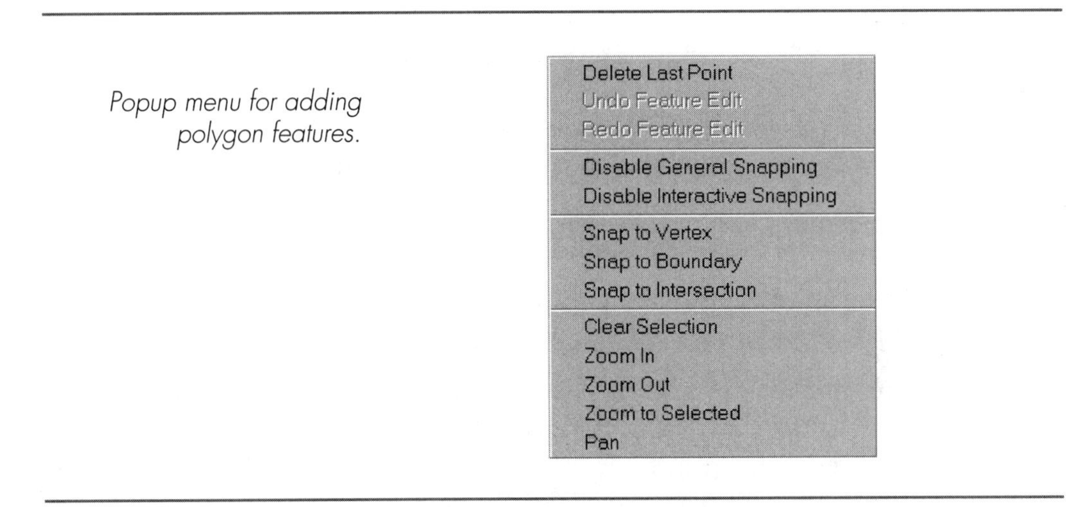


*Popup menu for adding polygon features.*

If a polygon created with the Polygon tool overlaps another polygon, the two polygons remain discrete, and do not topologically intersect. To split a polygon, select the Polygon Split tool from the Draw tool pulldown list. Add a line across the polygon for defining the boundary on which the polygon is to be split.

The Append Adjacent Polygon (AutoComplete) tool, available from the Draw tool pulldown list, can be used to append a new polygon to an existing adjacent polygon. The polygon is described by digitizing only the new portion of the polygon you are creating, using the segments of the existing polygons to complete the new polygon. Begin digitizing the new portion of the polygon boundary just inside the existing polygon, adding vertices with single clicks. Complete the new portion of the polygon by double-clicking to end the line, again just inside an existing polygon. (You can undershoot slightly if you have General Snapping enabled, provided that your ending double-click and the existing polygon are within the snapping tolerance.)

# Aggregating Polygon Features

Four functions are available for aggregating selected polygon features into new features: Union Features, Combine Features, Subtract Features, and Intersect Features. Each function handles overlapping polygons differently during the creation of the output shape. To perform any of the functions above, a minimum of two features must be selected.

Union Features merges the area of all selected features into a single feature. If selected polygons share a common boundary, the common boundary is removed. If selected polygons overlap, all boundaries of overlap are removed. If the selected polygons are not touching, the result is a single shape with multiple parts.

Combine Features merges the area of all selected features into a single feature. It differs from Union Features in that all areas of overlap between selected polygons are removed from the output polygon. In this manner, Combine Features can be used to create a hole in an existing polygon.

Subtract Features uses the area of overlap from one polygon to remove the corresponding area from the underlying polygon. You can envision this by picturing the overlying polygon taking a "bite" out of the underlying polygon. To reverse the order of operation, and have the underlying polygon take the bite out of the overlying polygon, hold down the <Shift> key while selecting Subtract Features from the Edit menu.

Intersect Features returns only the area of overlap between overlapping polygons. The remainder of the original polygons is removed.

The following four illustrations demonstrate the output from each aggregating operation.

*Rectangle with an overlapping circle before (left) and after (right) selecting Union Features.*

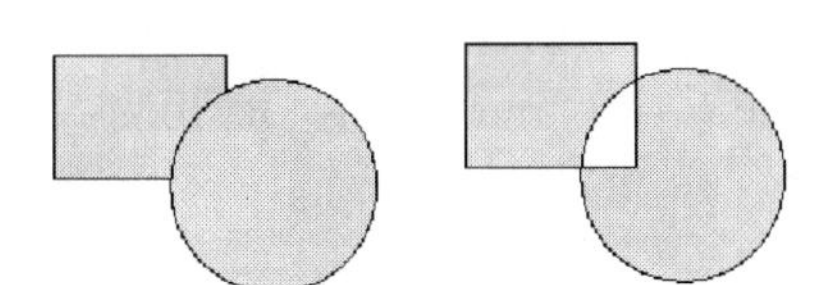

*Same rectangle with an overlapping circle before (left) and after (right) selecting Combine Features.*

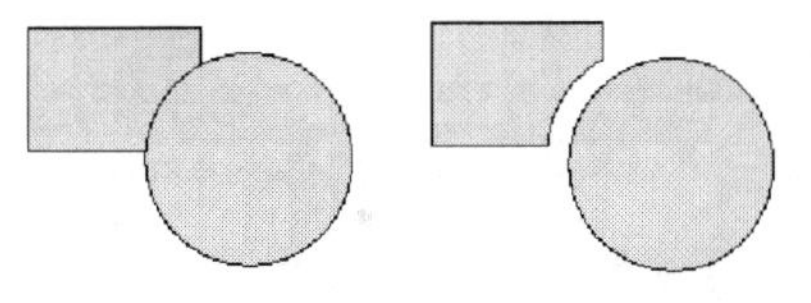

*Same rectangle with an overlapping circle before (left) and after (right) selecting Subtract Features. The resultant shapes have been pulled apart for clarity.*

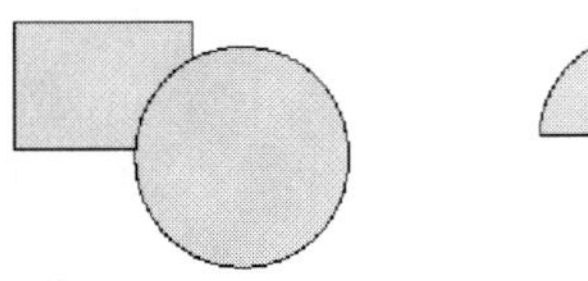

*Same rectangle with an overlapping circle before (left) and after (right) selecting Intersect Features.*

## Polygon Editing and Attribute Handling

Despite the additional options available, attribute handling for polygon editing is very similar to that for line editing. In all editing operations, polygons are either split or merged (unioned). By default, when polygons are split, field values for the original polygon are copied to the resultant polygons. When polygons are unioned, the attributes from the first record in the table are copied to the resultant polygon. The rule for handling attributes can be changed so that the values in the new fields are left blank. Additionally, for numeric fields, rules can be set such that the new value for the field is proportional to the area of the new polygon. For unioned polygons, the values for numeric fields can be added or averaged as well. The specific rule governing attribute handling can be set for each field in the feature attribute table.

# Using a Digitizer

Although heads-up editing on the screen with the mouse is adequate for routine feature editing, at other times a more precise method of feature input is needed. A digitizing tablet provides a precise method for feature input and editing.

**NOTE:** *Digitizing is not currently supported on UNIX or Macintosh.*

Before digitizing in ArcView, the driver for your digitizer must be loaded and configured. The driver for your digitizer must be WinTab-compliant. (If you do not have the correct driver, check your digitizer manufacturer's web page.)

To configure your digitizer, access the digitizer driver control panel. This is typically located in the Main group in Program Manager. In a typical configuration, the digitizer puck's left button—0 or 1—is typically configured to represent a left mouse button click. The right button—3 or 4—should be configured to represent a right mouse button click. Additionally, one button should be configured to represent a left double-click, which is used to end line and polygon features. Another button should be configured to represent a middle double-click. This is used to toggle the digitizer between *absolute* (digitizer) and *relative* (mouse) mode.

## Preparing the Map

To register a map, you need to know absolute coordinates for at least four points on the map.

1. Locate these points on your map, and write the coordinates next to each point.
2. Next, the map should be taped to the digitizing tablet so that it does not move while digitizing.

Before registering the map, ArcView's Digitizer Extension must be loaded. The Digitizer Extension is part of the core ArcView product.

1. To load the Digitizer Extension, make the project window active and select Extensions from the File menu.
2. In the Extensions dialog window, click on the check box next to Digitizer to turn it on, and click on OK.
3. Next, open the view that contains the theme—existing or new—in which you wish to digitize.

You may digitize in either absolute units or projected units. However, if a map projection is specified for the view, the digitized coordinates will be stored as geographic coordinates—even if the map is registered using real coordinates. This is what is referred to as digitizing in *projected coordinates.* Because the underlying coordinates for the theme are stored in geographic coordinates, the theme can be projected with different projections, if desired.

If no projection is specified for the view, the digitized coordinates will be stored as *absolute coordinates.* Because these coordinates are already projected, you will not be able to change the theme to another projection at a later date.

1. To register the map for the first time, select Digitizer Setup from the View menu. This brings up the Digitizer Setup dialog window.
2. In the Error Limit field, specify the maximum allowable error. (The default is 0.004 inches.)
3. Next, click on the Digitizer Puck icon to begin digitizing control points.

4. Click on a control point on the map. Type the X and Y coordinates from the map into the table in the window. Continue until you have input at least four control points.

Once you have input four control points, the RMS error (root mean square error of all points) will be displayed, along with the error associated with each point. If the error is smaller than your Error Limit, you can register your map. If the error is larger than your Error Limit, you must either re-digitize one or more control points on the map, or increase the error limit. (Note: Setting the error limit too high may result in inaccurate digitizing.)

*Digitizer Setup dialog window.*

Digitizer Setup

6426
4412
Load... Save...
Units Meters

| | Id# | X Coordinate | Y Coordinate | Error |
|---|---|---|---|---|
| * | 1 | 100000.000000 | 100000.000000 | 549.9705 |
| * | 2 | 100000.000000 | 400000.000000 | 551.8906 |
| * | 3 | 400000.000000 | 400000.000000 | 555.4831 |
| * | 4 | 400000.000000 | 100000.000000 | 553.5631 |

Error Limit: 0.008 in Error Report... Register
Calculated Error: 0.007 in Cancel

When four or more control points have been digitized, and the error is less than your error limit, you may register your map. If you intend to use the control point ground locations for more than one editing session, you may save the registered points for later use by clicking on Save before clicking on Register. This lets you save the registered points to a file. You will be prompted for the location and file name to store the registered points.

1. Click on the Register button in the Digitizer Setup dialog window to register the map and close the window.

2. To use previously stored control points, open the Digitizer Setup dialog window and select Load. Navigate to the proper directory, select the file containing the stored control points, and click on OK.

3. Highlight each point in the table, and digitize the corresponding point on the map. When at least four points have been digitized, compare the reported error with the Error Limit, and register or adjust as needed.

4. The map has now been registered to the digitizer. To digitize features into the theme, make the desired theme active and select Start Editing from the Theme menu. If desired, you may also create a new theme for digitizing.

5. When inputting features with a digitizer, the same Edit and Draw tools are used as when editing with the mouse. Toggle the digitizer puck from Absolute to Relative mode to select tools from the ArcView tool bar. (If you have not configured a button on your puck to toggle from relative to absolute mode, you may switch modes by pressing the F2 function key.)

6. The right mouse button, which you have mapped (ideally) to the right button on your digitizer puck, will bring up the Editing popup menu. This can be used to delete points, undo and redo edits, and pan and zoom as needed.

7. When you are finished digitizing, select Stop Editing from the Theme menu. You will be prompted to save your changes to the edited theme.

## Creating a New Theme

Most new themes are created by selecting features from existing themes, creating a new shapefile from the selected set, and editing as necessary. Occasionally, however, you will have no choice but to start entirely from scratch. When this occurs, rest assured: whether you need to draw a few polygons to delineate sales territories, or a few lines to describe alternative freeway corridors, ArcView provides the tools to accomplish this task.

1. To create a new theme, select New Theme from the View pulldown menu.

2. Select the feature type—point, line, or polygon—and specify the name and location for the new theme. After choosing a feature type, click on OK.

3. The theme is added to the Table of Contents. By default, the theme is editable and ready for you to begin adding shapes.

*New Theme dialog window.*

4. To add attributes to the new theme, make the theme active. Select the Open Theme Attribute Table icon from the button bar. The attribute table for the new theme is opened.

5. This table contains the Shape field and a record that corresponds to each added feature. By default, the table is editable. Additional fields can be added as needed, and attributes can be entered for each record.

If you need a refresher on editing tables, see "Working with Tables" in Chapter 11.

# Exercise 10: Editing Shapefiles

This exercise is going to be rather freeform! Since we cannot look over your shoulder to see what you are doing, we must trust that you will experiment until your results match the process we are describing. Remember—you learn by blowing things up, then looking at the pieces.

1. Begin by opening the project you saved at the end of Chapter 9. Actually, any project will do, including a new project, but the units associated with the CAD themes lend themselves to easily interpretable snapping tolerances.

2. First you need shapefile themes to edit. If you have not already done so, select the line and polygon themes.
3. Convert them to shapefiles by selecting Convert to Shapefile from the Theme menu. This will give you plenty of shapes to edit if you tire of creating your own. (If you would prefer to skip this step, you may open *ch10.apr* in the */INSIDEAV/projects* directory. This contains two themes derived from shapefiles, *line56.shp* and *poly56.shp,* which are ready to edit.)

## Line Editing

1. Begin by displaying a line shapefile theme and highlighting that theme in the Table of Contents to make it active.
2. From the Theme menu, select Start Editing. A dashed box appears around the theme's check box, indicating that it is presently being edited.
3. Pan and zoom in on a blank area adjacent to some existing line features. This will give you an expanse of blank canvas, as well as some line features to manipulate should you so desire.
4. Select the Line tool and add some polylines. (Remember that initial and intermediate vertices are placed with a single click; a double-click ends the polyline.)
5. Next, use the Pointer tool to select a line feature. Resize the line using the corner and side selection handles. (Resizing can be quite useful when copying features between themes where the features have been input at different scales or projections.)
6. Make the Vertex tool active and reshape the same line by adding, deleting, and moving vertices.
7. Next, turn on General Snapping by accessing the Editing section of the Theme Properties dialog window. Place a check in the box next to General for Snapping.

8. Back in the view, click on the Snap tool and drag a circle to describe the snapping tolerance. (Use a circle that approximates a tenth of an inch on the display.)

9. Add some new lines using the Line tool. Observe how lines are adjusted when new vertices fall within the snap tolerance of existing lines. Notice that, by following the vertices of an existing line (select the line using the Vertex tool to reveal the vertices), you can make a new line feature coincident with an existing line.

10. Return to the Editing section of the Theme Properties dialog window and enable Interactive Snapping. Now you can control the snapping that applies to the next vertex by using the right mouse button to call up the Editing popup menu. From there you can elect to snap the next vertex to an existing vertex, to a line segment (boundary), or to a line endpoint.

11. Switch to the Line Split tool and add additional line features. Digitize new lines to cross existing lines. Observe how both lines are split at the point of intersection.

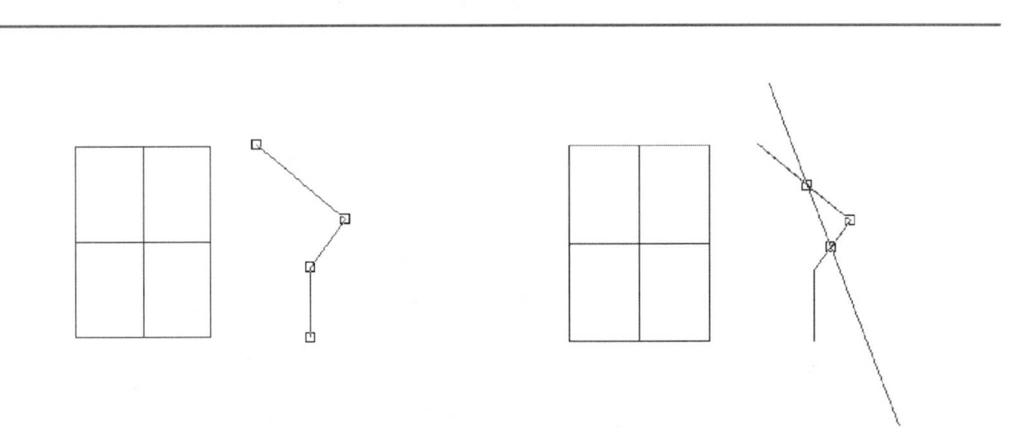

*Polyline with four vertices before (left) and after (right) being split with a single line using the Line Split tool. Note the new vertices created at the points of intersection.*

12. Finally, switch back to the Pointer tool. With General Snapping enabled, move a line segment adjacent to another line feature. Notice that vertices of the selected line are snapped to existing vertices within the snapping tolerance. General Snapping is enabled when moving or resizing features, not just when digitizing or editing vertices.

# Polygon Editing

1. Begin by displaying a polygon shapefile theme and highlighting the theme in the Table of Contents to make it active.
2. From the Theme menu, select Start Editing. A dashed box appears around the theme's check box, indicating that it is presently being edited.
3. Pan and zoom in on a blank area adjacent to some existing polygon features.
4. Select the Polygon tool and add some polygons. (Remember that initial and intermediate vertices are placed with a single click; a double-click closes the polygon.) Add some polygons that overlap existing polygons; note that new polygons overlap, rather than intersect, existing polygons.
5. Next, use the Pointer tool to select a polygon feature. Resize and reshape the polygon using the corner and side selection handles.
6. Switch to the Vertex tool; the vertices of the currently selected polygon will be displayed. Reshape the polygon by adding, deleting, and moving vertices as desired.
7. Next, enable General Snapping by accessing the Editing section of the Theme Properties dialog window. Place a check in the box next to General for Snapping.
8. Back in the view, click on the Snap tool and drag a circle to describe the snapping tolerance. (Use a circle that approximates a tenth of an inch on the display.)

9. Add several new polygons using the Polygon tool. Observe how polygons are adjusted when new vertices fall within the snap tolerance of existing polygons. Notice that, by following the vertices of an existing polygon (select the polygon using the Vertex tool to reveal the vertices), you can make the new polygon boundary coincident with that of an existing polygon.

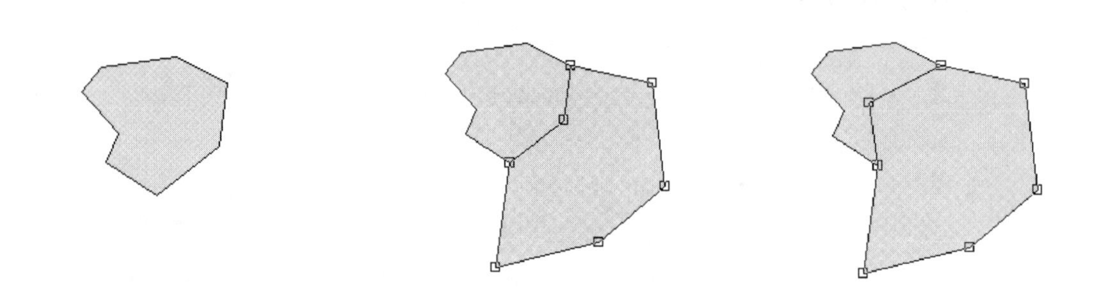

*Three stages in the process of digitizing coincident polygons: (left) the initial polygon is digitized; (center) the second polygon is digitized, using Vertex Snapping to ensure the new vertices are snapped to the vertices of the first polygon; (right) a common vertex is moved to demonstrate that the resultant vertex is shared by the two polygons.*

10. Continue by accessing the Editing properties from the Theme Properties window and enabling Interactive Snapping. Now you can control the snapping that will apply to the next added vertex by using the right mouse button to call up the Editing popup menu. You can elect to snap the next vertex to an existing vertex or polygon boundary.

Digitizing polygons using the Polygon tool, with or without snapping enabled, requires you to digitize the complete polygon. Two other tools allow you to create polygons by inputting only the new portion of the polygon boundary: the AutoComplete tool and the Split Polygon tool.

Selecting the AutoComplete tool is done from the Draw tool pulldown list. Once done, you can add polygons that share a common boundary with an existing polygon by starting and ending the line that describes the new portion of the polygon boundary within the boundary of an existing polygon. The line describing the new polygon boundary can fall outside an existing polygon boundary only if General Snapping is enabled and the point falls within the snapping tolerance of the existing boundary. By using the AutoComplete tool, new polygons can be added without having to retrace existing polygon boundaries for all new polygons.

Existing polygons can be split by selecting the Split Polygon tool from the Draw tool pulldown list. Digitize a line across the existing polygon that will be the boundary on which the polygon is to split, taking care that the start and end points of the line intersect the polygon boundary. You can undershoot the polygon boundary if Interactive Snapping is enabled, and if the start and end points of the line fall within the snapping tolerance of the polygon boundary.

# Editing Common Features

The ArcView shapefile, as discussed previously, utilizes a nontopological data structure to store features. In this data structure, each feature is discrete, and elements located in common between features, such as nodes and vertices, are stored separately for each feature, rather than stored in common between features. Through such functionality, such as feature snapping within ArcView and conversion to shapefiles from ARC/INFO coverages, shapefiles can be created which maintain features in coincident locations. Functionality has been added to allow simultaneous editing of coincident elements shared by two or more features. In this manner, a vertex that is shared by two polygons can be selected and moved; both polygons will be reshaped by the new vertex location simultaneously. If only one of the two polygons were selected, moving the common vertex of the selected polygon would cause it to pull away from the adjacent polygon.

1. Continue editing the polygon theme by making the Vertex tool active and clicking directly on a vertex shared by two polygons.

   Notice that when you select a shared vertex, ArcView draws the vertex as a hollow box, and the adjacent vertices of each polygon as circles. These circles serve to represent the anchor points and indicate how the polygons will be reshaped when the vertex is moved.

2. Select and drag the vertex to reshape the common polygons.

3. Next, click on a segment that is shared by two or more polygons, but away from the polygon vertices.

4. The interior vertices of the common segment will be drawn as squares, with the vertices at the ends of the common segment drawn as circles. Any editing operation you perform on this shared segment—adding, deleting, or moving vertices—will be performed on all polygons sharing the segment.

*Results of clicking on a line segment held in common by two polygons: the common vertices are squares, and the anchor vertices at either end of the common boundary are circles.*

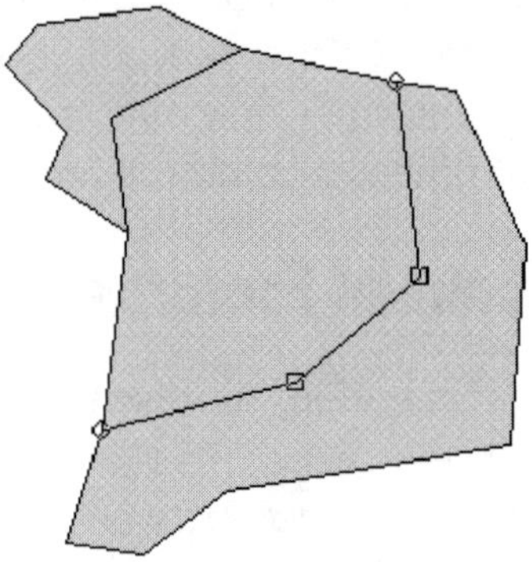

## Saving Your Edits

When you are done editing the theme, select Stop Editing from the Theme menu. You may save or discard your edits at this time. Alternatively, you can save your edits incrementally by selecting Save Edits from the Theme menu as you work.

# On Your Own

If you have experimented with each of the steps covered in the exercise above, you should now have a good feel for the tools that are available for editing shapefiles. We encourage you to experiment with changing the attribute rules to examine how attribute handling is performed when splitting or merging lines and polygons.

## Chapter 11

# Beyond the Basics

When writing this chapter, our greatest difficulty was identifying the basics. Some might agree that geocoding a table of street addresses to create a point theme is central to many projects, but certainly it is not as basic as adding a theme from an ARC/INFO coverage. Others might argue that overlay operations such as spatial join and theme-on-theme selection, which permit examination of the relationships between themes, constitute part of the core functionality that should be present in any desktop mapping software, and as such should be considered basic as well.

Ultimately, we elected to cover topics that involve the *manipulation* of components already present in an ArcView project. These components are organized into the following categories:

- ❒ Overlay Operations
- ❒ Advanced Classification
- ❒ Hot Links
- ❒ Working with Shapefiles
- ❒ Working with Tables

## Overlay Operations

Overlay operations, or spatial join and theme-on-theme selection, were discussed in previous exercises. These operations involve relating the features of one theme to the features of another. The operation can be temporary, as in the case of theme-on-theme selection, or permanent, as in the case of spatial join (in which the attributes of the source theme table are joined to the attribute table of the destination theme).

## Theme-on-Theme Selection

Theme-on-theme selection uses the selected features from one theme (the selecting theme) to define the selected set of features from one or more active themes (the target themes). Examples include identifying building permits issued within a historic district, or locating properties within a mile of a proposed freeway alignment. The criteria used to make this selection are determined by specifying the spatial relation type. The following six spatial relation types are available for feature selection:

- *Are Completely Within.* Features in the target theme fall entirely within features of the selecting theme.
- *Completely Contain.* Features in the selecting theme fall entirely within features of the target theme.
- *Have Their Center In.* Features in the target theme have their center within features of the selecting theme.
- *Contain the Center Of.* Features in the selecting theme have their center within features of the target theme.
- *Intersect.* Features in the target theme have at least one point in common with features in the selecting theme. This includes features in the target theme that are totally contained by features of the selecting theme.
- *Are Within Distance Of.* Features in the target theme are within a specified distance of features in the selecting theme. Included are features in the target theme that are totally contained by features in the selecting theme. Specifying a selection distance effectively creates a buffer around the features in the selecting theme, although the actual buffer polygon is not visible.

Theme-on-theme selections are performed by choosing Select By Theme from the Theme menu. You then choose the selecting theme and the spatial relation type from pulldown lists. If Are Within Distance Of is selected, enter the selection distance in map units.

> **NOTE:** *You can assign the same theme to be the selecting theme and the target theme. If Intersect is used, this procedure can be used to locate features in a theme "adjacent" to the selected features. If Are Within Distance Of is used, the procedure locates features within a buffer zone of the selected features.*

When the themes and spatial relation types have been specified, you must then choose a selection method. The following selection methods are available:

- ❒ *Add to Set.* Adds the newly selected set of features to the currently selected set.
- ❒ *New Set.* Creates a new set of features from all candidate features.
- ❒ *Select from Set.* Selects new features from the currently selected set.

*Select By Theme dialog window.*

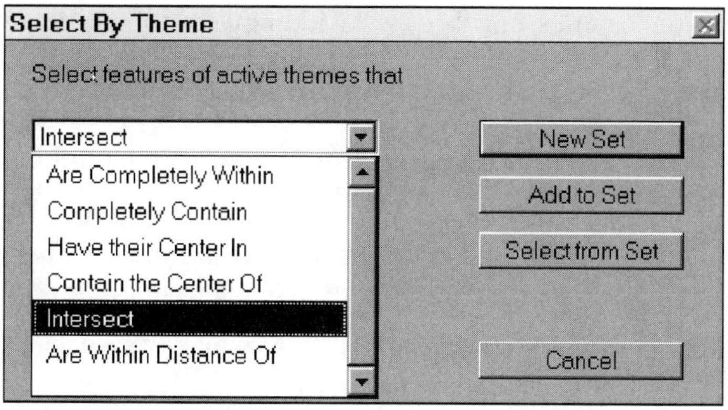

By choosing from the different selection sets, you decide whether to treat each selection separately, to keep a running tab, or to isolate features according to whether they meet a series of conditions. For example, target mailing lists might be generated by keeping a running list of zip codes that meet a certain criteria. Site selection might best be served by performing *Select from Set* operations with increasingly stringent criteria until a single zip code or two remain selected.

## Spatial Join

A spatial join functions much like an attribute join on two tables, with one important difference: an attribute join is based on attribute equivalency in the specified field, whereas a spatial join is based on spatial location equivalency.

The following spatial joins are supported in ArcView:

- ❐ Point to Point
- ❐ Point to Line
- ❐ Point in Polygon
- ❐ Line to Point
- ❐ Line in Line
- ❐ Line in Polygon
- ❐ Polygon in Polygon

Spatial queries can help you identify customers in specified service areas, zip codes within sales territories, roads within school districts, and a variety of other data sets related to "mixed" geographies.

1. To perform a spatial join, open the attribute tables for the two themes.
2. Highlight the Shape field in both tables. The table of the theme to receive the attributes resulting from the spatial join should be the active (or destination) table.

   For example, you may want to associate psychographic codes with customers, a process that is typical of market research. You would associate polygon attributes with point features that are located within those polygons; the point theme (customers) would be the destination table.
3. To perform the spatial join, select Join from the Table menu.

Three types of spatial relation types are used; which one depends on the theme feature types. Point in Polygon, Line in Polygon, and Polygon in Polygon joins are performed using the Are Completely Within spatial relation type. For example, assume you wish to join two polygon themes, one for block groups and one for zip code boundaries. Block groups is the destination attribute table. The only block groups to receive zip code attributes following the spatial join are those contained entirely within a single zip code area.

A Line in Line spatial join uses Intersect as the spatial relation type, whereas Point in Point, Point in Line, and Line in Point spatial joins use the Nearest spatial operator. In joins that use the Nearest relation type, a Distance field is added to the joined table. The Distance field contains the distance measured in map units between the joined features.

> ✘ **WARNING:** *Due to the manner in which a saved project is reopened, accessing a theme with a spatial join in a restored project can cause a fatal error. To prevent this error, it is necessary to remove all spatial joins from a project before saving. To preserve the result of the spatial join for future use, save the target theme as a new shapefile before removing the spatial join. Add the resultant shapefile to the view as a new theme.*

## Spatial Aggregation

Consider a map that depicts voting precincts. Each precinct is identified by a unique name and number. If you desire, you could apply a unique classification to a precinct theme, with each theme subsequently shaded in a unique color.

Additionally, each precinct is also identified by the legislative district it falls in. Although it is possible to classify this precinct theme on legislative district—by using a unique color to identify each group of precincts with their corresponding legislative district—it is also possible to edit the precinct shapefile and remove the lines between precinct shapes that share the same legislative district. In such a manner, the precinct shapefile could be manually aggregated to create a legislative district shapefile.

The term *spatial aggregation* refers to merging features by attributes. Through this process, a theme's shapes that share a common attribute value are merged together to form a single shape. In the ARC/INFO environment, this process is termed a *dissolve;* when dissolving an ARC/INFO polygon coverage, lines between polygons that share common attribute values are removed. Due to the nontopological structure of an ArcView shapefile, a merge by features that is done on a polygon shapefile may or may not behave in the same manner as an ARC/ INFO dissolve.

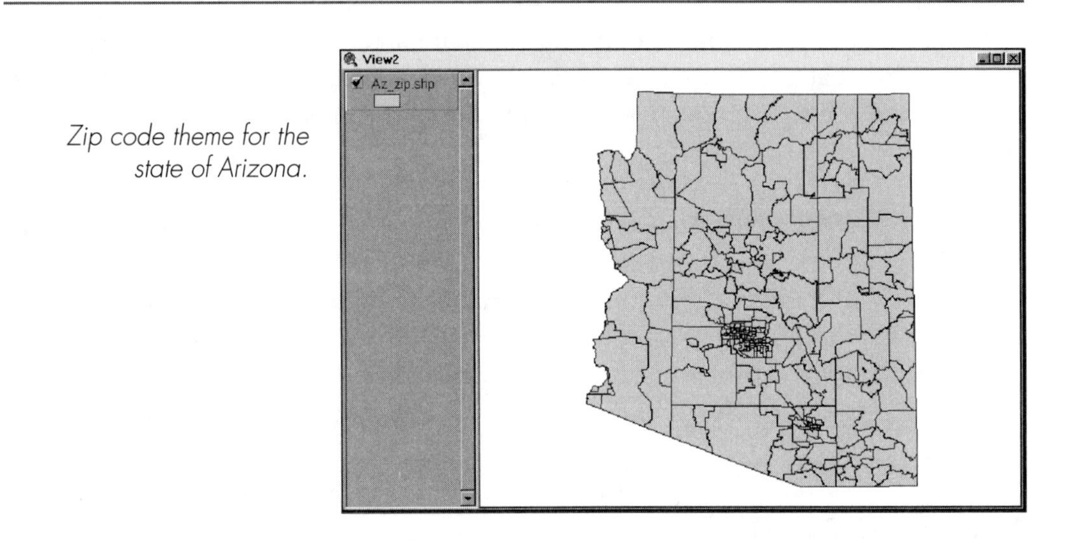


*Zip code theme for the state of Arizona.*

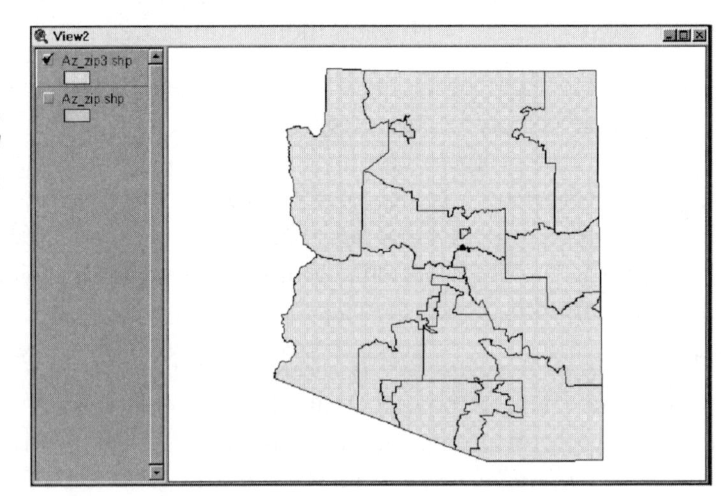


*Same zip code theme, this time merged on the first three digits of the zip code.*

The specific rules for merging polygons in ArcView shapefiles are listed below:

- *Adjacent polygons*—If two polygons share a common boundary, this boundary is removed in the output polygon.
- *Disjunct polygons*—If two polygons share a common attribute, but do not intersect, the set of polygons sharing the same attribute will subsequently treated as a single shape. For example, individual shapes representing each of the Hawaiian Islands could be merged to create a single shape representing the State of Hawaii.

These first two examples are intuitively easy to follow. The remaining three are not as intuitively clear. They do, however, predictably follow the same rules regarding intersecting set behavior.

- *Intersecting polygons*—If two polygons partially overlap but are not wholly contained by one another, the resultant merged shape contains the area of both original polygons, excluding the area of overlap that the original polygons shared. For example, if the two shapes of Equador and Peru overlap and are merged, the resultant shape would exclude the territory claimed by both countries.
- *Island polygons*—If one polygon is wholly contained by another polygon, merging the two shapes results in a hole that represents the inside island's removal from the enclosing polygon.
- *Nested island polygons*—If polygons are successively nested, for example, a polygon for California containing a polygon for Mono Lake containing the island in Mono Lake, the alternating nested shapes are combined into the merged shape. In this case, a single shape would result, one that contained both the land mass for California and the land mass for the island, excluding the water mass for Mono Lake.

The ability to merge shapes based on attributes is one of the most powerful tools in your GIS arsenal. While the list above may be intimidating initially, if you experiment with merge operations on a shapefile until the concepts click, your time will have been well spent.

**NOTE:** *When existing shapes from a shapefile theme are merged into a new shape, the records in the theme attribute table for each theme will be deleted and replaced with a single new record. The fields making up this new record will be blank rather than arbitrarily summarized. If necessary, you can subsequently edit the table to enter new values for the new record.*

1. To merge features by attribute values (spatial aggregation), open the attribute table for an active theme and highlight the field on which to base the merge.

   For example, you might choose to merge a zip code theme by an attribute identifying the salesperson assigned to each zip code, thereby creating a sales territory map from the original zip code theme. In this case, you would highlight the salesperson-ID field on the zip code theme attribute table.

2. Next, open the Summary Table Definition dialog window by selecting the Summarize tool from the button bar.

3. For the Field, select Shape. For Summarize By, select Merge. Select Add to add this field to the summary list, and specify the name and location for the output file.

   Note that in the case of a merge operation, although you are entering a name for the resultant output *.dbf* file, the name you provide will be used for the resultant shapefile as well.

4. When you are done, select OK; the merge operation will proceed and, upon completion, you will be prompted to add the new theme to the view.

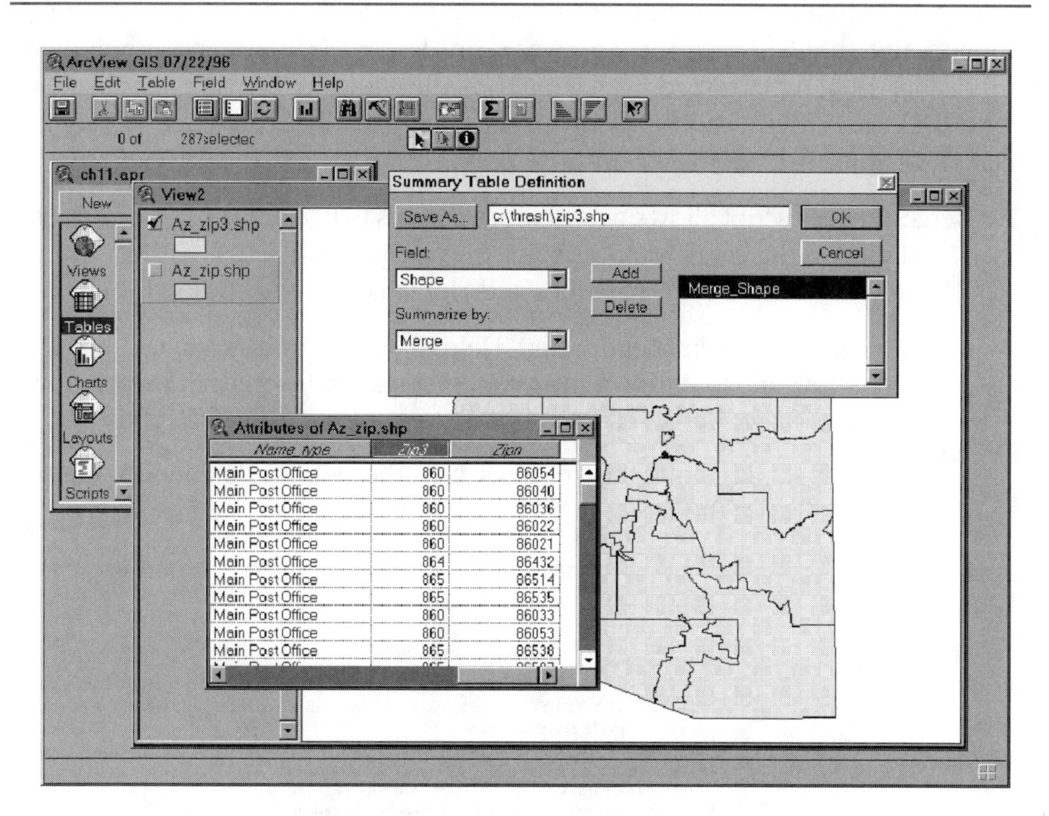


*Summary Table Definition dialog window for a spatial merge on the attributes in the Zip3 field.*

The resultant output theme will contain new shapes whose boundaries correspond to unique values for the merge field from the input theme. The merge operation, most commonly associated with polygon themes, can be performed on line themes as well.

## Merge Themes

Now you know you can perform spatial aggregation based on a single theme and on common attributes, but what do you do when you need to combine shapes from two different themes and shapefiles? The ArcView feature—one that mirrors the Append command in ARC/INFO—that lets you do this is available through a sample script supplied with the ArcView script library. Loading and running this script is not as straightforward as making a menu selection, but its utility makes it worth the extra effort. (The topics of ArcView scripts and Avenue are covered in detail in Chapter 14, "ArcView Customization.")

The Merge Themes script allows two or more themes that share common attributes to be combined into a single output theme and shapefile. The use of this script is demonstrated in Exercise 11, "Shapefiles and Hot Links," at the end of this chapter.

1. Under the Help menu, select Help Topics. In the Index tab, scroll to *Scripts,* and then down to *samples.* Display this topic, and scroll down to the *View.MergeThemes (ArcView Sample Data/Scripts)* entry.
2. Double-click on this entry to display the help topic entitled "Merges two feature themes."

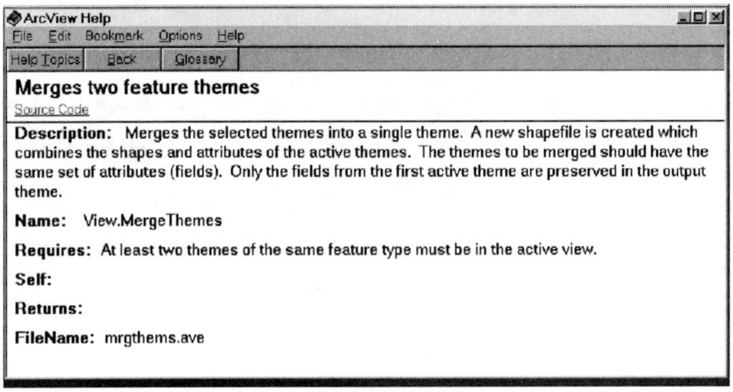

*ArcView Help topic that describes the View.MergeThemes sample script.*

ArcView Help
File Edit Bookmark Options Help
Help Topics | Back | Glossary

**Merges two feature themes**
Source Code

**Description:** Merges the selected themes into a single theme. A new shapefile is created which combines the shapes and attributes of the active themes. The themes to be merged should have the same set of attributes (fields). Only the fields from the first active theme are preserved in the output theme.

**Name:** View.MergeThemes

**Requires:** At least two themes of the same feature type must be in the active view.

**Self:**

**Returns:**

**FileName:** mrgthems.ave

3. The "Merges two feature themes" help topic contains a thumbnail description of the script. Follow the *Source Code* hypertext link. This opens another help window, this one containing the source code for the *View.MergeThemes* script. To copy the script to the Windows clipboard, click on the Options button and select Copy.

**4.** Click on the Scripts icon in the project window and then on New to open a new script document window. The GUI now contains menus and buttons appropriate for working with script documents.

**5.** Select Paste from the Edit menu to paste the copied script into the Script Document window.

**6.** Click on the Check button in the button bar to compile the script. Note that the running man icon, which was grayed-out previously, is now black, indicating that the script has been compiled and is ready to be run.

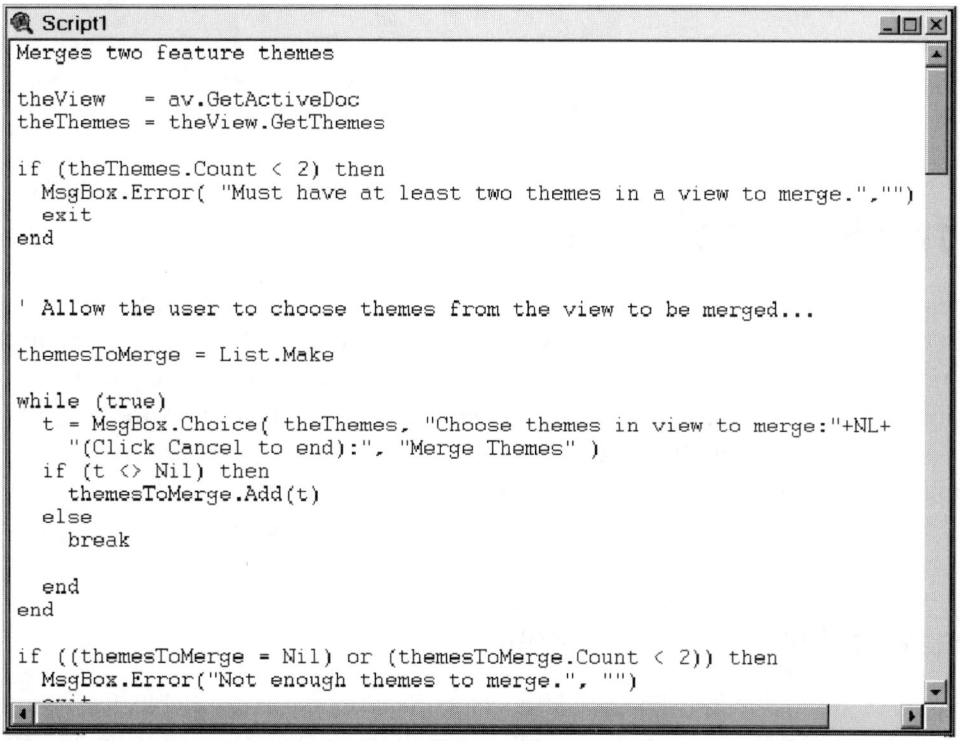

Script1

```
Merges two feature themes

theView    = av.GetActiveDoc
theThemes = theView.GetThemes

if (theThemes.Count < 2) then
  MsgBox.Error( "Must have at least two themes in a view to merge.","")
  exit
end


' Allow the user to choose themes from the view to be merged...

themesToMerge = List.Make

while (true)
  t = MsgBox.Choice( theThemes, "Choose themes in view to merge:"+NL+
    "(Click Cancel to end):", "Merge Themes" )
  if (t <> Nil) then
    themesToMerge.Add(t)
  else
    break

  end
end

if ((themesToMerge = Nil) or (themesToMerge.Count < 2)) then
  MsgBox.Error("Not enough themes to merge.", "")
```

*View.MergeThemes sample script pasted into a Script document and ready to run.*

7. With the view window open behind the script window, click on the running man to execute the script.

8. From the Merge Themes pulldown list, select the themes to merge, highlighting and clicking on OK after each selection. When you have finished selecting themes to merge, click on Cancel to close the window. You will be prompted for a name and location for the resultant shapefile, and asked if you wish to add the shapefile to the view.

   The resultant shapefile and theme will contain all the features and attributes from the input themes. It should be noted, however, that only those fields found in the first active theme will be preserved in the output theme.

If you have followed the steps above independently in your own exercise, you have now had your first exposure to *Avenue*, ArcView's scripting language. You will delve further into Avenue in Chapter 14.

# Advanced Classification

Thematic classification was introduced in Chapter 5. The basics of legend types and classification types, editing class values, assigning symbology, and using thematic classification as a data exploration tool were discussed. It may surprise you to learn that there is still more to cover! However, thematic maps are at the heart of ArcView, and the ability to customize a thematic map by applying a classification based on theme attributes is central. Now that you understand the basics, you will learn about additional functionality that will enhance your ability to perform thematic classification.

## Classification Types Revisited

Chapter 5 discussed the basics of the classification types available for thematic classification. In this section you will learn how to select a classification type appropriate to your data, manipulate the classification and data to best reveal the patterns contained in your data, and refine the symbology to best display these patterns on the map.

Consider the case of population data that has been gathered by state. A quantile classification could be applied to the raw data, resulting in a classification of, say, four classes, each containing an equal number of states. The resulting class breaks, as displayed on the map, could serve to illustrate some patterns regarding population distribution in the country. This map could even be significant, in the event that the absolute ranking of states by population became a criteria for determining funding of programs; however, such a map would do nothing to correct the inherent inequity introduced by the fact that certain states are much larger than others.

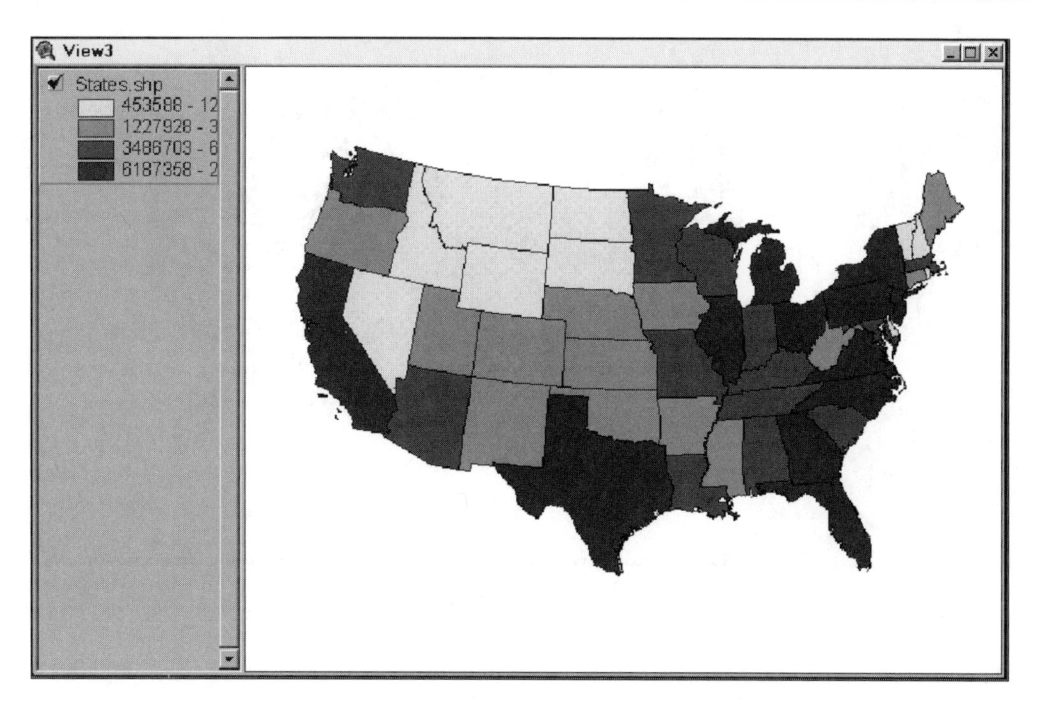


*States theme classified on Pop1990 using a four-class quantile classification.*

To better compare population data, you may wish to display population on a per unit area basis. One way to accomplish this is to switch the legend type from Graduated Color to Dot Density. In a dot density map, a value is assigned to the dot that represents a specific number of units in your data. In the population map, each dot can be set to represent 100 people. When the map is generated, the total number of dots needed for each state will be distributed uniformly within that state. In this way, the density of dots on the map provides a visual representation of the actual density of population throughout the country. In effect, creating a dot density map on raw population counts has the effect of normalizing your population data by area.

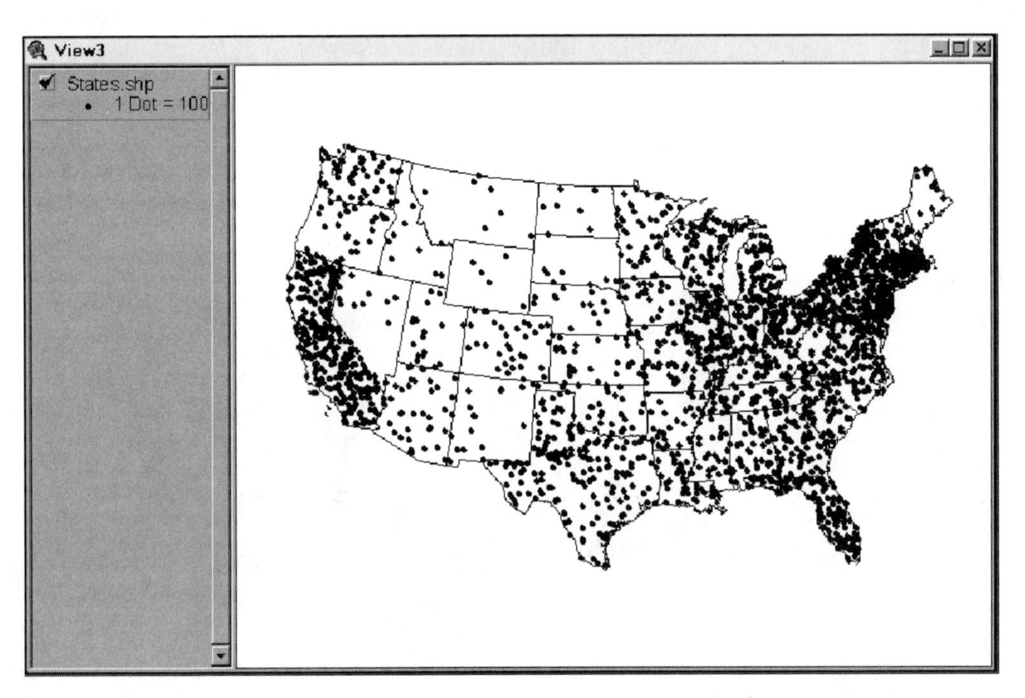


*Same States theme classified on Pop1990 using a dot density classification where one dot equals 100,000 people.*

Normalization of data, in which the value of one field is displayed in relation to the value of a second field, is a powerful tool for effectively displaying data on a map. Displaying total sales by territory, for example, may not be as effective as displaying average sales per customer, which is the result of normalizing the sales field by the customer count.

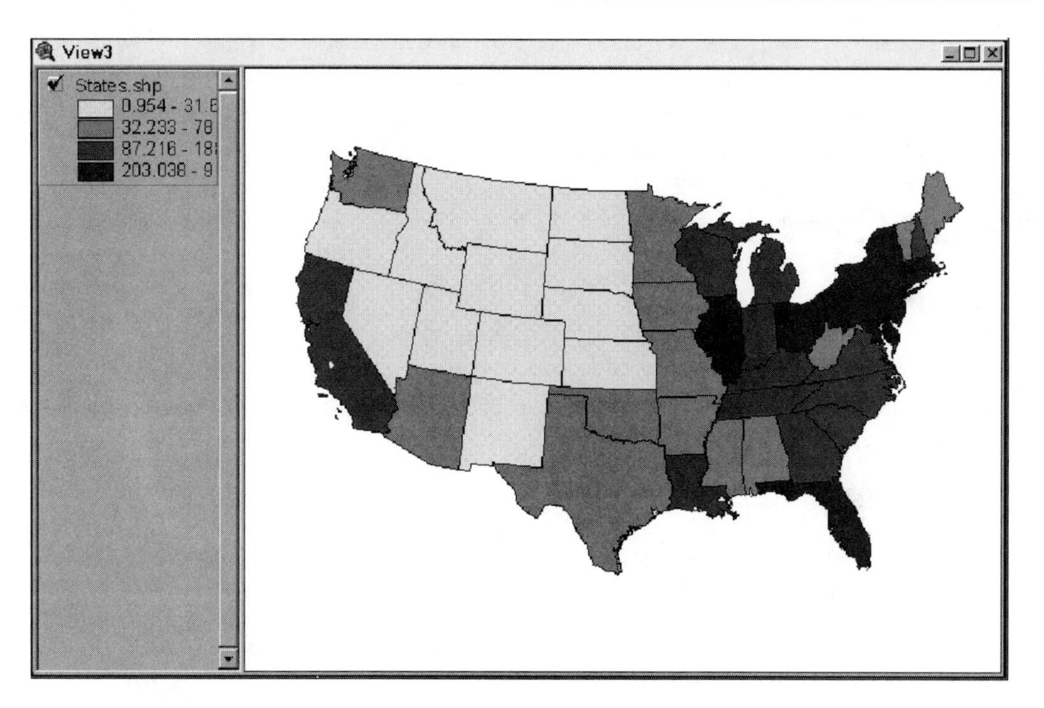


*States theme classified on Pop1990 using a four-class quantile classification and normalized by Area.*

One common method for normalizing data is to describe that data as a percentage of the whole. Using the previous example of population data, the population for each county could be expressed as a percentage of the state's population, providing an easily interpretable value that indicates the relative amount each county contributes to the total state population.

> **NOTE:** *If your data is already normalized, as in the cases of percentage or density data, further normalizing this data is likely to produce misleading results.*

Normalized data is well suited for display using Graduated Color or Graduated Symbol legend types. If you have several normalized fields you wish to display, a Chart legend is a good choice. In a Chart legend, field values can be displayed as a column chart or a pie chart. Additionally, a size field can be set that will dictate the relative size used to display each chart when drawn on the map. This can serve as another way to normalize raw data, such as charting 1980 and 1990 population for each county, and scaling each chart symbol based on the area of the county. Chart symbols can be applied to point, line, and polygon themes.

## Advanced Symbolization

Chart symbols are but one of the advanced symbolization capabilities available in ArcView 3.0. The initial discussion of thematic classification reviewed the Graduated Color and Graduated Symbol legend types; it is feasible to apply both types to a single theme, further accentuating the class breaks in the process. The following method is recommended:

1. Select the symbol.
2. Apply the Graduated Color legend type to construct the desired color ramp.
3. Manually change the symbol size for each class (using the line symbol palette) to represent the classes using graduated symbol size.

When symbolizing line or point features, ArcView offers additional symbolization options, such as scaling the symbol size automatically as you zoom in or out on a view. In this manner, the symbol size can be scaled to visually approximate the actual size of the feature, as with roads or borrow pits. Additionally, symbols for line features can be assigned an offset distance. In this manner, features can be shown next to their actual location, enabling features such as pavement test sections to be displayed as offset from the full street net. Also, point features can be rotated, using a value stored in a user-designated rotation field. This enables plotting of such features as wind roses, in which the symbol is rotated to indicate wind direction.

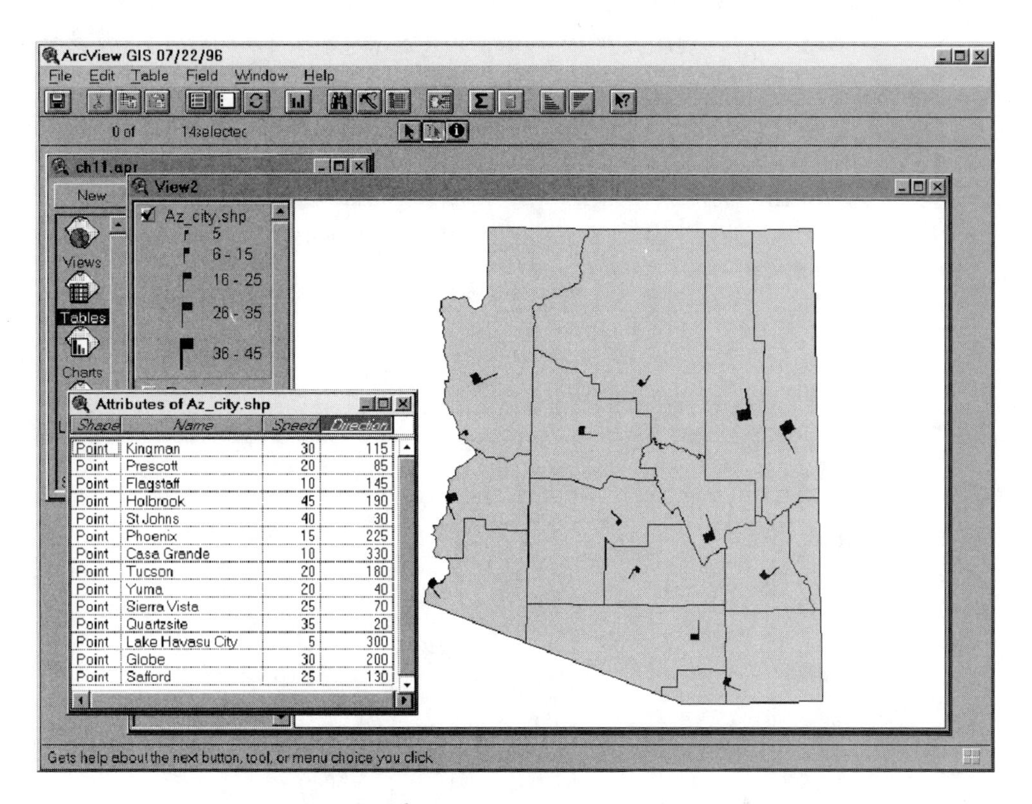


*Az_city point theme, classified on Speed with graduated symbols that have been rotated using the amount of rotation angle in the Direction field.*

*No scaling* is the default setting for line and point symbols, meaning features are drawn with the same size symbol as you zoom in and out on a view. In contrast, text labels that are created with the label tool and attached to a theme do scale by default, meaning the text grows larger and smaller as you zoom in and out on a view. Through the Text Labels option, which is part of a theme's properties, text labels can be set to *not* scale, meaning text will be drawn at the same size regardless of the extent to which you zoom in or out.

✓ **TIP:** *ArcView provides additional flexibility in a way that may not be readily apparent: through thematic classification on polygon themes based on ARC/INFO coverages. If the options available for classification on polygon themes are not sufficient, add the theme again to the view. Open the cover folder and select the labelpoint entry. The resultant theme contains all attributes associated with the polygons, this time displayed as point features. In this manner, legend types specific to point features, such as Graduated Symbols, can be applied.*

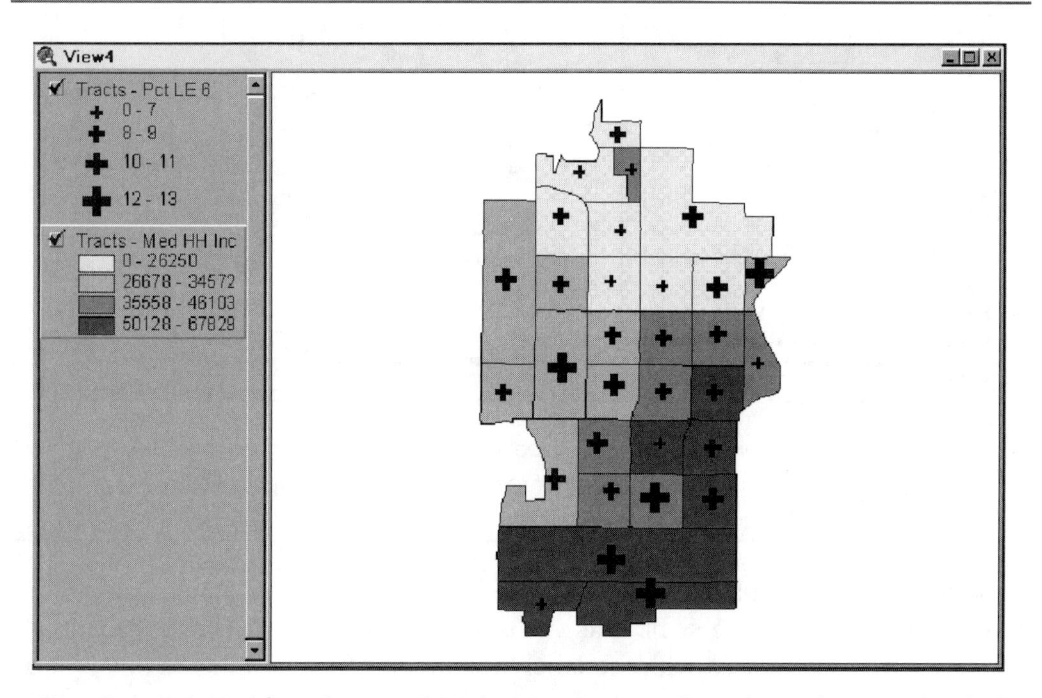


*Two themes derived from the same ARC/INFO coverage. The polygon theme is classified on median household income (Med HH Inc), and the labelpoint theme is classified on percent age six or less (Pct LE 6).*

# Massaging Your Data

As mentioned previously, the ability to classify and symbolize thematic data is central to GIS functionality, and ArcView provides a robust toolkit for this purpose. With an awareness of all the classification options available and the ability to preview new classifications and revert to the prior state via the Undo button, ArcView becomes a powerful data exploration tool. The following steps can be used as a guideline for optimizing thematic classification:

1. Select the legend type that best fits the data. Even before the classification type has been selected and the class breaks refined, one legend type should emerge as the one best suited to portray your data.

**NOTE:** *Not all legend types are available for every data type. Graduated Color and Graduated Symbol classifications cannot be applied on character fields, even if those fields contain numeric data. Similarly, a Unique Values classification cannot be applied on numeric fields containing decimal values, even if the decimal value is an artifact of how the data is coded (such as census tracts) rather than an instance of true floating point data. In these instances, it may be necessary to add a new field to the table and use the Avenue request AsString or AsNumber to convert the values to the appropriate data type before performing the desired classification. See the section on working with tables later in this chapter for additional information.*

2. Apply different classification types until the trends in your data begin to emerge. Changing the number of classes is a good way to reveal patterns that may not be immediately apparent. Consider normalizing your data if no trends are evident.

3. Fine-tune the classification by manually adjusting class breaks and assigning values to the No Data symbol as necessary.

4. To further bring out patterns in the data, assign symbology. For example, a strong midpoint break in the distribution of values may indicate a good candidate for a dichromatic color ramp, with the classes below the breakpoint ramped from blue to white, and the classes

above the breakpoint ramped from white to red. Anomalous classes can be handled by applying a symbol that is different from the one used for the rest of the classification.

5. Lastly—experiment! Take full advantage of the Undo function to explore your data. If nothing works, start over. Remember—it is your map and, to be successful, it must communicate the message you wish to convey.

# Hot Links

*Hot Link tool.*

A "hot link" allows you to associate an action with a theme's feature. The action could be to display an image or text file, to open an ArcView project document such as a view or chart, or to link to an external application via an Avenue script. If you have ever seen links to spreadsheets or imagery in other applications, you know that this is a powerful analysis and presentation capability.

To execute a hot link, make a theme that contains one or more hot links active. Select the Hot Link tool from the button bar, and then click on a feature in the active view. If a hot link is associated with the feature, the defined action will begin. If no hot link has been defined for a theme, the Hot Link tool will remain grayed-out while the theme is active.

To define a hot link, make a theme active. Choose the Hot Link icon from the Theme Properties dialog window to access the Hot Link dialog window. Choose a field in the theme attribute table to be the one that will contain the name of the file or ArcView document to access, and the action to be performed. Select either a predefined action (choose from Link to Text File, Link to Image File, Link to Document, and Link to Project) or the execution of an Avenue script.

The ability to hot link to an image file allows you to display documents that contain additional information about a feature, such as a digital photograph or scanned blueprint, when that feature is selected. ArcView supports the following image file formats:

- ❒ GIF (Graphics Interchange Format)
- ❒ MacPaint (on Mac systems)
- ❒ Microsoft DIB (Device-Independent Bitmap)
- ❒ TIFF (Tag Image File Format)
- ❒ TIFF/LZW compressed image data
- ❒ X-Bitmap
- ❒ XWD (X Windows Dump Format)
- ❒ Sun raster files

One particular type of useful hot link is the link to a specific ArcView document. This document could be a different view within the same project, which makes it possible to construct a "view of views." In such a project, the initial view can serve as an index to a series of views, each containing information about a specific region in greater detail.

## Working with Shapefiles

The shapefile is ArcView's native format for storing spatial data. Shapefiles can be created from scratch, from an ARC/INFO coverage, or via translation from another desktop mapping format. Individual shapefiles correspond one-to-one with a particular set of geographic features. Shapefiles contain both graphics and associated attributes for geographic features.

The advantages to using the shapefile format for storing spatial data in ArcView are summarized below:

- ❒ Shapefiles display more rapidly on a view.
- ❒ Shapefiles can be edited within ArcView.
- ❒ New themes can be created in ArcView using the shapefile format.
- ❒ Features can be merged or dissolved based on common attribute values.
- ❒ ARC/INFO can convert ArcView shapefiles back to ARC/INFO coverages.
- ❒ Shapefiles use an open format, allowing them to be read or written to by other software applications.

# Converting to Shapefiles

Any theme based on an ARC/INFO coverage or a CAD drawing can be converted to the ArcView shapefile format. To carry out the conversion, make the theme active and select Convert to Shapefile from the Theme menu. If a set of active features has been selected from the theme, only the selected features will be converted to the new shapefile. You will be prompted for the name and location of the shapefile, and asked whether you wish to add the shapefile to the view.

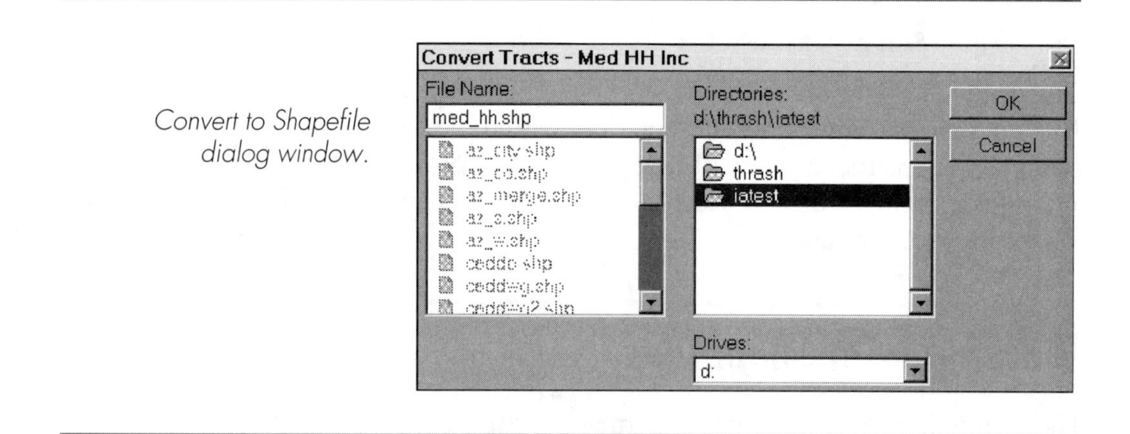


*Convert to Shapefile dialog window.*

# Strengths of Converting to Shapefiles

There are many advantages to converting your spatial data to the shapefile format—even if your data is already stored as an ArcView shapefile! Some of the advantages are highlighted below:

- The Convert to Shapefile choice from the Theme menu converts any theme to the shapefile format, even if the source data is from a CAD drawing or ARC/INFO coverage. The resultant shapefile can be subsequently edited within ArcView—an important consideration for sites not possessing the software that created the original data.

- Only the currently selected set of features is exported to the new shapefile. If the selected set results from a spatial overlay, exporting to a new shapefile can preserve the selected set for subsequent modeling and analysis.
- If a logical query has been applied to a theme via Theme Properties, exporting the theme to a shapefile and replacing the original theme with one based on this new shapefile will eliminate the need to perform the logical query on the theme. This can help reduce the initialization time when the project is opened.
- If a theme contains tabular joins, the exported theme will preserve the joined fields and records as part of the primary theme attribute table. This can eliminate the need to recreate table joins on project initialization, again reducing the time required to open the project.
- If a theme containing relatively few features is joined to a source table containing many records, exporting this theme causes only those records from the source table that match records in the primary table to be written to the new shapefile. In this manner, the size of the data sets required for a custom project can be reduced, facilitating subsequent project distribution.
- If a project contains a theme that resulted from a spatial join, this theme can be exported and added to the view, and the spatial join on the original theme removed. This will prevent the fatal error that can occur when a project containing a spatial join is reopened.

For detailed coverage of editing shapefiles, see Chapter 10.

## More Work with Tables

Chapter 6 covered the basics of tables. This section is dedicated to the discussion of advanced topics, including editing tables, converting field types, generating summary statistics, and exporting tables.

# Editing Tables

ArcView can access tables in several formats: dBase, INFO, delimited ASCII, and RDBMS tables such as Oracle or Informix via an SQL connection. Only dBase and INFO files, however, may be edited within ArcView. To edit other tables within ArcView, the table must first be *exported* (converted) to dBase or INFO format, and then added back into the project. (To export the table, select Export from the File menu.)

To begin editing a table, make the table active and select Start Editing from the Table menu. If you do not have write access to the table, the Start Editing selection will be unavailable. When the table has been opened for editing, the editable field names will be displayed in a standard, rather than italic, font.

> **NOTE:** *If a table is the result of a Join operation, only fields in the destination table can be edited. To edit joined fields, open the primary table and perform the edits. Use the Refresh selection from the Table menu to update the joined table after editing is complete.*

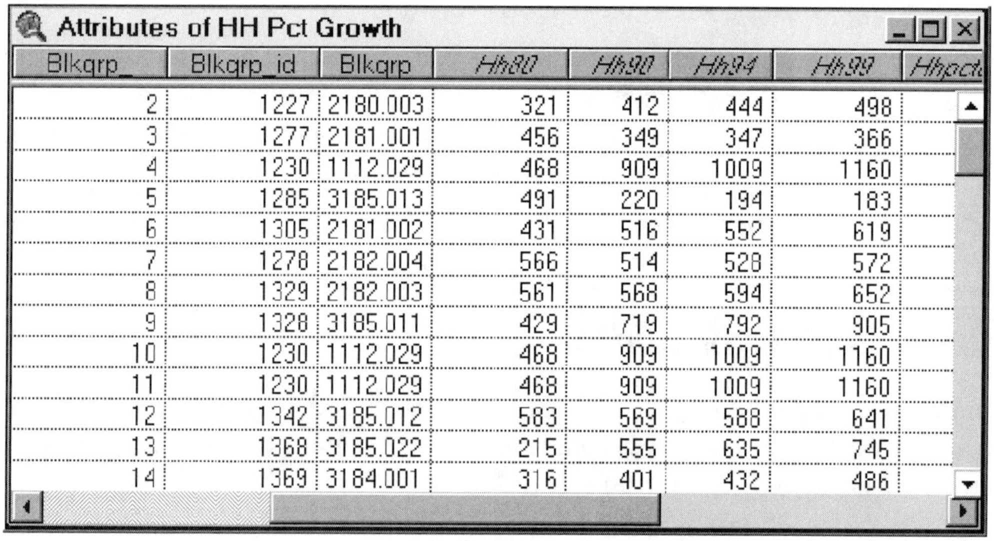

---

Attributes of HH Pct Growth

| Blkgrp_ | Blkgrp_id | Blkgrp | *Hh80* | *Hh90* | *Hh94* | *Hh99* | *Hhpct* |
|---|---|---|---|---|---|---|---|
| 2 | 1227 | 2180.003 | 321 | 412 | 444 | 498 | |
| 3 | 1277 | 2181.001 | 456 | 349 | 347 | 366 | |
| 4 | 1230 | 1112.029 | 468 | 909 | 1009 | 1160 | |
| 5 | 1285 | 3185.013 | 491 | 220 | 194 | 183 | |
| 6 | 1305 | 2181.002 | 431 | 516 | 552 | 619 | |
| 7 | 1278 | 2182.004 | 566 | 514 | 528 | 572 | |
| 8 | 1329 | 2182.003 | 561 | 568 | 594 | 652 | |
| 9 | 1328 | 3185.011 | 429 | 719 | 792 | 905 | |
| 10 | 1230 | 1112.029 | 468 | 909 | 1009 | 1160 | |
| 11 | 1230 | 1112.029 | 468 | 909 | 1009 | 1160 | |
| 12 | 1342 | 3185.012 | 583 | 569 | 588 | 641 | |
| 13 | 1368 | 3185.022 | 215 | 555 | 635 | 745 | |
| 14 | 1369 | 3184.001 | 316 | 401 | 432 | 486 | |

*Table resulting from a join—note the editable fields in non-italic typeface.*

---

To edit field values, select the Edit tool from the tool bar, and then click on the cell you wish to edit. Key in the new value, followed by either <Tab> or <Enter>. When editing cells, the following keyboard selections are allowed:

- <Tab> makes the cell to the right active.
- <Shift> + <Tab> makes the cell to the left active.
- <Enter> makes the cell below active.
- <Shift> + <Enter> makes the cell above active.

When you are done editing the cell value, pressing <Tab> or <Enter> will commit the new value to the cell. If you change your mind, Undo Edit and Redo Edit are available from the Edit menu.

> **NOTE:** *Unlike most software applications, pressing the <Esc> key in ArcView commits the edited value to the cell, just as the <Tab> and <Enter> keys do. If you find yourself editing the wrong cell and wish to back out of the edit, press one of the keys above to commit the edit, then select Undo Edit from the Edit menu to restore the prior value to the cell.*

On occasion you may need to add or delete a field. To add a field, select Add Field from the Edit menu. You will be prompted for the field name and the field type (Number, String, Boolean, or Date). In addition, for String fields you will be prompted for the field width; for Number fields, the field width and number of decimal places. Keying in zero (0) for decimal places will make the field an integer. To delete a field, highlight the field in the table header, then select Delete Field from the Edit menu.

> **NOTE:** *The Undo Edit selection from the Edit menu will not restore a deleted field to a table. If you delete a table field that you wish to restore, select Stop Editing from the Table menu, then answer No when asked whether to save changes. Any editing you have done on the table will be lost—but you will have saved your deleted field as well!*

When you have finished editing, select Stop Editing from the Table menu. You will be asked if you wish to save your changes or quit without saving. Alternatively, you may also elect to save the edited table as a new table by initially selecting Save

Edits As from the Table menu. A new table will be created, one that contains your edits. The original table will be preserved in its initial state.

## Converting Field Types

Periodically, you may want to change a field type in order to perform a specific operation; for example, to change a number field to a string field so that you can locate features with the Find tool. A change may be required by an ArcView operation that demands a certain field type, such as a number field for performing a quantile classification. Changing a field type may also be required to join two tables on a common field, because the field must be the same type in both tables.

The primary method for changing a field type, without manipulating the table in dBase or INFO, is to add a new field with the desired field type. Then convert the field as a value is calculated. To perform the conversion, begin editing the table and highlight the target field. Select the Calculator tool to access the Field Calculator window. Form an expression from the source field, using the AsString or AsNumber request to translate the source field to the desired format.

*Field Calculator dialog window showing a calculation that moves the numeric field Fips to the character field Fips_C.*

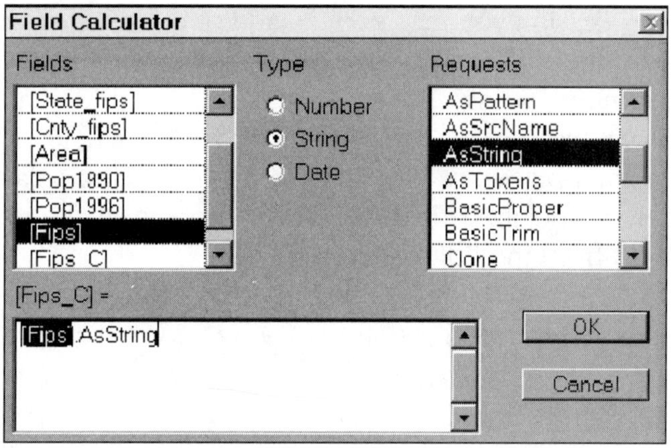

# Statistics

Statistics can be generated for any numeric field in an ArcView table. To generate statistics, select Statistics from the Field menu, or select the Summarize tool from the Tables button bar.

Selecting Statistics from the Field menu will create a window of statistics for the selected records in the highlighted field of the active table. If no records are selected, statistics are calculated for the entire table. The statistical output is made up of the field's sum, count, mean, maximum, minimum, range, variance, and standard deviation.

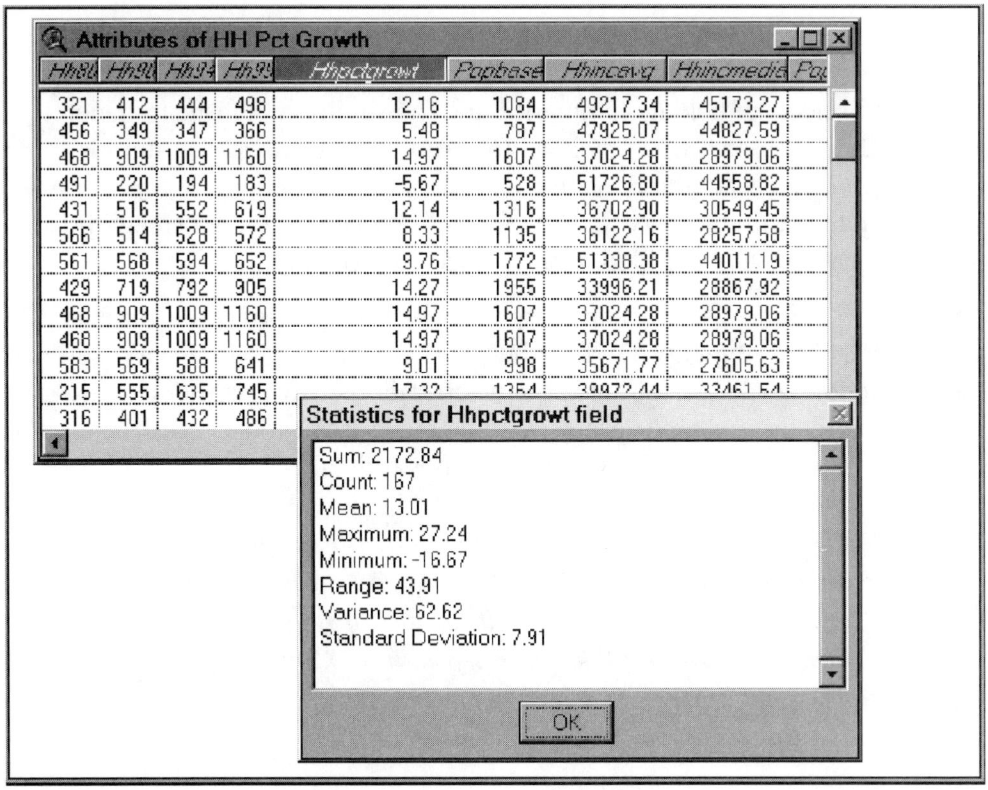

Attributes of HH Pct Growth

| Hh8 | Hh9 | Hh94 | Hh9 | Hhpctgrowt | Popbase | Hhincavg | Hhincmedia | Po |
|---|---|---|---|---|---|---|---|---|
| 321 | 412 | 444 | 498 | 12.16 | 1084 | 49217.34 | 45173.27 | |
| 456 | 349 | 347 | 366 | 5.48 | 787 | 47925.07 | 44827.59 | |
| 468 | 909 | 1009 | 1160 | 14.97 | 1607 | 37024.28 | 28979.06 | |
| 491 | 220 | 194 | 183 | -5.67 | 528 | 51726.80 | 44558.82 | |
| 431 | 516 | 552 | 619 | 12.14 | 1316 | 36702.90 | 30549.45 | |
| 566 | 514 | 528 | 572 | 8.33 | 1135 | 36122.16 | 28257.58 | |
| 561 | 568 | 594 | 652 | 9.76 | 1772 | 51338.38 | 44011.19 | |
| 429 | 719 | 792 | 905 | 14.27 | 1955 | 33996.21 | 28867.92 | |
| 468 | 909 | 1009 | 1160 | 14.97 | 1607 | 37024.28 | 28979.06 | |
| 468 | 909 | 1009 | 1160 | 14.97 | 1607 | 37024.28 | 28979.06 | |
| 583 | 569 | 588 | 641 | 9.01 | 998 | 35671.77 | 27605.63 | |
| 215 | 555 | 635 | 745 | 17.32 | 1354 | 39972.44 | 33461.54 | |
| 316 | 401 | 432 | 486 | | | | | |

Statistics for Hhpctgrowt field

Sum: 2172.84
Count: 167
Mean: 13.01
Maximum: 27.24
Minimum: -16.67
Range: 43.91
Variance: 62.62
Standard Deviation: 7.91

OK

*The statistics for the Hhpctgrowt field.*

The Summarize tool is used to produce summary statistics for additional fields based on the value of a specified summary field. The results of the summarize operation are written to a new table. Summarize creates one summary record for each unique value in the summary field. The record will include the value for the summary field, a count of the records containing this value, and any summary statistics on additional fields which were requested. Available summary statistics include count, first, last, sum, average, maximum, minimum, variance, and standard deviation.

For example, a table of census demographic data could contain three fields: block number, population by block, and tract number. (A tract number identifies the census tract associated with each block.) The Summarize tool, or summary field, could be used to obtain the total population for each census tract.

To produce summary statistics, select the Summarize icon from the button bar. This brings up the Summary Table Definition dialog window. Two pulldown lists allow you to select the field to summarize and the summary statistic to generate for this field. After each selection, click on the Add button to add the summary item request to the list of summary statistics to generate. When the list is complete, specify the name and directory for the output file. Click on OK to create the table. The resultant view will be added to the project and opened for use.

One common way to use the summary table is to re-join it to the primary table using the specified summary field as the join field. In this manner, the summary statistics become available for subsequent operations, such as thematic query or classification. In the example above, a quantile classification could be used to identify the most populous block groups in the study area.

✓ **TIP:** *To produce a summary table with statistics on a single field, such as those produced in the Statistics window, summarize on a field containing the same value for all records. If no such field is available, a new field can be added and left blank to serve as the specified summary field.*

# Exporting Tables

Exporting a table is a convenient means of manipulating tabular data for later use in ArcView or other applications. The Export function will write all visible fields in the selected records of the active table to dBase, INFO, or delimited text format. To export a table, select Export from the File menu of the Table menu bar and specify the export format with the name and location of the output file.

Many uses for the Export function may not be immediately apparent. Noteworthy Export capabilities include the following:

- Allows you to create a custom output file that contains only fields and records of interest. By using Table Properties to control which fields are visible, an exported table can be confined to just a few fields and records extracted from a larger table.
- Allows the results of joining tables in ArcView to be preserved for subsequent use. In this manner, additional attribute information or summary statistics can be permanently associated with records from a primary table.
- Allows data to be converted to a new format. Tables from an INFO database can be output in dBase format for subsequent use in another application. The results from an SQL query on an external database, such as Oracle or Informix, can be stored locally for future use.
- Allows data to be converted to a format that allows subsequent editing. Tables obtained from a delimited text file or an external SQL database can be written locally to dBase or INFO format and subsequently edited.

If you plan to work with tabular data, it would be helpful to become familiar with the Export function.

# Exercise 11: Shapefiles and Hot Links

In this exercise you will use sample data supplied with ArcView, along with custom data sets, to demonstrate selected properties of shapefiles and hot links.

1. Open a new project, and then open a new view window in that project.
2. Select the Add Theme icon from the button bar. Navigate to the *\esri-data\usa* directory at the location where the ArcView sample data resides on your system. Add the *counties* theme. When added to your view, this theme will appear with the default name of *Counties.* Display the theme and you will see all U.S. counties, including counties in Alaska and Hawaii.

**NOTE:** *If you have not previously loaded the ArcView sample data on your system, you will need to install the data at this time. An alternative is to access the ArcView sample data directly from the ArcView CD-ROM. In order to access the CD-ROM directly from the Add Theme dialog window, you must type at least the partial path in the Directory input box, as opposed to selecting the CD-ROM drive from the Drives list. In this case, typing* f:\esri\esridata\usa *(assuming your CD-ROM has been mounted as the f: drive) will result in a listing of available data sources, including the desired* counties *theme.*

You are viewing an ARC/INFO coverage. Because the coverage is stored in geographic coordinates, it appears flat and distorted. With such a vast geographic area on display, your first step will be to experiment with projections.

Counties theme in geographic coordinates.

3. Select Properties from the View menu and then click on the Projection button to access the Projection Properties dialog window. From the Categories list, select Projections of the United States. Several choices will be available in the Type list.

4. Experiment with these choices, but return to Projections of the United States and Lambert Conformal Conic (North America) when you have finished.

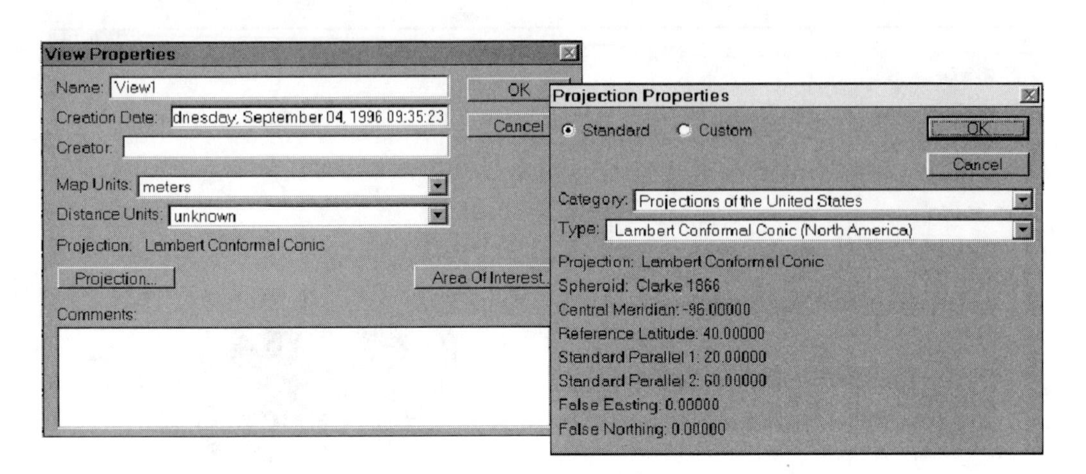


*Projection Properties dialog window as applied to View1.*

Congratulations—you have changed the world! Let's continue the exercise by zooming in on Arizona to demonstrate shapefiles.

**5.** Select the Query Builder tool from the View button bar to access the Query Builder dialog window.

**6.** Construct the query *State_name = Arizona* and choose New Set. When the process is complete, the selected counties in Arizona will be highlighted on the view. Use the Zoom to Selected Set icon to zoom in on the selected features.

**7.** You can now experiment with changing these selected features to an ArcView shapefile. From the Theme menu, select Convert to Shapefile. You will be prompted for the name and directory in which to store the completed shapefile. Use the name *Az_co.shp,* and store the file in your working directory. Click on Yes when you are prompted to add the shapefile as a theme in your view.

8. The new theme will appear in the Table of Contents. Turn it on and display the results. Make the *Az_co* theme active and zoom to the extent of the active theme. Notice that Arizona is skewed to the right—the result of the view still being projected to Lambert Conformal Conic. Access the View Properties dialog window, change the projection category to UTM, and change the Type to Zone 12. Click on OK to apply.

9. Next, click on the *Cnty* theme in the Table of Contents and select Delete Themes from the Edit menu. Because your selection is complete, you will no longer need the previous theme. (The incremental project has been saved as *ch11a.apr*.)

Now that you have created the *Az_co* shapefile, you are ready to investigate different types of data manipulation. Let's begin by examining spatial aggregation, which is accomplished with the Merge operation.

1. Select the Add Theme tool from the View button bar, and navigate to the directory that contains the data supplied with this book. Add the theme *Az_zip.shp,* the zip code theme for Arizona, from the ArcView sample data. The theme contains an additional field, *Zip3,* the three-digit zip code class.

2. For a quick look at the distribution of the *Zip3* attribute, double-click on the *Az_zip.shp* theme to bring up the Legend Editor.

3. Classify the *Az_zip.shp* theme on the unique values for *Zip3*. Select *Zip3* from the Field list. From the Classify window, select Unique Value. Apply the classification. Your map will display the *Zip3* areas across Arizona.

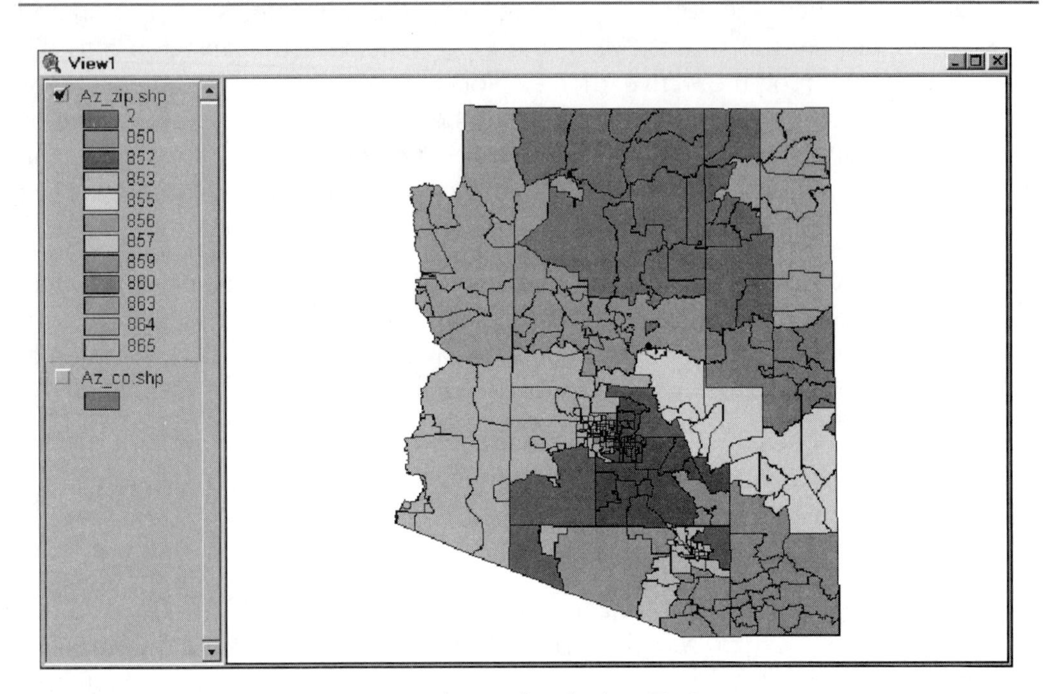


*Az_zip theme classified on Zip3.*

4. You are now ready to perform a merge on the *Zip3* field. Make the *Az_zip* theme active and open the theme's attribute table. Highlight the *Zip3* field in the table.

5. From the Table button bar, select the Summarize tool. In the Summary Table Definition dialog window, select Shape from the Field list, and Merge from the Summarize By list. Click on Add to add *Merge_Shape* to the summary definition list.

6. Specify the name and directory in which to store the output. Note that although you are prompted for the name of a file that has a *.dbf* suffix, the Merge operation creates a new shapefile. The name you supply will be applied to the new shapefile as well. Use the name *Az_zip3.dbf* for the output file name.

Summary Table Definition
Save As... d:\insideav\data\az_zip3.dbf
OK
Cancel
Field:
Shape
Add
Delete
Merge_Shape
Summarize by:
Merge

*Summary Table Definition dialog window for the Merge operation.*

7. Click on OK to begin the merge. When it is complete, you will be asked if you wish to add the new theme to the view; click on OK. Display the theme to confirm your results. (The incremental project has been saved as *ch11b.apr.*)

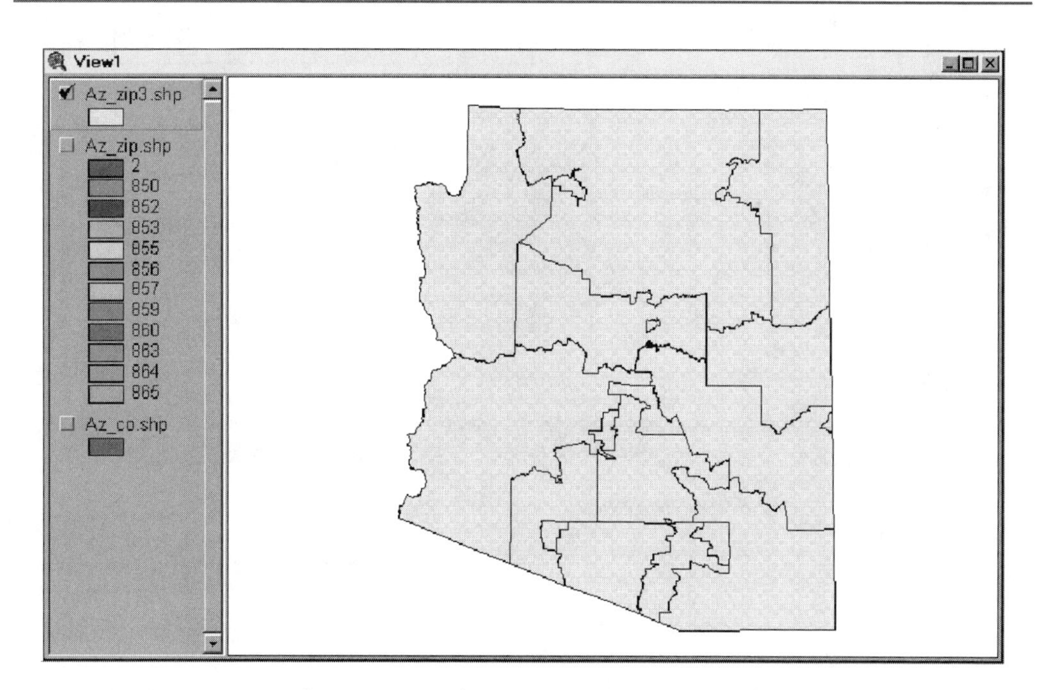


*Resultant Az_zip3 theme merged on the Zip3 field.*

You have just aggregated all the Arizona zip code areas into larger units, known as Arizona Zip3 areas. You will now use this theme and the *Az_co* theme to illustrate the Merge Themes script.

1. Make *Az_zip3* the active theme. Access the Query Builder by clicking on the Query Builder icon in the button bar, and form the query *([Zip3] = 853)*. Select New Set. One feature, comprised of two discontinuous shapes, is highlighted in the southwest portion of the state.

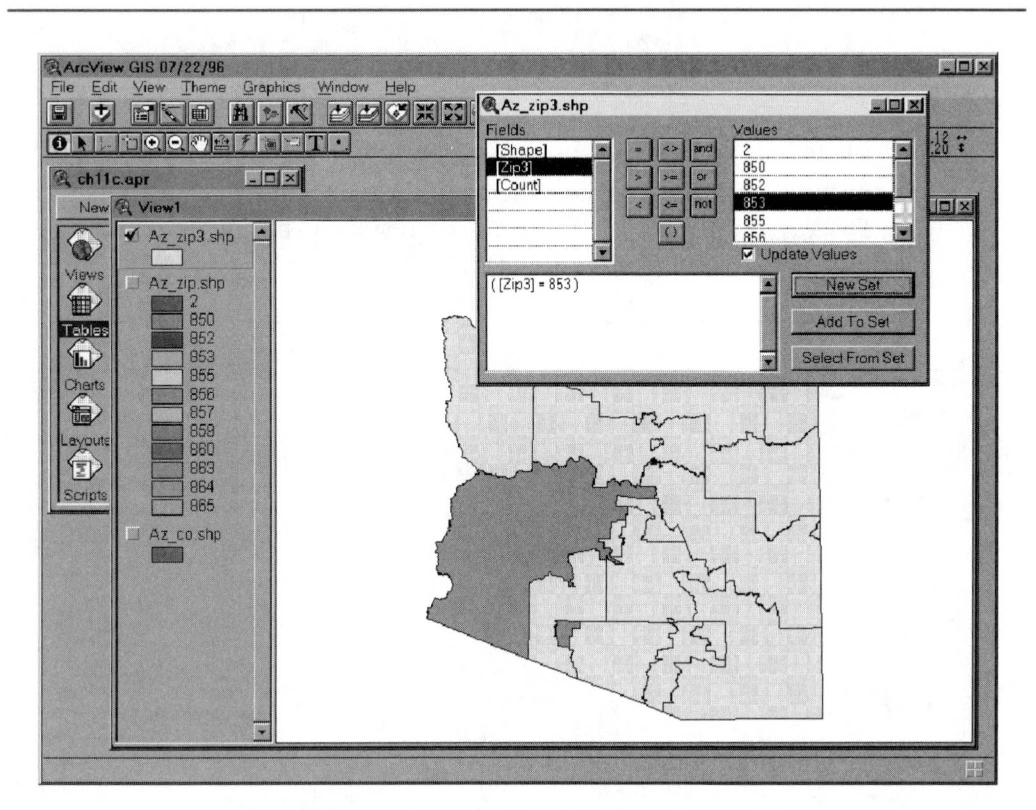


*Selected feature derived from the query ([Zip3] = 853).*

2. Dismiss the Query Builder dialog window, and make the *Az_co* theme active.

3. From the Theme menu, choose Select by Theme. In the Select by Theme dialog window, select features of active themes that intersect the selected features of *Az_zip3.shp*. Click on New Set.

4. Turn off the *Az_zip3* theme and turn on the *Az_co* theme to display the results. Eight counties are highlighted in the *Az_co* theme.

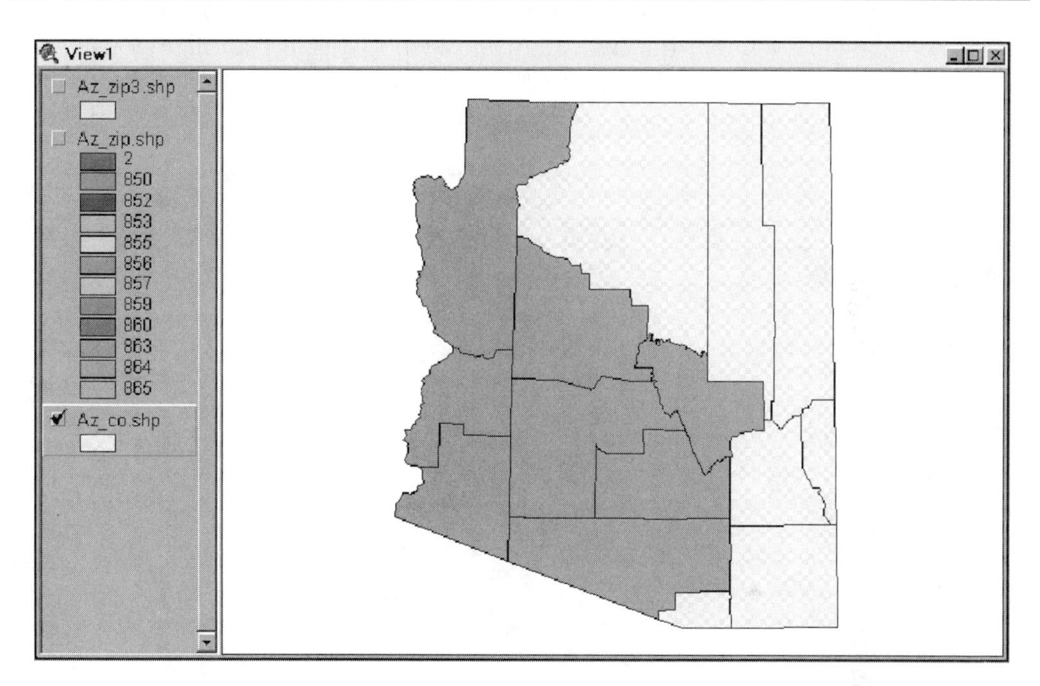


Selected counties from the Az_co theme.

5. Save these selected features in a new shapefile. From the Theme menu, select Convert to Shapefile. Navigate to your working directory and save the theme as *Az_w.shp*. Add the theme to View1.

6. Turn off the *Az_co* theme. Turn on the *Az_zip3* theme, and make it active. Clear its selected features by clicking on the Clear Selected Set icon from the button bar.

7. Make the *Az_co* theme active, and clear its selected features.

8. Repeat steps 1 through 3, this time forming the query *([Zip3] = 852)* and using the selected features to do a Select By Theme from the *Az_co* theme. Five counties should be highlighted in the *Az_co* theme. Save these features in your working directory as *Az_s.shp*.

9. Turn off all other themes and turn on the *Az_w* and *Az_s* themes. Access the Legend Editor and change the symbols to diagonal hatch patterns with transparent background. Select the hatch color and angle such that the two themes, when displayed simultaneously, will show the area of overlap. There should be four counties common to the two themes. Save your project before proceeding to the next step.

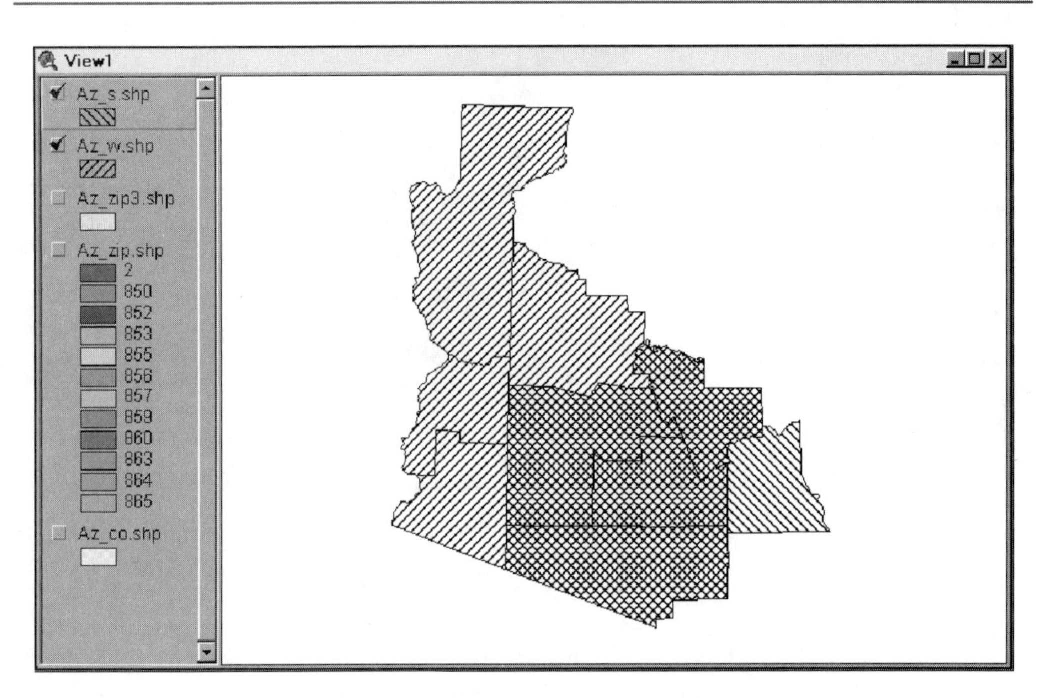


*Az_s and Az_w themes showing the area of overlap.*

You are now ready to run the Merge Themes script.

1. Make the project window active, click on the Scripts icon, and select New to open a new script window. A blank script window with the title Script1 is opened.
2. From the Help menu, select Help Topics. Click on the Index tab, and type *Scripts* into the search box. From the list of Scripts topics, double-click on Samples. Scroll down the alphabetical list of sample scripts to the *View.Merge Themes* entry (in *scripts* subdirectory of ArcView sample data).
3. Click on the Merge Themes link. From the Merge Themes help page, click on the Source Code link. The source code for the Merge Themes script will be displayed in a separate window. Click on the Options button, select Copy, and close the help windows. The Merge Themes script has now been copied to the system clipboard.
4. Make the Script1 window active and select Paste from the Edit menu. A copy of the Merge Themes script appears in this window. Resize the window to examine the script, if desired.
5. Click on the Check icon in the button bar to compile the script. After compiling, the Check icon will be grayed-out, and the running man icon (the Run icon) to the right will become black. This indicates that the script has compiled successfully and is ready to run.
6. Click on the view window to make it active, then click on the script window to return to the script. Click on the Run icon to execute the script.
7. From the pulldown list in the Merge Themes dialog window, select *Az_s*, click on OK, then select *Az_w* and click on OK. When the process is finished, click on Cancel to dismiss the dialog window.
8. Save the output merged shapefile to your work directory with the name *Az_merge.shp*. Add the output theme to View1.
9. Close the script window, and turn on the *Az_merge* theme. Notice that the theme overlaps with the combined areas of the *Az_w* and *Az_s* themes. A count will show nine counties in the *Az_merge* theme.

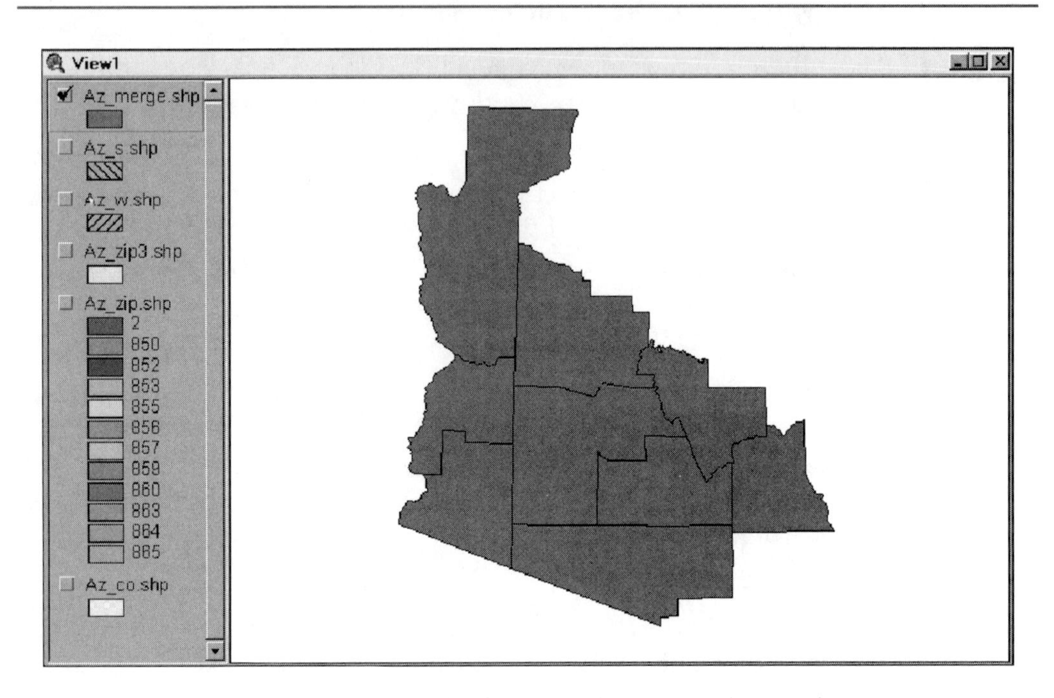


*Az_merge theme, the result of merging the Az_s and Az_w themes.*

**10.** Make the *Az_merge* theme active and click on the Open Tables icon in the button bar to open the attribute table for the active theme. Notice that the status line in the tables tool bar reports 0 of 13 selected. Scroll through the records in the table. You will notice that Gila, Maricopa, Pima, and Pinal counties are all duplicated in the table.

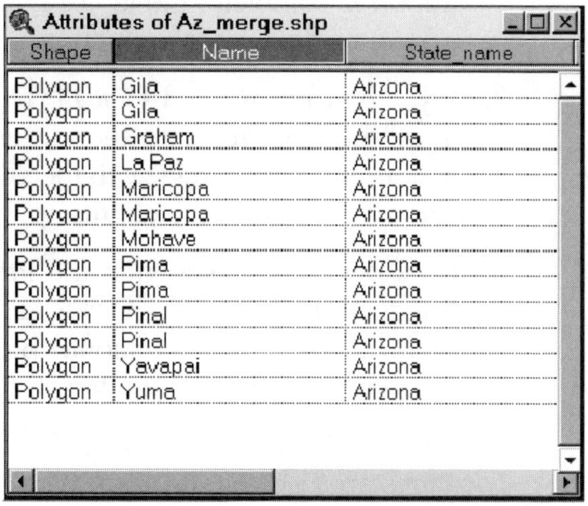

*Attribute table for the Az_merge theme showing the duplicate entries for some counties.*

Attributes of Az_merge.shp

| Shape | Name | State_name |
|---|---|---|
| Polygon | Gila | Arizona |
| Polygon | Gila | Arizona |
| Polygon | Graham | Arizona |
| Polygon | La Paz | Arizona |
| Polygon | Maricopa | Arizona |
| Polygon | Maricopa | Arizona |
| Polygon | Mohave | Arizona |
| Polygon | Pima | Arizona |
| Polygon | Pima | Arizona |
| Polygon | Pinal | Arizona |
| Polygon | Pinal | Arizona |
| Polygon | Yavapai | Arizona |
| Polygon | Yuma | Arizona |

**11.** Close the table and click on the Identify tool to make it active. Click on one of the counties in the south-central portion of the map. Notice that two records are listed in the Identify window for each of the four south-central counties in the theme.

The Merge Themes script has successfully merged the shapes and attributes from the input themes into a single output theme. It does *not,* however, check to see if any features are duplicated across themes. Given that the ArcView shapefile format is a nontopological data structure, overlapping features are perfectly legal—and, in many cases, desirable. It is up to the user to properly interpret the results from any merge themes operation. (The incremental project has been saved as *ch11c.apr.*)

In the next part of this exercise, you will create a hot link.

**1.** First, you need to import the project you saved at the end of the exercise in Chapter 8. Make the Project window the active window, and select Import from the Project menu. Choose *Project (*.apr)* from List Files of Type, and navigate to the directory in which you save the exercise project files. Select *ch8.apr* as the file to import. When the import is complete, the tables and views from this project should be added to the current project.

2. If you had a view titled *View1* in the project from Chapter 8, notice that it will be imported into the current project with the same title. This will result in two views with the title *View1* in the same project. Duplicate view titles can be confusing, but you are about to change that.

3. Click on the Views icon in the Project window. Three views will be listed: *View1, View1,* and *View2.* Click on the second *View1* entry and select Open. You should be able to view the exercise as you saved it at the end of Chapter 8.

4. From the View Properties window, change the name of this view from *View1* to *TEMPE* (all caps). Close the view, and return to the *View1* window you have been working in for this exercise.

5. In the original *View1,* one of the fields in the *Az_zip* attribute table was *Name,* which contained the city in which the zip code was located. Make the *Az_zip* theme active and use the Query Builder to select the polygons for which *Name = TEMPE.*

6. Four polygons should be selected. Use the Zoom to Selected tool to zoom in on the selected set. Note that the city of Tempe contains four zip codes.

You are now ready to add the hot link. A hot link is the means by which an action can be associated with a geographic feature in a view. This action can be displaying a table or image, opening another view or project, or starting up an entirely new application. In this example, you are going to create a hot link that will open another view when you click on any of the four zip codes in Tempe.

*Query Builder ready to select polygons for which Name = TEMPE.*

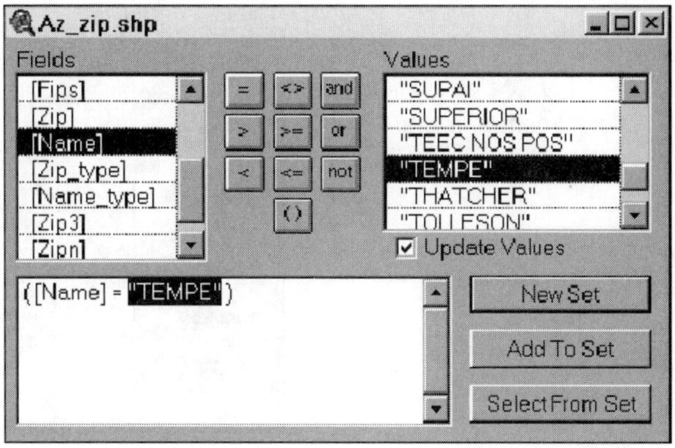

1. From the Theme Properties window, click on the Hot Link icon to call up the Hot Link dialog window.

2. From the Field list, select Name, and from the Predefined Action list, select Link to Document. This will define what happens when you use the Hot Link tool to click on a feature for which a hot link is defined. By selecting Link to Document as the action and Name as the field to be used for this action, you will establish a hot link in which clicking on a zip code polygon will open a document with the corresponding city name from the Name field.

3. Click on OK. The Script box will show the entry *Link.Document.* Click on OK to add the hot link to the theme.

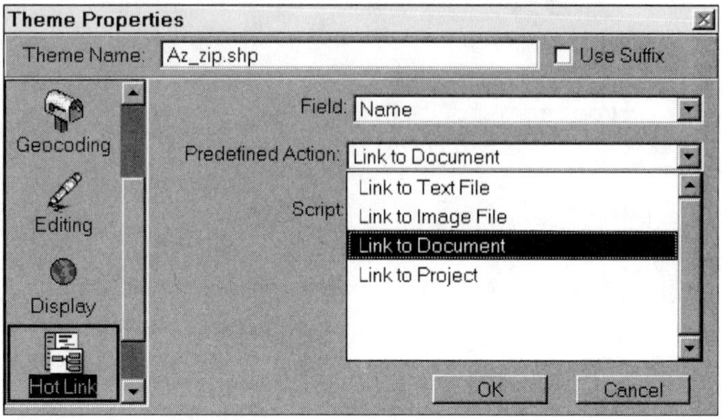

*Hot Link portion of the Theme Properties dialog window.*

4. Click on the Hot Link tool. The cursor changes to a lightning bolt.
5. Click on one of the four zip code polygons for Tempe. The view for *TEMPE* will automatically be opened. (The incremental project has been saved as *ch11fin.apr*.)

You have just created a hot link! This particular exercise demonstrated a view of views. By using this technique, a master coverage can serve as an index to additional themes that contain portions of a more detailed study area.

## Chapter 12

# Advanced Functionality

In the first eight chapters of this book, we took you through the basics of ArcView. Even though we were dealing with basic functionality, it was already apparent that much could be done with regard to viewing and modeling spatial data. In the last three chapters, we have taken you beyond the basics. We have addressed CAD and image themes, the editing and digitizing of shapefiles, and advanced overlay functionality—all of which further extend the ArcView data model, allowing you to better design your ArcView project to address your needs. However, we are not done yet! There is yet another level of functionality that extends the power of ArcView into areas that were previously the sole domain of powerful workstation-based GIS software.

The ArcView model of modular functionality is realized through loading additional ArcView extensions. These extensions allow you to add advanced functionality as needed. In this manner, enhancements to ArcView can be provided without re-engineering the core product.

Three extensions to the ArcView data model will be addressed in this chapter: Network Analyst, Spatial Analyst, and the Spatial Database Engine (SDE). In addition, we will also discuss a separate product, the Data Automation Kit, which brings additional data editing and conversion functionality to the Windows 3.*x*, Windows 95, and Windows NT platforms.

# Network Analyst

Consider the following problems:

- A truck leaves a warehouse with packages to deliver to ten businesses before returning to the warehouse. What is the optimal order and route for the driver to follow when delivering these packages?
- A city is served by three hospitals. An accident occurs at 10th and Main. In terms of response time, which hospital is closest to the accident location?
- A school district currently contains three elementary schools. A fourth school is scheduled to open in the fall. To optimize travel distance and school enrollment, how should students be allocated among the four schools ?

These problems share a requirement: the need to get from one point to another using a linear delivery system, the city street network. These situations also hold something else in common: the problem can be modeled and solved using the ArcView Network Analyst extension.

To solve these problems, Network Analyst requires a street network comprised of connecting line segments, each of which is coded with the cost associated with traversing that segment. This cost-coding schema can be simple or complex, depending on the problem at hand. To allocate students to schools based on their distance from that school, for example, the network requires only one field: the length of each segment.

When assigning an accident to a hospital based on response time, a more complex data schema is required. In addition to the travel time required to traverse each segment, additional information is needed about the time to traverse the intersections, or nodes, between segments. Typically, the time required to travel straight through an intersection is less than that required to turn left or right. This information is stored in a special file called a *turntable*.

Additionally, the travel time of line segments often differs with respect to the direction traveled, depending upon whether travel is with or against the dominant flow of traffic. Within a sophisticated network, separate fields may be used to model travel costs for the morning commute, the midday hour, and the afternoon commute. Using Network Analyst, you can specify which fields to use for computing travel cost, based on the time of day for anticipated travel. This makes Network Analyst a very powerful engine for solving complex network problems.

To determine the optimal route for a delivery van, the steps to take are as follows:

1. Create a file that contains the addresses of delivery stops.
2. Use the ArcView Geocoding Editor to create an event theme of point locations for each delivery stop.
3. Begin the routing process by making the street network theme active and selecting Find Best Route from the Network menu.
4. Specify the starting point for the route by using the Add Location by Address button to enter the address.
5. Load the stops from the delivery stop event theme.
6. Specify that the truck needs to return to the warehouse at the end of the route.
7. Specify that the Network Analyst is to find the best order when assigning stops.
8. Click on Solve to calculate the optimum route. The route is displayed in the view as a new theme, and the stops and cumulative travel time are displayed in the Best Route dialog window.

That's all there is to it! The best route solution is a new theme based on an ArcView shapefile, and as such can be manipulated as any other ArcView theme, with tabular joins, overlay operations, and the like.

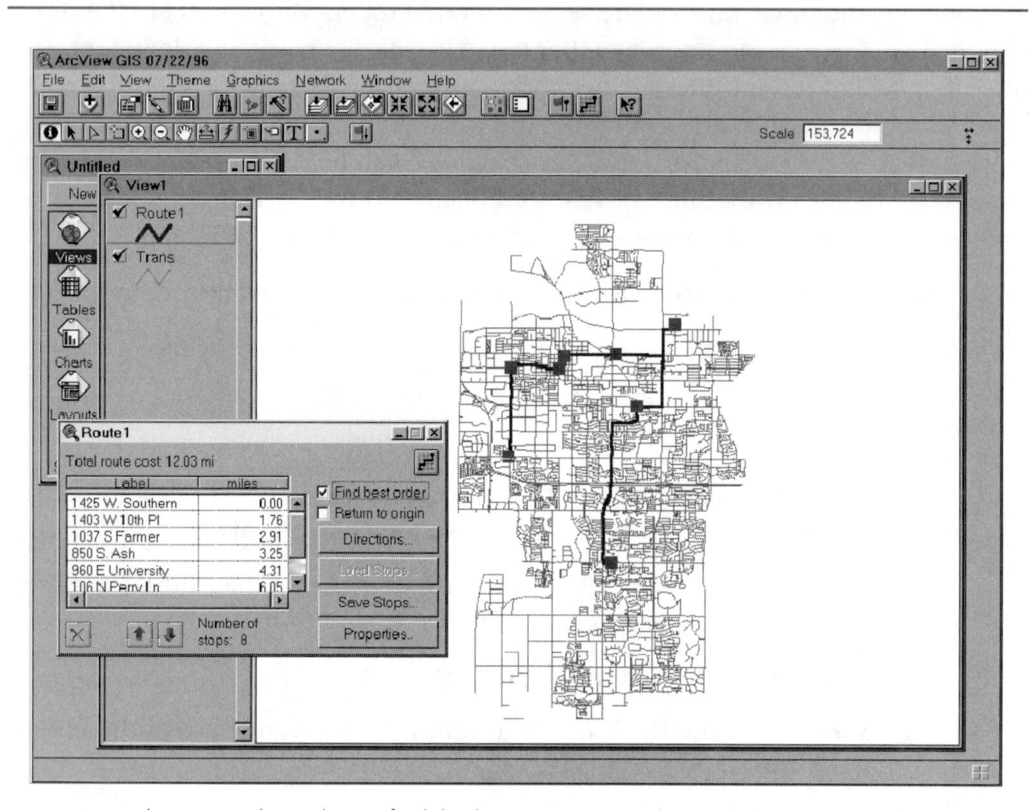


*Using the Network Analyst to find the best route—one that includes a series of stops.*

The most difficult part of this process—setting up a properly coded street network—can be expedited by obtaining a ready-to-use street network from a third-party data provider. The ability to solve routing and allocation problems is not common to all GIS applications, but when you need it, you *really* need it!

ArcView Network Analyst constitutes a separate product, and is not included with the core ArcView product. Contact your ESRI representative for additional information.

# Spatial Analyst

The ArcView data model as provided in the core product is based on a vector data structure. ArcView's native shapefile format, as well as ARC/INFO coverages and CAD drawing files, are all comprised of line segments. As you have seen, much can be accomplished using a vector data structure; however, there is another GIS data format: *raster* (grid cell) data.

The raster data model, in which data is represented by a two-dimensional matrix of uniformly spaced rows and columns, is well suited for depicting certain types of geographic data and solving certain types of spatial problems. Data that represent such properties as elevation, slope and aspect, and density and proximity are types for which the grid cell model is appropriate.

ArcView Spatial Analyst uses the ARC/INFO grid data structure to store and model raster data. Additionally, most of the modeling capability found in Grid's Map Algebra has been extended to Spatial Analyst.

Spatial Analyst is suitable for the following applications:

- Generating a proximity map based on distance from major roads.
- Generating a temperature contour map based on temperature readings at point locations.
- Combining separate resource data sets, such as slope, soils, and vegetation, to construct a land use suitability map.

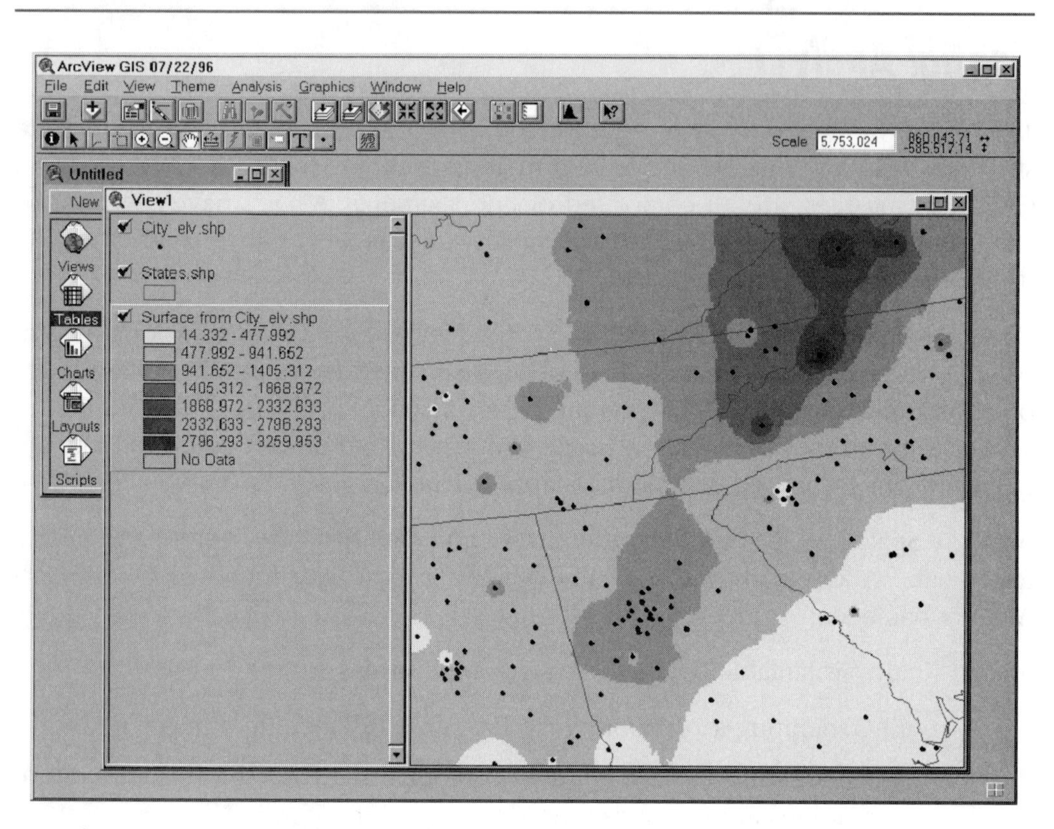


*Surface elevation derived from the Elevation field associated with the point theme City_elv.*

ArcView Spatial Analyst also lets you convert back and forth from an ArcView shapefile to a grid, providing a smooth means of exchange between Spatial Analyst and core ArcView applications.

Using Spatial Analyst, you could prepare a range site suitability map via the following steps:

1. Use a line shapefile that depicts streams as an input data source to create a continuous output buffer grid showing distance to water. Classify this output grid on two categories: one for distances of less than one mile to water, and one for distances of greater than one mile to water.

2. From an input elevation grid, derive a slope map; extract from this output grid those areas of less than two percent slope.
3. From an input polygon shapefile that depicts soil mapping units, construct a discrete category grid, assigning a soil mapping unit value to every grid cell.
4. Using the Map Calculator, combine the above three grids to create a single output grid; classify the resultant grid to display those areas of soil mapping units that (1) have sandy surface textures, (2) are less than one mile to water, and (3) have less than two percent slope.

The resultant map will display areas of prime range suitability, as dictated by your input criteria. If you do not like the results, change the input criteria and generate the map again. The ease with which output grids can be generated from multiple input themes makes this a valuable tool for examining "What if?" scenarios.

As with Network Analyst, ArcView Spatial Analyst is a separate product, and is not included with the core ArcView product. Contact your ESRI representative for additional information.

## Database Themes

The Database Themes extension in ArcView allows you to access data maintained by ESRI's Spatial Database Engine (SDE). The Spatial Database Engine stores geographic features and feature attributes in a relational database, such as Oracle. SDE supports extremely large data sets—in excess of one million features—with spatial processing that is independent of database size.

SDE stores spatial data in the RDBMS using an object-oriented data structure comprised of simple geographic elements, such as polylines, polygons, and donuts. The purpose of SDE is to provide spatial data resulting from a logical or spatial query. It relies on the served application—in this case, ArcView—to perform the spatial processing required to support geographic modeling and analysis.

This client–server-based approach to managing spatial data results in several fundamental differences in how feature themes and database themes are modeled and manipulated. Database themes, which result from a set of features served from

RDBMS, are based on a cursor model (one record at a time), rather than the spreadsheet model used for feature themes (all records presented in a table view). As a result, some differences are readily apparent. First, there is no linked attribute table. Second, there are also constraints on the thematic classification available through the Legend Editor. A little thought about the logistics involved in performing, for example, a standard deviation classification on a million-plus record data set, points out some of the unavoidable difficulties involved in the design of database themes.

Although there is no accompanying attribute table associated with a database theme, you can export the current set to a table and add the resultant theme to the view. In this manner, the full set of ArcView functionalities, including theme classification and table joins, is available.

When evaluating the positives and negatives of database themes, you should keep in mind the type of data sets for which SDE is appropriate. SDE is designed to support large, mature databases, containing upward of a million features (for example, parcel databases or large hydrologic databases). In such instances, you will probably work with standard queries and a standard legend predefined by the database administrator. SDE and database themes are not optimized to support a data exploration type of database. However, as previously mentioned, it is possible to export a selected set to an ArcView shapefile and table for subsequent query and analysis.

The Database Theme extension is included in the core ArcView product; however, to be functional, it does require an SDE database on the client side.

## Data Automation Kit

Strictly speaking, the Data Automation Kit is not an ArcView extension. It does, however, extend functionality to PC platforms in the areas of data creation and editing, data conversion, and map projection.

The data creation and editing module is the full version of PC Arcedit. This allows creation and editing (including digitizer support) of PC ARC/INFO coverages and associated dBase attribute tables. Also included is the PC version of Clean and Build, which gives you full control over the creation and maintenance of cover topology.

The data conversion module allows for two-directional exchange between Arc-View shapefiles and the following formats:

- MIF (Mapinfo exchange format)
- BNA (Atlas exchange format)
- E00 (ARC/INFO exchange format)
- ARC/INFO coverages

It also allows one-way transfer from TIGER94 and .dlg to ARC/INFO coverages, and from ARC/INFO coverages to .dxf.

The map projection module supports 46 map projections, 30 spheroids, and 200 datums, thereby supporting the transformation of PC ARC/INFO coverages into virtually any map projection.

It should be pointed out that the Data Automation Kit is based on the PC ARC/INFO data module. As such, it does not support double precision ARC/INFO coverages, ARC/INFO coverages containing more than 5,000 vertices in a single polygon, or the extended ARC/INFO data model [routes, regions, node attribute tables (NAT), or text attribute tables (TAT)]. It does, however, allow the creation and manipulation of ARC/INFO coverages for users without access to workstation ARC/INFO.

# Chapter 13

# Optimizing Project Design

Previous chapters have demonstrated that ArcView provides a powerful tool for the modeling and analysis of spatial data. You probably have numerous ideas about projects you want to undertake. The only remaining issue might be one of how to begin, or how to organize project design.

We suggest that your first step be to determine the nature of your project. Is the project intended to answer a specific question, or to support a more general process? The former can be described as one-shot, or disposable applications, and the latter as reusable applications. This chapter addresses the design of reusable applications. Keep in mind, however, that even if you are using ArcView to obtain a specific answer to a problem you do not intend to address again, the principles of project organization are relevant.

Proper database and workspace management can make it easier to work through complex assignments. These management techniques allow you to more easily choose different approaches if your first efforts go wrong, and they make it easier to resume project design after interruption.

## Data Organization

A successful desktop mapping project involves establishing a link between spatial data—the geographic features on your map—and the tabular or attribute data associated with these features. How the tables are structured is highly correlated with the ease with which the data can be manipulated to answer your questions.

# Spatial Data

Typically, you will have less control over attribute data than spatial data. If you obtain spatial and attribute data from a commercial vendor, the spatial data has been determined for you based on the attribute or tabular data via a spatial or locational "hook." The hook may be a census block group or zip code. While you may have some ability to aggregate data to larger units, such as census tracts, you have little choice but to obtain the vendor's preformatted digital maps. The only situation in which you can ensure that the hook in your tabular data fits your existing spatial hooks is to generate your own tabular data via primary research.

For example, assume you obtain ten tabular data sets, and all are referenced by census tract. Meanwhile, the only spatial data set (digital map) you have represents census blocks. In this situation, you would be forced to obtain a census tract spatial data set. The underlying assumption is that replication of your tabular data set is unlikely, because the data is either a snapshot in time or would be prohibitively expensive to replicate. By contrast, the proper spatial data set can likely be obtained from a vendor or government agency or, as a last resort, digitized from scratch.

Wherever possible, verify that a one-to-one relationship exists between your mapping units and the corresponding units in your tabular data. In many instances your map may be coded with more than one locational identifier. For example, a block group theme may also be coded with attributes that associate each block group with a census tract or zip code. While it is possible to link a table of data gathered by census tract to this theme, a more efficient procedure is to first prepare a new census tract theme using the Merge operation. Then link this table to the new theme. In this way, the one-to-one relationship between geographic features and attribute records is preserved. Otherwise, links to the block group theme could produce errors when you model attributes gathered by census tract.

# Tabular Data

It is also important that you organize your tabular data. Whenever possible, construct your tables so that no more than one join is required; namely, the join to the theme attribute table. Multiple joins increase the likelihood of an incomplete table affecting the resultant theme, particularly when files are chained and the constraining table occurs in the middle of the chain.

The benefits of data organization also extend to bringing data into ArcView. A theme attribute table or joined table will often contain numerous fields pertaining to a variety of attributes. Through thematic query and classification, it is possible to produce a single theme that displays a combination of attributes. However, it is usually preferable to begin with each theme representing a single attribute, and to use thematic query and overlay for exploring the relationship between themes.

## Project Organization

By maintaining separate documents for storing themes, charts, tables, and layouts, ArcView imposes a certain amount of organization on any project. In addition, ArcView lets you further organize thematic data by allowing multiple views.

The use of multiple views allows you to group similar data together. In this manner, environmental data could be grouped into one view; demographic data grouped into a second view; and administrative boundaries and infrastructure made common to both views. Because duplicate themes point to a single data source, there is no storage redundancy. Moreover, distributing themes in this manner reduces the size of each view's representation in the Table of Contents, thereby potentially creating an ideal situation in which the contents can be viewed without scrolling.

Using multiple views to organize a project can be aided by the use of hot links. As discussed in Chapter 11, hot links can be established between a reference theme and additional views to create a view of views. With hot links, clicking on a feature in a reference theme can open a view that displays additional information associated with that feature. By removing the need for the user to know how to navigate between views, greater use can be made of multiple views as a means of organizing thematic data. If the ability to navigate between multiple views is not sufficient, hot links can also be established between projects.

## Project Optimization

Once the thematic data organization is determined, you should also pay attention to speed and maintenance issues. In particular, when designing an ArcView application that will be used repeatedly by many people over time, these factors are just as important as data organization to the effectiveness and usefulness of the project.

## Optimization for Performance

The ArcView data structure is designed to be efficient with regard to retrieval and display. Given adequate hardware and a moderately sized project, ArcView will deliver very acceptable performance. However, projects have a way of growing in scope and complexity, and there may come a point when performance becomes an issue.

In ArcView, performance issues center around initial project start-up time and in-use performance. Optimization for in-use performance can result in improved project initialization as well.

### Start-up Time

Optimizing project initialization involves reducing the time necessary to read the thematic data into ArcView. To determine the type of optimization that can be accomplished, it is first necessary to understand how an ArcView project is saved.

Because an ArcView project is dynamic, the project file stores *references* to the data displayed rather than the data itself. Consequently, changes to the data made subsequent to the last ArcView session will be reflected when the project is reopened. The disadvantage of this is that ArcView has to repeat all the steps that were taken to reach the point at which the project was last saved: all spatial and tabular data must be reloaded; all tabular joins and logical queries, reconstructed; and all supporting documents, such as charts and layouts, recreated. If your project and data sources are complex, start-up time can be lengthy.

The primary means for decreasing project start-up time is to reduce the number of joins and logical queries on themes and tables. To do this, spatial and tabular data can be customized specifically for use within ArcView. This technique challenges accepted tenets of database management, which stress avoiding redundant or duplicate data. However, duplicating data in order to reduce start-up time may be a worthwhile trade-off.

Logical queries on themes can be eliminated by performing the logical query once and then using the Convert to Shapefile option to save the selected set as a new shapefile. Adding this shapefile to the project will make the new shapefile and its associated attribute table available immediately when the project is reopened. This approach may be helpful when a logical query is required to produce a subset of features based on attribute values, or when an ARC/INFO coverage contains island polygons of "no data." In ArcView these no-data polygons must be selected out to ensure that they are not subsequently shaded. If the selected set described above is saved as an ArcView shapefile, this query will not be necessary.

The same procedure is useful for queries on tabular data. If a logical query is performed in order to extract a subset of records from a master data file for use in a particular project area, the start-up time required to perform this selection can be eliminated by writing the selected records to a separate file prior to opening the project.

Overhead associated with joining tables is typically not as severe as that associated with logical queries. However, when tuning for performance, reducing the number of joins from single attribute tables to the feature attribute table can result in improved start-up time. If necessary, all feature attributes can be added directly to the primary feature attribute table. If you convert a theme containing tabular joins into a shapefile, your result is an output theme in which all attributes are located on the primary theme attribute table. Replacing the original joined theme with this output theme will result in improved project initialization, because the joins will not have to be re-established each time the project is opened. Because of the extra overhead that comes with coverage maintenance, however, this procedure should be considered only after all other options have been explored.

## In-use Performance

One basic method for accelerating data access and display is to *index* the data. An index can be created on one or more fields in a table. Indexing will improve operations based on values for these fields, such as simple queries, and will improve Join or Link operations using these fields.

To create an index on a table, open the table, highlight the fields to index, and then select Create Index from the Field menu. If an index already exists for the field, the menu choice will be Remove Index. The index is stored in the source data directory. If the user does not have write access to the source directory, a temporary index will be created and then removed at the end of the ArcView session.

➡ **NOTE:** *Indexing a field will improve the performance of a simple query involving the field, such as [County] = 'Lincoln'. Indexing will not improve queries that involve string matching, such as [Name] = "Sm*", or queries that involve comparison, such as [Age] << 18.*

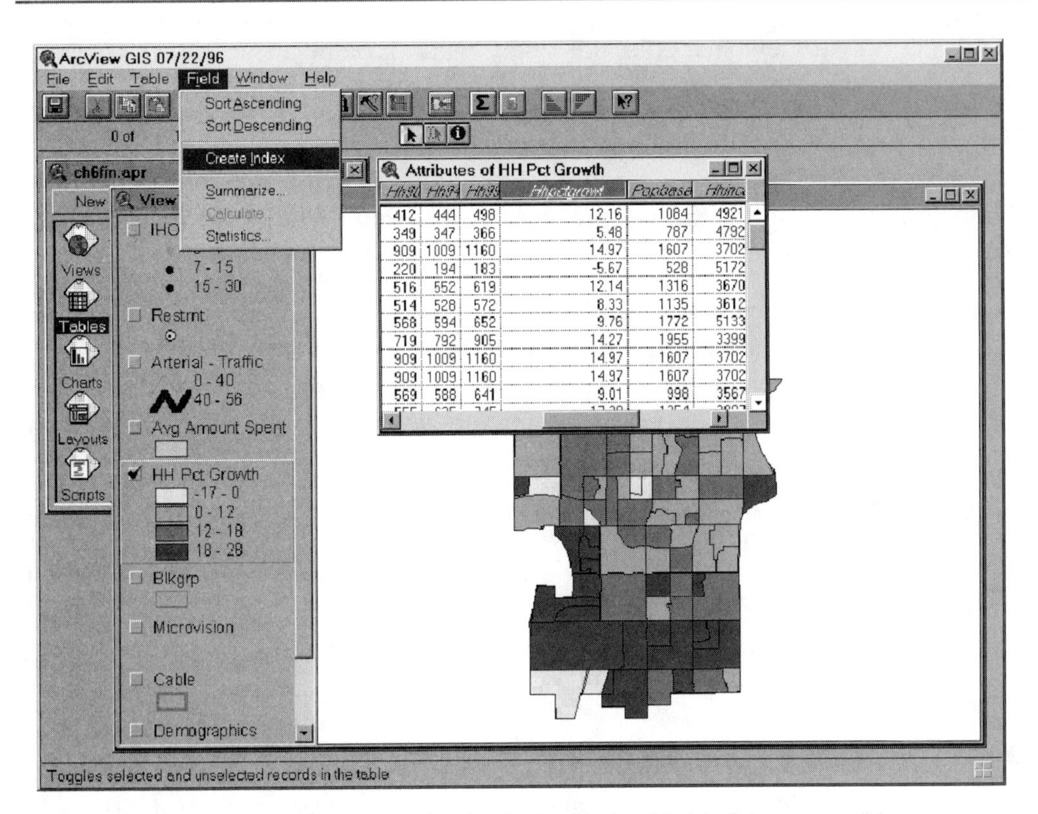


*Preparing to create an index for the highlighted field of the open table.*

If the Shape field is selected as the field to index, a spatial index will be created. Spatial indexes improve the following query and selection operations:

- ❒ Identify
- ❒ All drawing operations
- ❒ Spatial joins
- ❒ Theme-on-theme selection

When a spatial join or theme-on-theme selection is performed, a spatial index is automatically created. As with tabular indexes, spatial indexes are stored in the source data directory, provided the user has write access to this directory.

Another means of improving in-use performance is to convert any complex geographic data based on ARC/INFO coverages to the ArcView *shapefile* format. The shapefile format is a simpler, nontopological data format optimized for use in ArcView. A theme based on an ArcView shapefile will display more rapidly than a theme based on the same data from an ARC/INFO coverage. If necessary, a shapefile can also be converted back to an ARC/INFO coverage for maintenance, or for use in other operations.

In addition, some of the methods noted in project organization can improve performance. A "view of views" can eliminate the need to display detailed themes except when zoomed in on a specific area. Next, you can set a maximum display scale for detailed themes. This eliminates some re-drawing when zoomed out to the full project extent. Another way to keep performance at an acceptable level is to ensure that the spatial geography is not more detailed than the thematic data it represents.

Lastly, the Spatial Database Engine (SDE) can be used to further optimize ArcView performance for data sets comprised of a large number of features. Using the Spatial Database Engine, the spatial data is stored as primitive graphic elements directly within the RDBMS. This lets you work with a single database for storage of both feature and attribute data, and lets you tune the RDBMS to efficiently serve this spatial data to the ArcView application. Additionally, the cursor-based approach allows queries to be optimized and/or constrained so that an inappropriate request does not degrade performance. The Spatial Database Engine is a separate product, available from ESRI. It currently supports the Oracle RDBMS; other databases, such as Sybase and Informix, will be supported in the near future.

## Optimization for Ease of Maintenance

The design issues involved in optimizing a project for ease of maintenance can be complex. When an ArcView seat is part of a network (for example, when it connects to a server running ARC/INFO or a network of PCs storing a distributed database), design criteria are much different than when the ArcView seat is located on a stand-alone PC, separate from spatial and tabular data sources. Moreover, because the optimal design for maintenance purposes can conflict with the optimal design for performance, it may ultimately be necessary to balance the trade-offs.

When ArcView is used on a network and linked to external spatial and tabular data sources, and when performance is not an issue, the design criteria for maintenance are simple: as much as possible, keep data at the source in native format. In this way, the maintenance is done with the originating application, and the problems associated with keeping duplicate data sets in sync are kept to a minimum. If data restructuring is required, either in post-processing of spatial data coverages or in manipulation of tabular data, it can be performed using the originating application (including ARC/INFO), or the native DBMS software.

When data is to be used on a stand-alone ArcView installation, data format and the ability to move data on and off the PC must be taken into account. As discussed in the insert in Chapter 3, "Importing and Exporting ARC/INFO Data," it may be difficult to establish bidirectional spatial data transfer with Windows ArcView. Consequently, you must determine whether you need to update the data on the PC from the source only, or if you need to pass revised or archived data back to the host as well.

Although only two spatial data formats are supported by ArcView, ARC/INFO coverages and ArcView shapefiles, in reality the situation is more complicated. (Because they are limited with respect to attribute handling, themes based on CAD drawing files are not considered a primary spatial data source.)

ARC/INFO coverages come in three formats:

- PC ARC/INFO
- UNIX ARC/INFO 7.0
- pre-7.0 UNIX ARC/INFO

(See the insert titled, "Importing and Exporting ARC/INFO Data" in Chapter 3 for a detailed discussion of the formats.)

The discussion below focuses on the format constraints involved in the most difficult data transfer situtaion: moving data back and forth from a UNIX ARC/INFO source to a standalone Windows ArcView destination. The underlying assumption here is that you have decided you must take an active role in data maintenance.

If periodic data updates are to be provided from UNIX ARC/INFO to Windows ArcView, and if the entire workspace will be updated with each delivery, then the workspace can be maintained as an ARC/INFO 7.0 workspace and copied to the PC as a unit. If updates will be performed at irregular intervals, with single coverages updated rather than the entire workspace, then adjustments must be made to allow single coverages to be replaced on the PC without affecting the integrity of the remaining data.

If data is to be maintained in ARC/INFO 7.0 format, then each coverage can be transferred to the PC as a separate workspace. Alternatively, the 7.0 cover can be converted to ARC/INFO's Export format and then restored on the PC using the Import utility. When imported, the ARC/INFO coverage is restored on the PC in the PC ARC/INFO format. However, as discussed in the "Importing and Exporting ARC/INFO Data" insert, not all ARC/INFO 7.0 coverages are supported by the Import utility. Another alternative is to convert the ARC/INFO 7.0 cover to ArcView's shapefile format using ARC/INFO's ARCSHAPE command, and then copy the resultant shapefile to the PC.

If updates must be passed from Windows ArcView back to UNIX ARC/INFO, there are two alternatives:

1. If only attributes are to be edited, the data can be maintained as either ARC/INFO coverages or ArcView shapefiles. If maintained as ARC/INFO coverages, the entire workspace must be copied back to the workstation as a unit. The decision regarding whether a workspace will contain a single coverage or multiple coverages will rest on whether the covers require individual or group editing.
2. If both features and attributes are to be edited, the only option is to maintain the data in ArcView shapefile format, which is the only format that allows geographic features to be edited. As indicated in the "Importing and Exporting ARC/INFO Data" insert in Chapter 3, the Import utility creates a cover in PC ARC/INFO format. The most common way to import this cover back to UNIX ARC/INFO is to convert the cover to Export format using PC ARC/INFO. An alternative is to convert each feature class from the PC ARC/INFO cover to an ArcView shapefile. Subsequently, you must convert the shapefiles back to ARC/INFO coverages using UNIX ARC/INFO.

When compared to the complexity of configuring ArcView's spatial data for optimum maintenance, managing tabular data for optimum maintenance is straightforward. ArcView supports three tabular data formats: dBase, INFO, and delimited text files. All formats can be transferred easily between the PC and UNIX workstation environments, and most database and spreadsheet software can read and write data in at least one of these three formats. The key here is to follow standard database management procedure; that is, keep your key fields unique, and keep your data files discrete and modular.

# Theme and Project Locking

After working on an ArcView project, it may be useful (and probably advisable) to protect the project design from being inadvertently altered. To this end, two tools are available: theme locking and project locking.

The option of theme locking is available for all themes in a View. Theme locking places a password on the Theme Properties window for the active theme, requiring subsequent users to issue the password before they can alter theme properties. This capability can be useful for protecting theme properties such as a logical query applied on a theme or a hot link associated with a theme. Theme locking does not, however, lock the theme to all customization; the Legend Editor can still be invoked in order to change the theme classification or display.

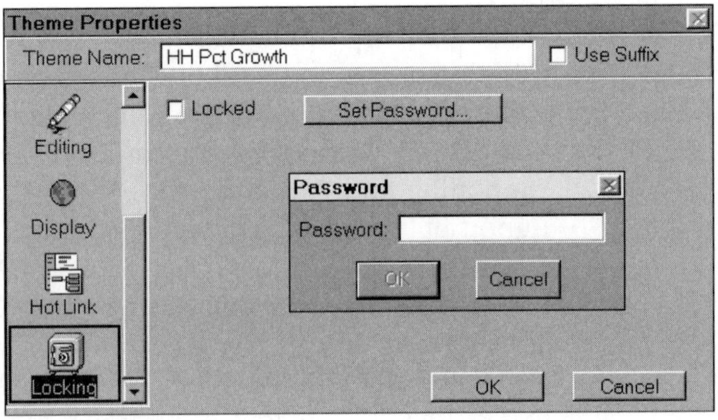

*Theme Locking portion of the Theme Properties dialog window.*

Project locking can be carried out at the system level by locking a project's *.apr* file. If you set the file properties to read-only, a user can open the project and make customizations but cannot save these changes back to the original project file. If the user wants to keep such changes, they must be saved to a new project file using the Save As option from the File menu.

## Looking Ahead

As suggested above, project design and management require careful consideration—even when you use ArcView "straight out of the box." If you require still more control, then customizing with Avenue is your next step.

Chapter 14

# ArcView Customization

## Why Customize?

Through 13 chapters and several detailed exercises, we have demonstrated the powers available in ArcView and how they can be used for real world problem-solving. Using ArcView "out-of-the-box" has been the focus of the preceding chapters. Despite ArcView's inherent strengths, you may find it necessary to customize an application to meet your needs. This is done via Avenue, ArcView's programming language. You do not have to be a programmer guru, however, to customize with Avenue. This chapter provides an overview of Avenue so that you will be aware of its strengths in the event that you need to use it.

## Customizing the User Interface

An ArcView project is a collection of documents, including views, tables, charts, and layouts. Each type of ArcView document has a unique graphical user interface (GUI). Each GUI is composed of three control groups: the menu bar, the button bar, and the tool bar. With Avenue, you have the ability to modify the GUI. Controls can be added, deleted, or rearranged, and the properties of each control can be modified. Avenue enables you to construct a customized interface that matches any user's needs.

Customizing the user interface differs from the rest of Avenue functionality in that you do not need to write an Avenue script to achieve results. Instead, you make modification through the Customize dialog window. This window is accessed by double-clicking on any blank area of the button or tool bar, or by selecting Customize from the Project menu.

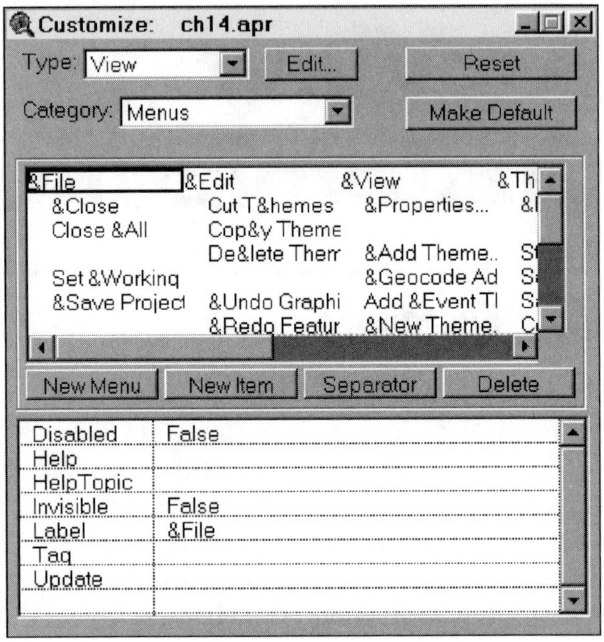

*Customize dialog window.*

The Customize dialog window is divided into three functional areas. The upper area contains pulldown lists for selecting the document and control type to edit, as well as one button that makes your changes the new default and one that returns the interface to the standard ArcView settings. The central area, the Control Editor, contains a representation of the control set, along with buttons that let you add or delete controls and separators. In the lower area, the Properties List presents a list of properties and settings for selected controls.

The following document choices are available in the upper area's Type list: *Appl* (Application), *Chart, Layout, Project, Script, Table,* and *View.* The Customize dialog window allows you to modify the GUI associated with each document type. Note that this window allows you to modify the GUI for each *class* of document rather than specific instances of document types. Thus, modifications to the View GUI, for example, will apply not only to the active view, but to all view documents in the project.

The Control Editor presents the controls for the current selection—the menu bar, button bar, tool bar, or popup menus—as they appear on the GUI. Controls can be repositioned by clicking on the element and dragging it to a new location. The New and Delete buttons allow you to add or delete controls; the Separator button allows you to add a separator to a control set. This translates to either a horizontal line for separating menu items or a space for separating buttons and tools.

When a control is selected, a box appears around the control on the Control Editor display. The properties of the control are displayed in the Properties List. The properties determine the behavior and appearance of a control. These are the properties that can be set in the Properties List:

- ❐ Displaying or hiding a control.
- ❐ Enabling or disabling a control.
- ❐ Selecting the icon associated with a button or tool.
- ❐ Selecting the cursor associated with a tool.
- ❐ Selecting the help string that will appear in the status bar.
- ❐ Selecting the help topic that will be associated with the control.
- ❐ Selecting the Avenue script that is executed by the control.

Through the control of menus, buttons, and tools displayed in the user interface and the Avenue scripts associated with each, you can carefully tailor the interface to fit a specific application. Associating customized Avenue scripts with specific controls by substituting them for default system scripts can greatly enhance the project environment.

## Document Properties and Virtual Documents

The previous discussion of customization centered on customizing standard ArcView document types: Views, Tables, Charts, Layouts, Scripts, and the encompassing Project and Application documents. As discussed earlier, customizing a document type will affect all existing documents of that type as well as all subsequent documents of that type added to the project.

As of ArcView 3.0, you can also modify the standard document types, either by modifying the default or by creating custom (*virtual*) documents. For example, the standard view document type could be copied and modified to produce a custom view as an index for exporting canned data sets or printing canned maps. This view document might contain additional custom map generation and data export tools and buttons. At the same time, other buttons could be removed that allow for opening attribute tables or performing queries. This new document type could be given the name *Canned Maps.* A new icon for this custom document type could be added to the project window. When selecting the Canned Map document type and choosing New, a new Canned Map document would open in the project, one that contained the customized controls and GUI. At the same time, the standard View document could remain available to the user, allowing new standard View documents to be added to the project as well.

In addition to customizing the document GUI via the Customize dialog window, you can modify the standard or custom properties of the document type. These properties include the type title; type icon; the scripts to run when the document is created, opened, or updated; and the display of the document type in the project window.

To access the type properties for an existing document type or to create a custom document type, click on the Edit button in the Customize dialog window.

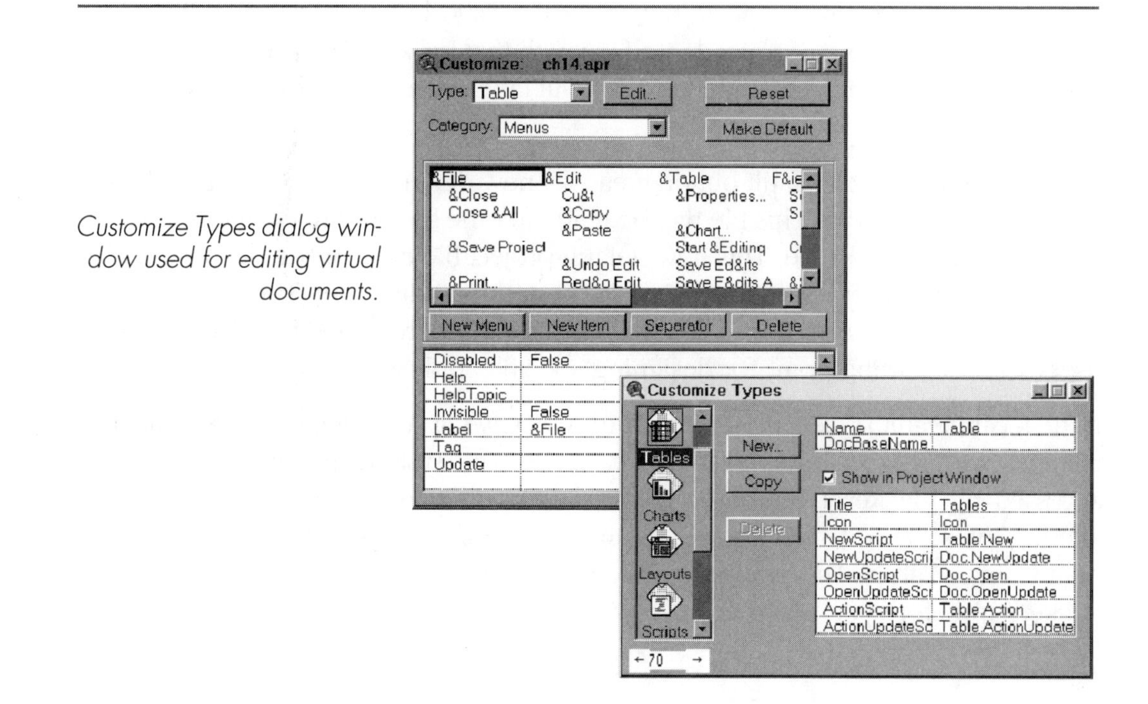


*Customize Types dialog window used for editing virtual documents.*

## Saving the Customized Interface

When ArcView is started, environment defaults are established by reading the project file titled *default.apr*. Customization has the effect of overriding these system defaults.

The final step to customizing the ArcView interface is to make your changes permanent. Three levels of change are associated with ArcView customization:

1. Changes can be saved for the specific project, and no other projects are affected.
2. Changes can be applied to all projects for all users that share the same home directory.
3. Changes can be applied system-wide.

Saving the project with the Save button or the Save menu selection will permanently associate your customized interface with that project. These changes will be restored when the project is opened, and will take precedence over any system or home defaults.

Clicking on the Make Default button creates a new default project file (*default.apr*) in your home directory. This file will be read after the system project file but before the application project file. This new default project file can apply to a specific user or to a user group, if those users share the same home directory. The home default project file can be used to set general ArcView properties for a specific class of users, or as a starting point for designing specific applications.

If the local default project file is copied to ArcView's system default project file, the changes will be applied to all ArcView users. The system default project file is accessed by the Reset button in the Customize dialog window. Consequently, altering the system default project file makes the initial ArcView defaults permanently unavailable. The ArcView system default project file should always be backed up before you make any changes.

# Avenue

You are now ready to dig a little deeper and explore Avenue. The fundamental building blocks of both ArcView and Avenue are called *objects*. ArcView is built with objects; Avenue is an object-oriented programming language.

## Fundamentals of Object-Oriented Programming

Object-oriented systems and programming languages (SmallTalk, for example) have gained popularity in recent years. Perhaps the best way to describe an object-oriented environment is to contrast it with the more traditional procedural programming environment. In procedural programming languages, such as Fortran or C, there is a major difference between the data and the actions taken on the data. Applications constructed with procedural tools often share a strong procedural focus. Thus, in many traditional applications, such as spreadsheets and database management systems, a clear distinction is maintained between the application and the data manipulated by the application.

In an object-oriented system, everything is an object. The data, the application, and even the user interface, are all objects in a unified environment. The distinction between the objects representing your data and the objects representing the tools used to model your data is made by defining the *properties* of each object.

The relationship between objects in such a system is defined by a formal hierarchy built on *classes*. A class is a template that defines common properties for a group of objects. Individual objects are referred to as *instances* of a class. For example, a project may contain several views, each of which is an instance of the View class.

Because classes are themselves objects, they can be associated with higher, or more general, classes. Thus, the View class is itself a member of the more general Doc (document) class. Other members of the Doc class include Chart, Layout, Table, and SEd (Script Editor). All members of the Doc class *inherit* common properties defined for that class. Individual members of a class build on inherited properties through the definition of additional properties that make them unique. Thus, while all members of the Doc class have a window and can be opened, a Table document can be sorted whereas a View document cannot.

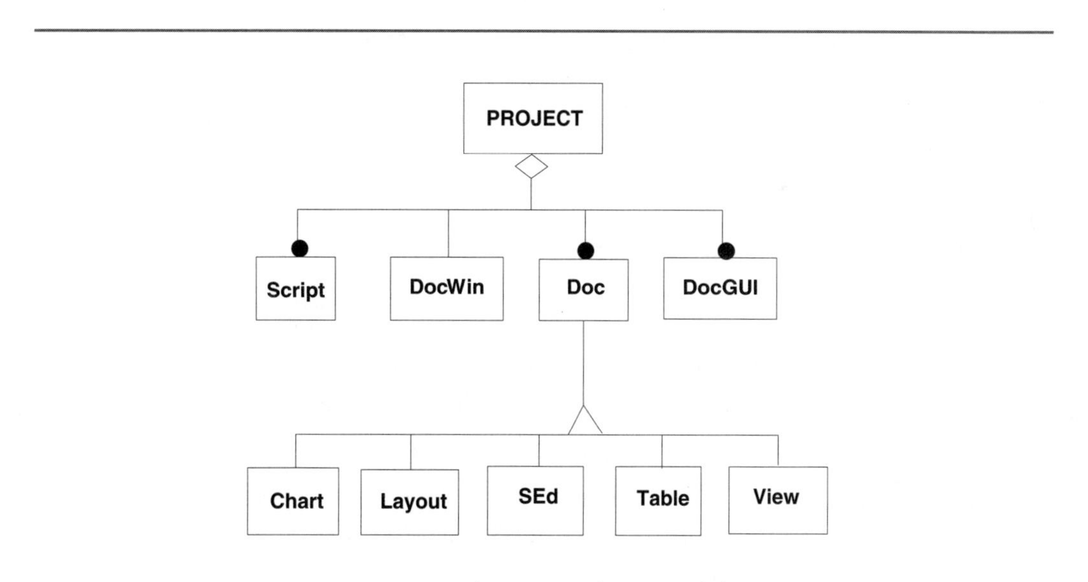


*ArcView documents object model.*

Objects interact by sending *requests* to each other. Requests are the mechanism by which actions occur in ArcView. A request can control an object, for example, by opening a view, or it can return information about an object, for example, by obtaining the name of an active view. A request sent to an object in Avenue results in the return of another object. Thus, when you ask ArcView to obtain the name of an active view, the result is the return of a string object that contains the name of the view.

An Avenue *statement* is comprised of objects and requests sent to those objects. The returned object can either be stored in a variable or passed on to another request. Certain requests will also accept *arguments,* which could be a theme with a specific name.

# Avenue Scripts

An Avenue *script,* then, is a structured series of Avenue statements that have been organized to accomplish a given task. Several examples of Avenue scripts are discussed below.

Let's start by examining the Zoom In function. The Zoom In selection (available from the View menu on the View menu bar) allows you to zoom in on the display of the currently active view by a determinate factor. The Avenue script associated with this control follows:

```
theView = av.GetActiveDoc
theView.GetDisplay.ZoomIn(125)
```

In the first statement above, the object that is returned by the *GetActiveDoc* request on the object *av* is assigned to the variable *theView.* (The object *av* is an Avenue-reserved word that references the overall ArcView application.)

In the second statement, the object returned by the *GetDisplay* request on the active document is the extent of the area of the screen that can be drawn to. This, in turn, is passed to the *ZoomIn* request. The *ZoomIn* request takes a numeric value as its argument, which is 125 in this example.

Thus, in a two-line script, the extent of the active view is determined and subsequently zoomed in on by a factor of 125 percent.

Alternatively, you can zoom in on the display by selecting the Zoom In tool from the View tool bar. You then either drag a rectangle that describes the display extent to zoom to, or click on a point to which the display will zoom by a set factor. The following Avenue script is associated with this tool:

```
theView = av.GetActiveDoc
r = theView.ReturnUserRect
d = theView.GetDisplay
if (r <> nil) then
d.ZoomToRect(r)
else
d.ZoomIn(125)
d.PanTo(d.ReturnUserPoint)
end
```

As with the previous script, the first statement returns the name of the active document and assigns it to the variable *theView.* In the second statement, the *ReturnUserRect* request, which takes a rectangle you drag as input, is made of the active view. The extent of the rectangle is assigned to the variable *r.*

Again, as in the previous script, the third statement uses the *GetDisplay* request to obtain the extent of the display for the active view. However, instead of the extent being passed to the *ZoomIn* request, as in the first script, the extent is instead assigned to the variable *d.*

The final statement of this script is conditional. First, the object assigned to the variable *r* is evaluated. If the value for *r* is not nil (i.e., if you have dragged a valid rectangle), then the rectangle is passed as an argument to the *ZoomToRect* request, which acts on *d*, the extent of the current display.

If you did not drag a valid rectangle (for example, if you single-clicked on the display), then the value for *r* will be nil, causing the *else* portion of the conditional statement to be executed. In this case, the *ZoomIn* request is used, again with the argument *125*, to zoom in on the current display (*d*) by a factor of 125 percent. Subsequently, the extent of the display is also centered or panned to the point you clicked on, by way of the *PanTo* request acting on the point returned by the *Return-UserPoint* request on the current display (*d*).

Thus, in four statements, the active document is determined and the display is zoomed in on by either a rectangle you define or a determinate factor at a specified point on the display. This example is a fine illustration of programming efficiency.

# Avenue Syntax

At this point, you most likely have made some observations about Avenue syntax. While the following is not a comprehensive discussion, it will serve as a quick guide to reading an Avenue script.

As mentioned previously, programming in Avenue involves sending requests to objects. The most common format for sending a request to an object is to follow that object with the request to it, and to join the two with a period or "dot" (e.g., *theView.GetDisplay*). Requests can also be chained together, as in the following statement from the script example:

```
theView.GetDisplay.ZoomIn(125)
```

Assignment of an object to a variable is performed with the equals sign (e.g., *theView = GetActiveDoc*). If a string is assigned to a variable, the string is enclosed in double quotes (e.g., *thePres = "Clinton"*).

Although Avenue is not case-sensitive, by convention, variable names begin with a lowercase letter. In contrast, objects and requests begin with an uppercase letter.

Parentheses are used to supply arguments to a request. Parentheses also enclose expressions to be evaluated as part of a conditional statement, as in the expression *if (r <> nil) then* from the script example.

Comments are designated by a single quotation mark, and can either be placed on a separate line or follow an Avenue statement.

```
' This entire line is a comment.
theView = GetActiveDoc ' Get the active view document.
```

# Creating an Avenue Script

Creating an Avenue script involves the following steps:

1. Enter the script.
2. Compile the script.
3. Debug the script.
4. Run the script.
5. Link the script to a control (optional).

To begin creating an Avenue script, open a script window. Click on the Scripts icon in the Project window and select New.

When a script window is the active document, the user interface changes to the scripts GUI. The scripts user interface contains controls for editing, compiling, and running Avenue scripts. You can type Avenue scripts directly into the script window. Alternatively, you can load a script from a text file or from an existing system script.

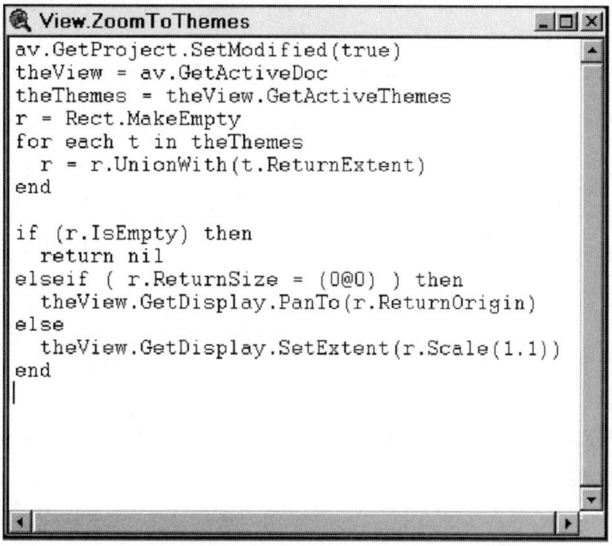

*Active script window displaying the View.ZoomToThemes script.*

View.ZoomToThemes

```
av.GetProject.SetModified(true)
theView = av.GetActiveDoc
theThemes = theView.GetActiveThemes
r = Rect.MakeEmpty
for each t in theThemes
  r = r.UnionWith(t.ReturnExtent)
end

if (r.IsEmpty) then
  return nil
elseif ( r.ReturnSize = (0@0) ) then
  theView.GetDisplay.PanTo(r.ReturnOrigin)
else
  theView.GetDisplay.SetExtent(r.Scale(1.1))
end
```

When a script is complete, you use the Compile button to convert the script to an executable format. At this point any errors in Avenue syntax are flagged.

Once a script is compiled, the next step is to run the script. If a script runs but produces unexpected results, several tools are available for script debugging. These tools include the ability to step through the script one object and breakpoint at a time, and the ability to set breakpoints to interrupt script execution. You use the Examine Variables button to examine the name, class, and value of all variables at any breakpoint.

Although a script can be run from the script window, typically a completed Avenue script is associated with either an existing ArcView control or a new one. Linking a script to an ArcView control is accomplished by calling up the Customize dialog window from the Project Properties window.

# Sample Avenue Script

To illustrate the creation of an Avenue script, we will choose a property that can only be accessed through Avenue and not through the view GUI: the ability to lock a theme so that it cannot be modified via the Legend Editor. While a theme's properties can be locked through the Theme Properties dialog window, this does not prevent access to the Legend Editor. A user can still change the symbology or classification, often with disastrous results. This sample script will enable you to block access to the Legend Editor for all themes currently active in the view.

By ArcView default, double-clicking on any theme's entry in the Table of Contents activates the script *View.EditLegend*. This script calls up the Legend Editor for the theme. To lock the theme so that it cannot be edited, you need to change the *LegendEditorScript* property for the theme so that no script is called when a user double-clicks on the theme.

To begin, get the active document for the project and assign this to the variable *d*:

```
d = av.GetActiveDoc
```

Next, check that this active document *is* a view. If it is, return the list of all active themes in the view, assigning this list to the variable *t*:

```
if (d.Is(View)) then
   for each t in d.GetActiveThemes
```

For each active theme in the list *t*, assign a null string ("") to the *LegendEditor-Script* property for the theme:

```
t.SetLegendEditorScript("")
```

The complete script, with all loops closed, is as follows:

```
d = av.GetActiveDoc
if (d.Is(View)) then
   for each t in d.GetActiveThemes
   t.SetLegendEditorScript("")
   end
end
```

For our current purpose of illustration, we have not fleshed out the error-checking to cover all contingencies. You could, for example, present the user with a list of available views if a view document is not currently active, or prompt the user to select a theme if no themes are currently active in the view.

To restore the ability to access the Legend Editor for a theme, the script should be modified as follows:

```
d = av.GetActiveDoc
if (d.Is(View)) then
   for each t in d.GetActiveThemes
   t.SetLegendEditorScript("View.EditLegend")
    end
end
```

This sets the *LegendEditorScript* property back to the default script, *View.EditLegend.*

## ArcView Object Class Hierarchy

To program in Avenue, you must first understand how ArcView works. This may sound like a glib generalization, but there is a reason. Because Avenue is an object-oriented programming language, knowledge of the relationships and the structure of the objects from which ArcView is built is essential. Fortunately, the Avenue manual and the Avenue online help contain detailed diagrams that depict the relationships among all object classes in ArcView.

# The Avenue Help System

A review of the ArcView class hierarchy can be somewhat intimidating. Apart from wallpapering your office with object diagrams, how is it possible to keep them all straight? The answer lies in the Avenue portion of ArcView's online help.

The online help system contains an entry for every object class and every request. All classes and requests have been indexed so you can access them easily using the Search utility.

Each class entry contains a description of purpose and function, several examples, and placement in the overall class structure. The overview is followed by a listing of the superclass the class inherits from, the subclasses the class is inherited by, and all available class requests and instance requests. Each list item is hyperlinked so that clicking on a name accesses the help entry for that item.

Each request entry contains a description of what the request does, the syntax of the request, and the class of object returned by the request. Examples are provided, as well as a listing of related requests.

In practice, Avenue programming involves keeping the help window open and accessible while writing Avenue scripts. Hyperlinks and the ability to search for class and request names allows you to effectively program in Avenue even though you have not committed the entire object structure to memory.

The *Lists* listing from the Avenue online help system is reproduced below:

# List

## Discussion

A List is an ordered collection of heterogeneous objects. An object in collection is called an element of that collection. Each element in a list is referenced through a list index number. The first index is always zero and the last index is always List.Count - 1.

Use lists when you need efficient sequential access to an unbounded number of objects, or when you need efficient random access based upon the list index number.

*Using Lists*

You can use the 'for each' loop to gain access to each element in the list. Note also that "{","}" serve as list delimiters.

```
myList = { "red", 34, av.GetActiveDoc, av.GetProject.GetActive }
for each i in mylist
  msgbox.info( i.AsString, "List")
end
```

Certain lists can be sorted. See the Sort request for details.

You can use the '+' and '-' operators to merge and remove groups of elements. The '+' operator produces a new list that is the operand lists concatenated together. The '-' operator produces a list that contains elements from the left operand list that are not in the right operand list.

### Inherits From

Collection

### Inherited By

Clipboard

GraphicList

GraphicSet

SymbolSet

Template

### Class Requests

### Creating a list

Make : List

### Instance Requests

### Updating Lists

Add ( anObj ) : List

Insert ( anObj )

Remove ( anIndexNumber )

RemoveDuplicates

RemoveObj ( anObj )
Set ( anIndexNumber, anObj )
Shuffle ( anObj, anIndexNumber )
Sort ( ascending )

### Examining Lists

Count : Number
Find ( anObj ) : Number
FindByClass ( aClass ) : Number
FindByValue ( anObj ) : Number
Get ( anIndexNumber ) : Obj
HasKindOf ( aClass ) : Boolean

### Operating on Entire Lists

+anotherCollection : List
-anotherCollection : List
AsList : List
Clone : List
DeepClone : List
Empty
Merge ( anotherCollection ) : List

---

See also
Dictionary
String

## *Building Your Own with MapObjects*

Unable to mold ArcView into exactly the application you want? Need just a little mapping in a larger application? You may be a candidate for MapObjects, ESRI's collection of mapping and GIS components targeted strictly for developers.

MapObjects is an OLE (object linking and embedding) Custom Control (OCX) and a set of OLE-compliant programmable objects. As such, it can be plugged directly into many of the popular programming environments, such as Visual Basic, Delphi, PowerBuilder, Visual C++, and others.

MapObjects lets you view and query spatial data. It supports thematic mapping and classification, attribute and spatial queries, and address-based geocoding. Supported spatial data formats include ArcView shapefiles, SDE database layers, and georeferenced images.

MapObjects is particularly suited to smaller, highly customized mapping applications, and to incorporating mapping functionality into non-mapping applications, such as spreadsheet or database applications. It relies on either the Microsoft Windows 95 or Windows NT operating system and a development environment that supports OLE 2.0 controls. With MapObjects, you, as the developer, have another viable choice for delivering mapping functionality to the user.

# *Communicating with Other Applications*

Avenue can be used for a broad spectrum of tasks, from modifying ArcView controls to creating a fully customized application. Communicating with other applications warrants specific mention in this chapter, in concept if not in detail.

Communication with other applications can take two forms: (1) sending a command string to execute a system command or to start another application, and (2) using *inter-application communication* (IAC) to communicate with another application through a client-server relationship.

The *System.Execute* request is a simple way to pass a request to the operating system. One obvious use for this request in the UNIX environment is the ability to execute an ARC/INFO AML (ARC Macro Language) from within ArcView. A syntax sample for this application follows:

```
System.Execute("arc \&run plotit.aml &")
```

Under Windows, a system command can be used to start another application and load a file associated with that application, such as a spreadsheet or database. For example, the following command starts Microsoft Excel and opens a spreadsheet named *sales94*:

```
System.Execute("c:\excel\excel.exe c:\work\sales94.xls")
```

IAC is the general term for a process in which two applications exchange information. In this arrangement, the process initiating the communication is referred to as the *client*, which requests data or calls functions from the responding application, referred to as the *server*. The actual protocol involved in IAC is platform-dependent. In Windows, the process involves Dynamic Data Exchange (DDE); in UNIX, Remote Procedure Calls (RPC); and on the Macintosh, AppleEvent and AppleScript.

It is possible not only for ArcView to execute a command or run an AML within ARC/INFO, but for ARC/INFO to inform ArcView when the process or AML has completed and, optionally, to return a value from the executed process to ArcView. Using the above example, ArcView could initialize an AML to buffer the ARC/INFO coverage for the active theme. Upon completion, ArcView could receive the name of the newly buffered ARC/INFO coverage from the AML, at which point the coverage could be added as a new theme to the view.

## Start-up and Shutdown Scripts

Every time ArcView is started, a system default file is read before any project files are opened. This *startup* file is located in the */etc* directory in which ArcView is installed. A file named *shutdown,* which is located in the same directory, is read when you quit ArcView.

The start-up and shutdown files are used to set the home directory for the user and to display the ArcView banner. In Windows, the *startup* file also starts ArcView as a DDE server. Upon exit, the *shutdown* file is used to halt the DDE server.

You can also create start-up and shutdown scripts for a specific project. For example, a project start-up script can prompt a user for the path to a particular workspace containing ArcView data, thereby enabling a project to be made portable across platforms. Users can be prompted for their group names and passwords, after which the ArcView project specific to their group is initialized. Such customization could enable an agency-wide application to be initialized uniformly at the system level, with a user's group name used to start up a custom project tailored to that group. Similarly, a shutdown file can be used to remove any indexes created during the session and, optionally, to remove all links and joins in the event a project needs to be made portable across platforms.

As is the case when modifying the default project file, modifying the system start-up and shutdown files has the effect of globally applying the changes to all users. When modifying any system file, you should make a backup before editing so that you can remove alterations easily.

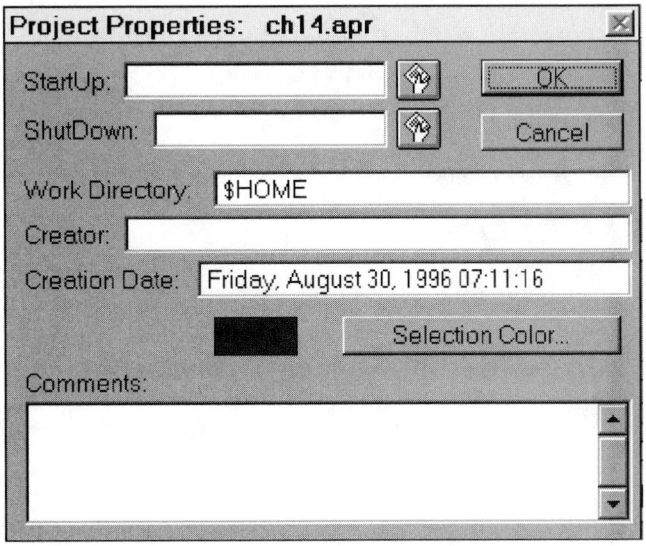

*Project Properties dialog window with fields for specifying start-up and shutdown scripts.*

# The Finished Application

Once you have completed an application, you need to make it permanent. The first step in this process may involve the creation of start-up and shutdown scripts, as described above. Additional steps could include embedding scripts, encrypting scripts, and locking the project to prevent further customization.

*Embedding* a script requires taking a script stored in the Script Editor and adding it to the list of scripts stored internally with the project. Script embedding reduces the number of external documents associated with the project, thereby making the project more self-contained.

An embedded script can be viewed by any ArcView user running Avenue. However, because the embedded script appears in the Script Manager window, the script could be loaded into the Script Editor and altered. If protecting proprietary Avenue scripts is necessary, the script can be *encrypted.* Encryption converts the script into a format that is machine-readable only. Because this process is not reversible, you should always store an unencrypted copy of the script in case additional changes are required.

Finally, an entire project can be locked by encrypting the project file. This action disables your ability to access the Customize dialog window or further modify the project by writing additional Avenue scripts. As with encrypting a script, encrypting a project file is not reversible.

# Extensions

If you have followed along up to this point, and if the degree of customization offered via virtual document types and Avenue scripting is not sufficient for what you need, you can go one step further by creating your own extension.

ArcView uses extensions as a way to modularize functions, particularly those specialized functions that are used infrequently or that have an additional resource requirement. Core functionality (found in CADReader, Imagine Image support, and SDE database themes) and optional functionality (such as Network Analyst or Spatial Analyst) are both added via extensions.

The ability to create custom extensions is available to any user as well. An extension is, at its core, an ArcView project. Custom scripts, controls, and documents are added to the script until it contains the desired functionality. When the customized project is complete, scripts are written to install and uninstall the extension. An Avenue script is written and executed to make the project into an ArcView extension. When complete, the extension is written out to a system file in the current directory, and is given an *.avx* file extension. It is then available to be loaded into a project by selecting Extensions from the Project document's File menu.

## In Summary

In keeping with the focus of this book, we have attempted to cover the basics of Avenue programming without getting too involved with details. The purpose of the discussion above is to provide you with enough information to determine if Avenue programming is required for your applications. We hope to have provided you with a basic understanding of the ArcView data structure as well as an appreciation of how the robustness of the object-oriented data model can serve to protect your investment in ArcView application development for years to come.

## Chapter 15

# ArcView in the Real World

After reading through the chapters leading up to this point, you may already have some ideas about your first few projects. Do not lose sight, however, of the fact that the applications and functions described in this text are just a few of a much broader set. Desktop mapping can influence a great many tasks in a wide variety of organizations. In the interest of helping you brainstorm about ways to use ArcView, this chapter is devoted to a few specific descriptions of how ArcView is being used to make contributions to projects in the "real world."

## ArcView in Epidemiology

In Maryland, the Secretary of Health and Mental Hygiene is responsible for the control of disease outbreaks. Health Officers are authorized by law to conduct investigations. Code of Maryland Regulations (COMAR) 10.06.01 delegates this responsibility to Health Officers. The Epidemiology and Disease Control Program (EDCP) has the overall responsibility for offering consultations to the local health departments. EDCP also assumes direct responsibility for outbreak investigations. Other program units such as the Office of Licensing and Certification have regulatory responsibility for the facilities in which outbreaks may occur. EDCP needs to keep those other units in the Department of Health and Mental Hygiene affected by outbreaks informed about any investigation.

EDCP has formulated a plan that involves GIS in a model surveillance program of the state's communicable diseases (excluding AIDS). These are the basic components of the surveillance program:

- ❐ **Passive surveillance**—The Department of Health and Mental Hygiene (DHMH) obtains surveillance data through notifiable disease reporting. Passive surveillance is inexpensive and useful when studying trends. In order to identify clusters and trends, geographic information systems have been proposed for increasing the analysis of collected data. At both the state and local level, this increased automation should allow analysis with greater emphasis on action that can be taken following the receipt of data. Collaboration with the University of Maryland to secure a long-term study of Lyme disease has been successfully implemented. For the past three and one half years, active surveillance of bacterial invasive disease has been conducted, in collaboration with Johns Hopkins University, through hospital infection control practitioners and laboratories.
- ❐ **Sentinel populations**—For emerging and re-emerging pathogens, sentinel populations offer a way to detect problems quickly. For example, efforts to identify vancomycin-resistant Staphylococcus aureus (VRSA) can be appropriately centered on hospitalized populations even though the first occurrences of this agent may be in nursing homes.
- ❐ **Population-based studies**—The future of Maryland's surveillance efforts will be heavily centered on population-based studies. The growth of managed care companies offers an easy partnership between Maryland's health department and Maryland's HMOs. This partnership's purpose is to develop adequate surveillance of the state.
- ❐ **Data collected for other purposes**—At the state level, the EDCP is currently spending a significant amount of resources to gain access to many available health care databases. Medicaid has given EDCP permission to use their database. Hospital discharge summary data for TB immunization and Sexually Transmitted Diseases (STDs) are currently being accessed.

## ArcView and Epi-Info

In epidemiology investigations, disease problems are generally characterized by person, place, and time, whether the problem concerns the emergence of a new disease, a change in the resistance pattern of a known pathogen, an emergency response to an outbreak, or a routine disease surveillance program.

The database that ties all this data together is the Maryland Electronic Reporting and Surveillance System (MERSS). The 24 local health departments (LHDs) upload their data to this system on a daily basis. The entire system is driven by Epi-Info software, jointly created by the Centers for Disease Control (CDC) in Atlanta, Georgia, and the World Health Organization (WHO) in Geneva.

In keeping with its objectives, EDCP is creating applications with ArcView 2.1 for use on the LAN and for use by the LHDs through dial-up access. Because a large part of the epidemiological world is heavily dependent upon Epi-Info, it was decided to integrate this software with ArcView. Epi-Info, then, becomes the database selection and statistical engine used to preprocess data before final spatial display by ArcView.

Aspects of the system are depicted in the images below.

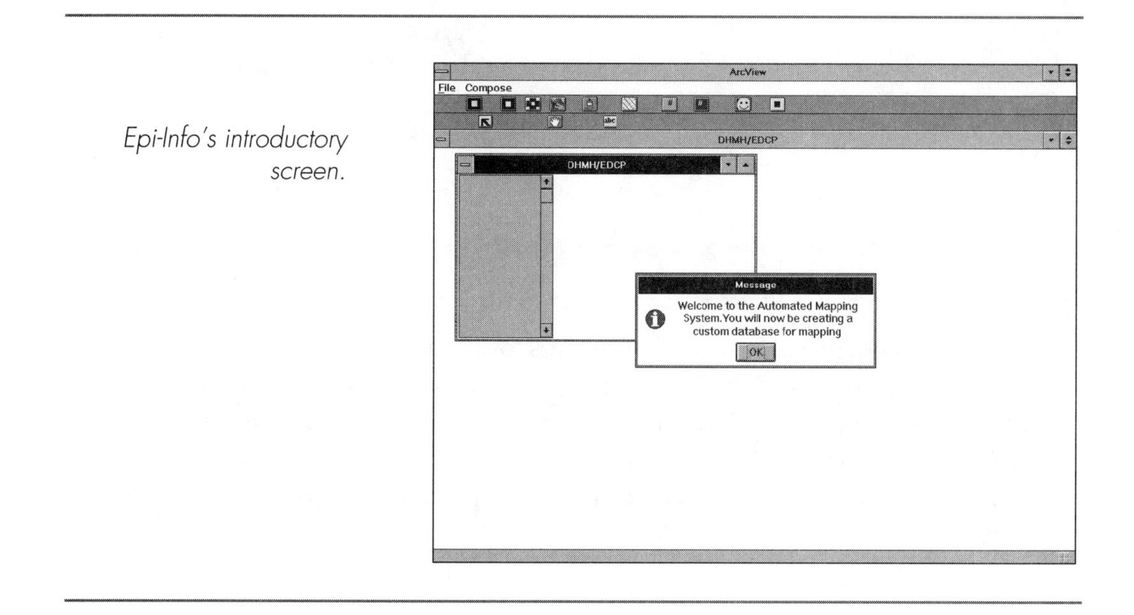


*Epi-Info's introductory screen.*

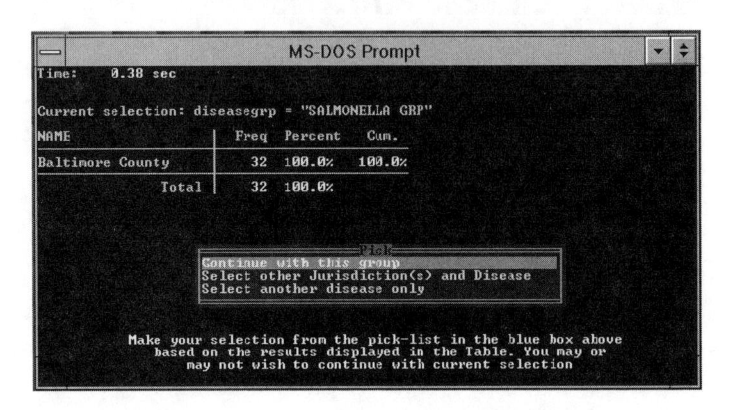


*Appearance of the DOS screen after selection of a database.*

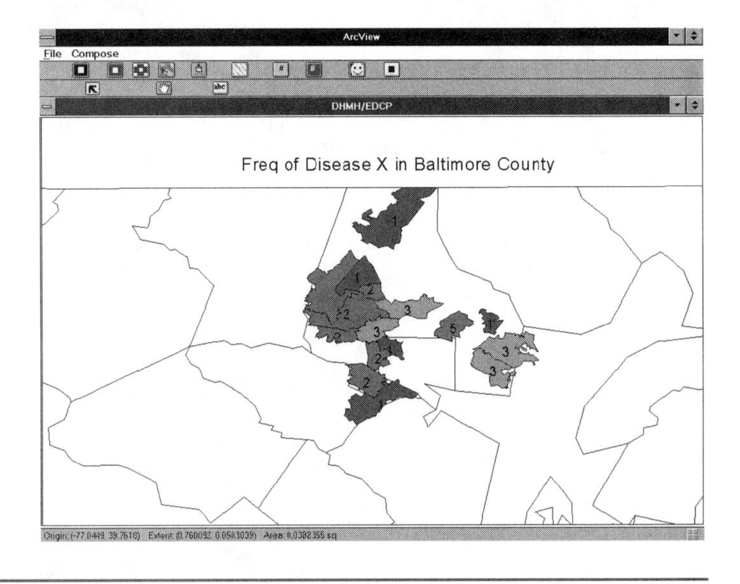


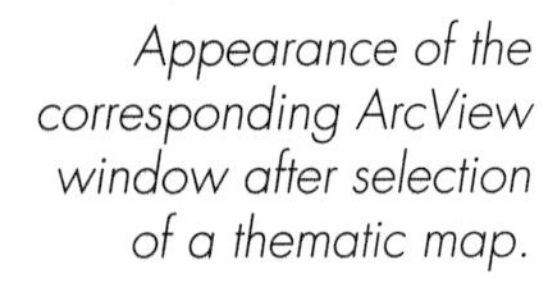

*Appearance of the corresponding ArcView window after selection of a thematic map.*

One can examine disease counts and rates by county or zip code boundaries. Because this will be the first time that most users see a GIS, the application has been kept very simple, with the absolute minimum number of buttons and menu options visible. In time, EDCP will provide several levels of user-chosen complexity so that users who are spatially inclined can use the fully functional version of ArcView.

# ArcView in Education

The Institute of Public Affairs (IOPA), at the University of South Carolina, in conjunction with the South Carolina Geographic Alliance, has developed an ArcView-based application called *The Siting Game*. This application is used as a teaching module for age groups that range from professionals to college undergraduates to high school students.

The game was developed to simulate the dilemmas faced by county officials and other decision-makers who choose sites for locally unwanted land use (LULU). The game's dynamic and interactive aspects provide a stimulating avenue for introducing GIS, teaching issues concerning environmental management, enhancing map cognition skills, and developing higher order thinking skills. Futhermore, the simulation allows future modifications and simplifications to be made to the game in order to make it suitable for middle school and elementary school children.

ArcView was chosen as the state-of-the-art software for displaying the geographic data used in the game and for introducing the basic concepts of GIS. ArcView serves as a good teaching tool for various reasons: the user-friendly GUI (which can be modified); the ability to import geographic data from a variety of sources to use as spatial, image, or tabular data; and the object-oriented scripting language and development environment of Avenue. Avenue, the same scripting language that allows users to modify the GUI and functions for the game's purposes, also enhances ArcView's overlay and query capabilities.

The Siting Game is a role play that takes place in fictitious Sandhills County. The players are all inhabitants of the county. The number of roles vary between 30 and 50 players, based on the number of expected players at the scheduled play time. The game is introduced by distributing information packets about each role to each player.

Once the role packets have been distributed, the players are introduced to Sandhills County and the dilemma at hand. The Supreme Court has ruled that the county's current landfill is almost full, and the county council must decide on another site for a new landfill at the next meeting or be fined an exorbitant amount of money. Also on the county council's meeting agenda are proposals to site a church campground, a microchip company, and a retirement home. The problem is that there are only three identified sites that have been inspected and approved for a landfill. The county council has two options: (1) siting the landfill and rejecting one of the other proposals, or (2) finding a fourth location that is suitable for the proposed development.

With the brief hint that information is power, the players are then dismissed to begin playing the game. There are three sources of information: the county planning office, the regional Department of Health and Environmental Control (DHEC) office, and the local DHEC office. The data for each office is available through the computers, using a modified form of ArcView. To simulate the tediousness and frustration often experienced in the real world when searching for information, differing data sets were dispersed among the three computers and intentional dead ends were scripted into the directed query system. In the search for information, the players interact, making deals and trading information. Mock public hearings and press conferences are held to express the community's concern to the county council members. Finally, the county council meeting is called to order and the game ends with the county council vote.

The computer modules developed for the Siting Game are more than just systems for map browsing; they are introductory GIS modules as well, offering overlay and spatial search capabilities. Menu options allow the user to display various maps, calculate distances, overlay maps, retrieve information about map features, locate a character's home, and search for a new site. The menus for calculating distances and retrieving map features information just toggle the appropriate ArcView tool. To locate a character's home, each user is prompted for his or her character's name. A string search matches the name in the feature table and displays the home as a dot on the screen. The display and overlay options, however, are designed to help non-GIS users visualize and understand the term *overlay* and how it works.

Under the display menu, users can display only one map at a time, which can be rather uninformative. The exception is that the Sandhills County base map, which includes the county boundary, the municipal boundaries of Wilshire Heights and Freetown, the interstates, the highways, and the three proposed sites, can be selected from the display menu and it will overlay the current map.

The overlay option is more specific in that it prompts the user to select any map as a base map and then select an overlay map. Then the user can overlay a map with not only the county base map, but with other maps as well. This separation of maps helps the players understand the overlay function and realize that they can retrieve more information using the overlay process than they can by viewing each map alone.

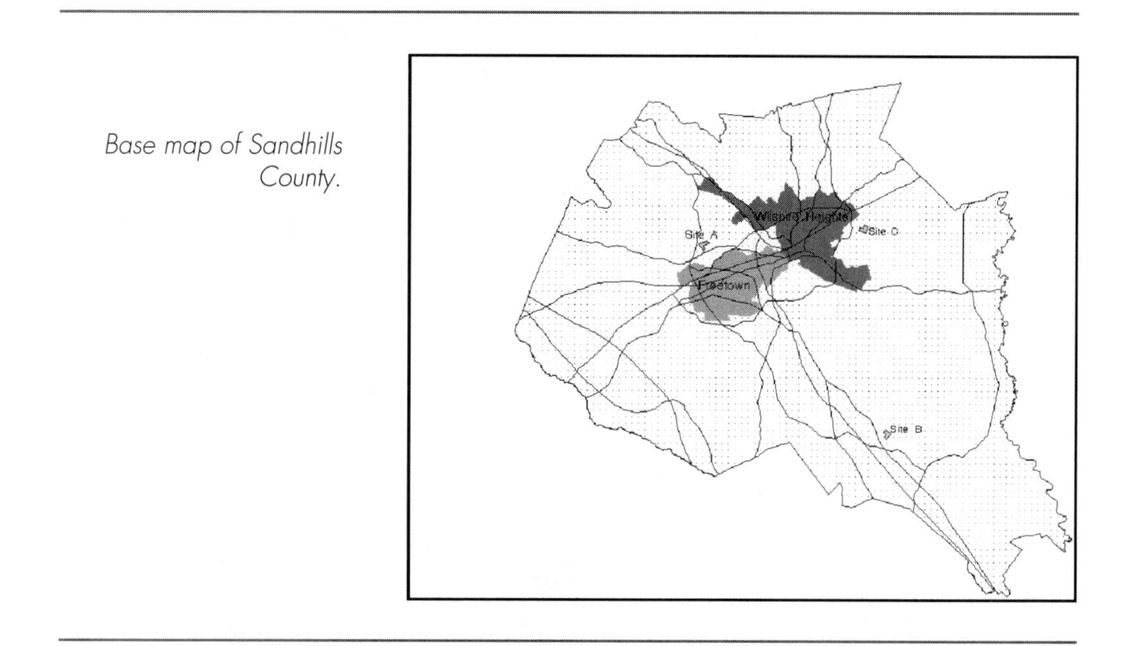


*Base map of Sandhills County.*

The Site Search menu option is a simplification and automation of the ArcView query process for the players. A grid layer of cells of one square mile was created to serve as a sort of raster buffer system for the spatial search option when locating a new site. The user need only specify search criteria, such as land use, census tract data category, and distance from particular map features. Based on those specifications, the corresponding grid cells are highlighted by obscuring the undesired grid cells and underlying map area. For a more direct and personal effect, each player is able to locate his or her character's home in order to determine what sort of impact the siting of different projects will have on that character's neighborhood.

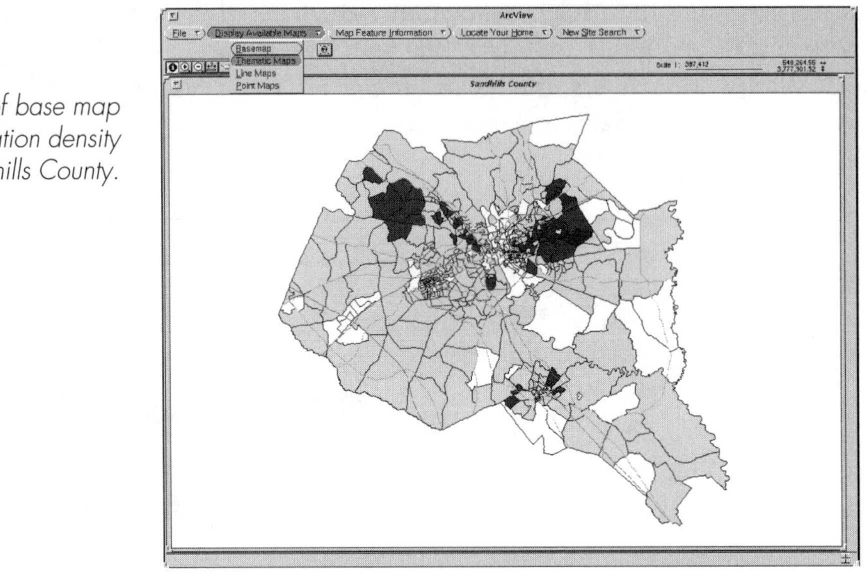


*Overlay of base map on the population density map of Sandhills County.*

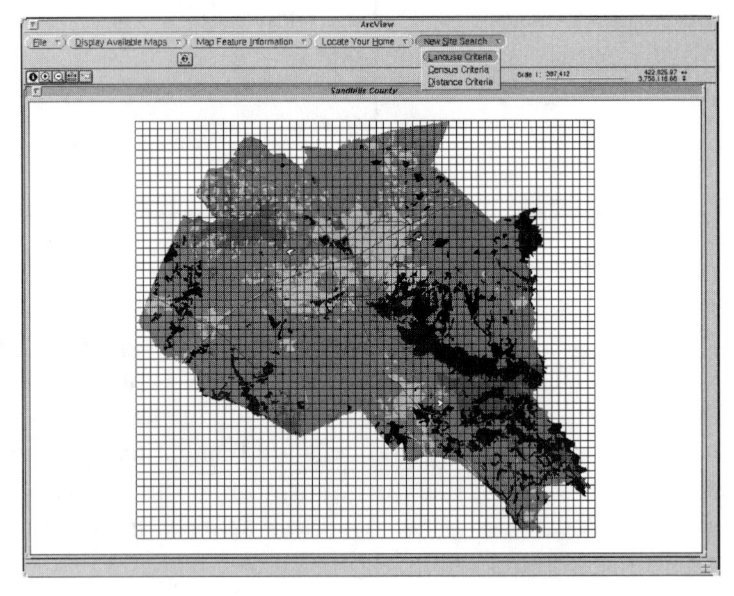


*Full land use map for Sandhills County showing the grid layer superimposed.*

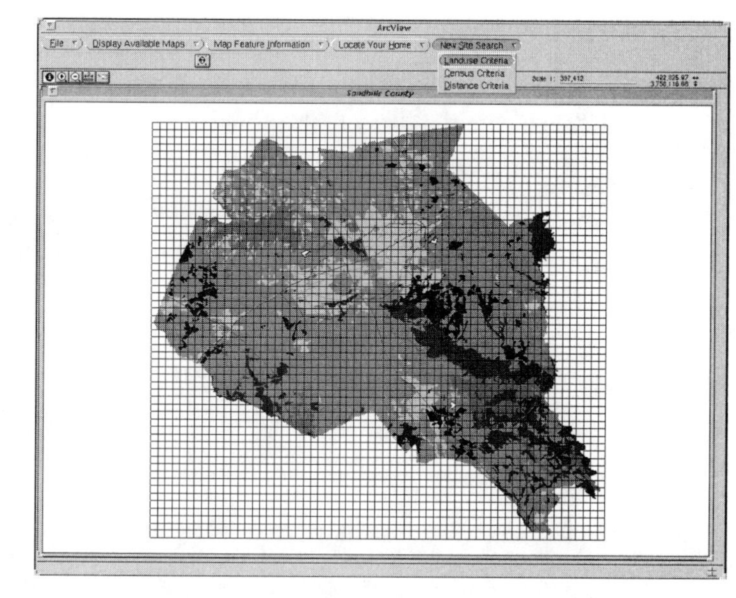


*Using the Site Search option, the user isolates all the grid cells with forested land use (the rest of the map is hidden).*

This research has been partially funded by a grant from the United States Department of Energy and the South Carolina Universities Research and Education Foundation. IOPA directs special thanks to the South Carolina Geographic Alliance for their support and aid in the development of the game.

## ArcView and Hydrography

In South Africa, the Department of Water Affairs and Forestry is one of the largest users of ESRI products. At the beginning of 1995, a major project was initiated. Its objective was to automate some traditional engineering procedures by using Digital Terrain Models (DTMs). Initially, ArcView 2.0 was used for presentation purposes only. When ArcView 2.1 was released, the department decided to use the Remote Procedure Call (RPC) functionality introduced in this version to transfer its system to the ArcView platform.

The major applications included in the system are the following:

- The extraction of surface cross-sections from a DTM and the calculation of their characteristics (area, wetted perimeter, hydraulic radius)
- The calculation of the evaporation area and volume of a dam reservoir for any gauge plate reading
- The calculation of stream characteristics (length, highest and lowest altitude, and mean slope) and the drawing of the corresponding profile
- The catchment delineation and associated characteristics (area, perimeter, length of the longest watercourse, mean slope, mean elevation, etc.)

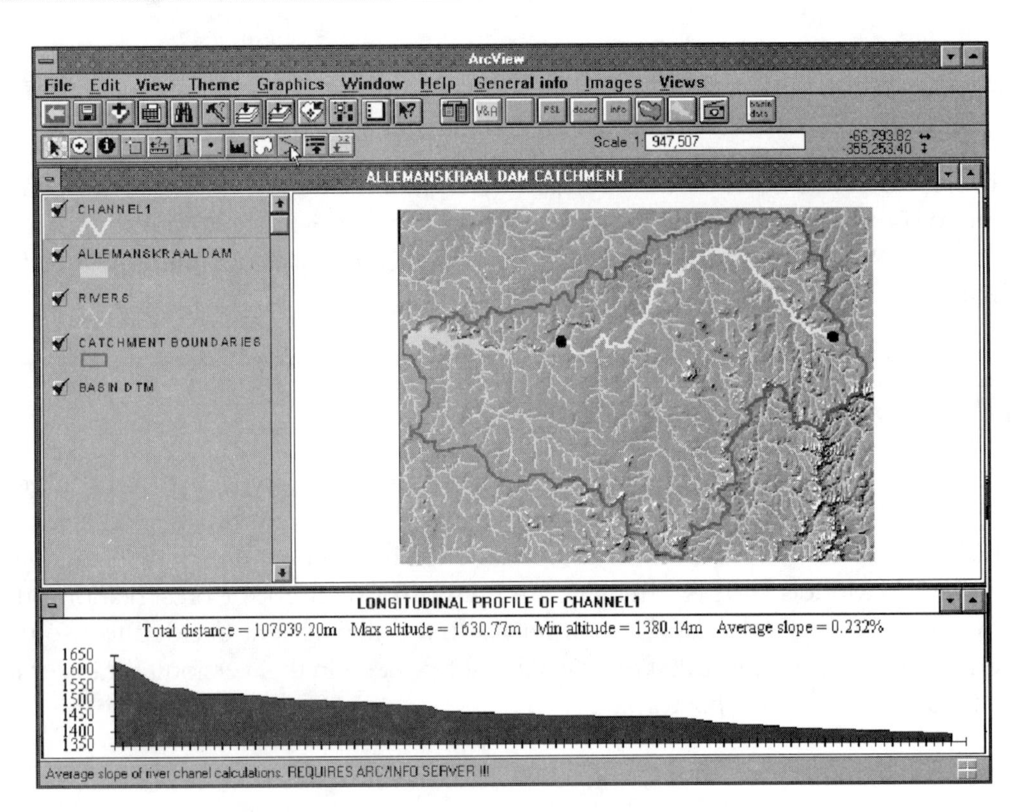


*Slope analysis.*

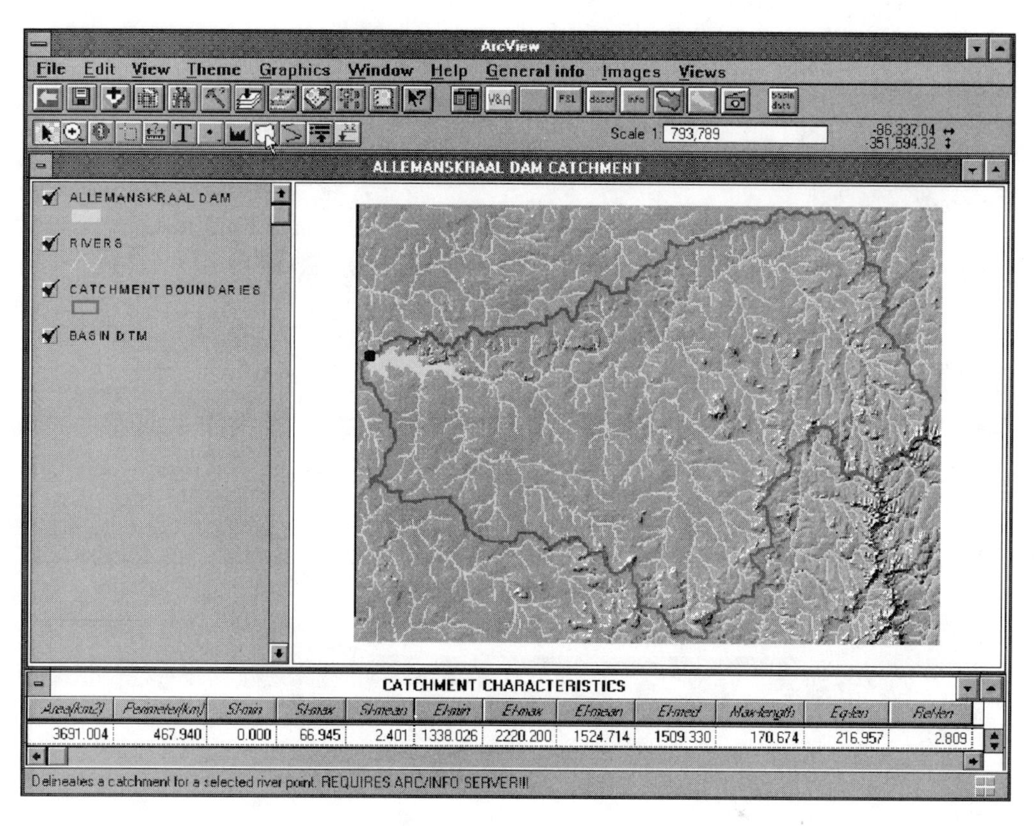


Catchment analysis.

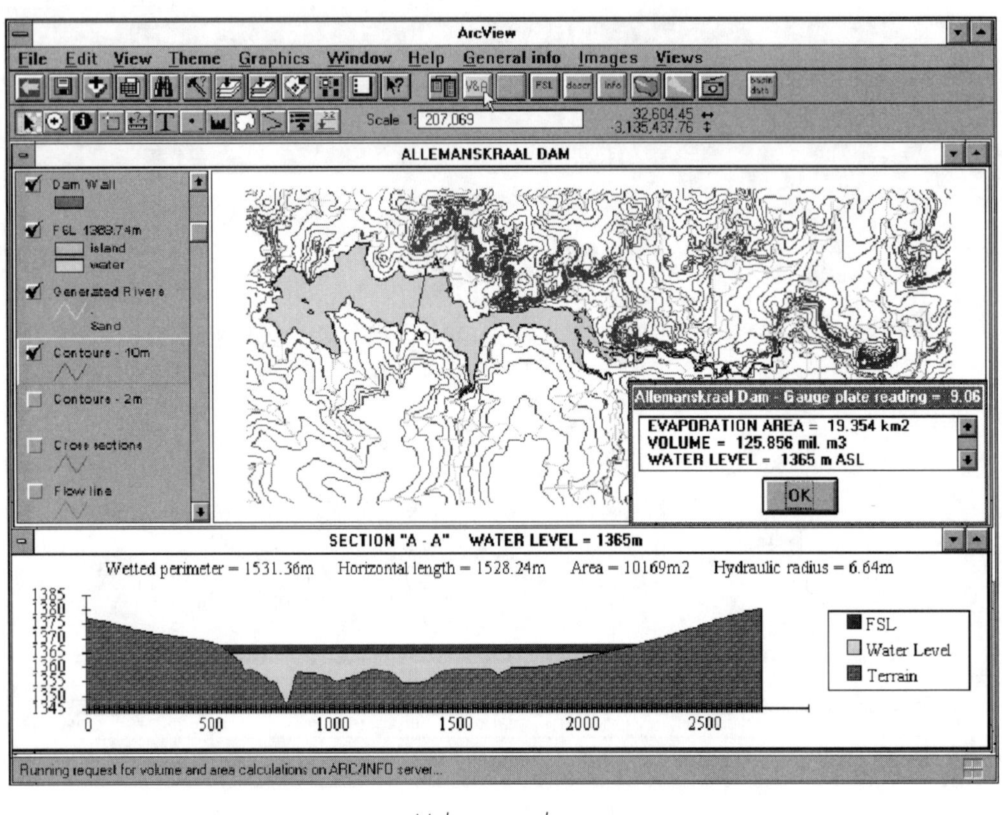


*Volume analysis.*

Also under development is an application that uses ArcView 2.1 in conjunction with ARC/INFO to prepare input files for the MIKE11 hydrodynamic module, which is used for floodline calculations. The application exports cross-sectional data extracted from a DTM to MIKE11. After performing floodline analyses, the resultant file is imported into ARC/INFO. The floodline coverage is generated and then presented in ArcView.

All applications are designed to use ArcView 2.1 to input user requests graphically. A request is sent to a UNIX-based ARC/INFO server, where the analyses are performed before the results (coverages, tables, variables, etc.) are returned to ArcView for presentation. The system combines the friendly user interface of ArcView with the powerful analytical tools of ARC/INFO, and provides the added benefit of allowing many users to send requests to the same ARC/INFO server from their PCs.

## ArcView and Environmental Analysis

*The following section is based largely on material contributed by Robert K. Greene, Robert Teller, Darrel Holt, Thom McVittie, Gerry Ramsey, Nancy Marusak (from Los Alamos National Laboratory), and Larry Batten (from ESRI).*

The Emergency Management & Response (EM&R) Team at Los Alamos National Laboratory (LANL) is presently incorporating ArcView into its system for mitigating accidental atmospheric releases of toxic and hazardous materials. EM&R currently uses the MIDAS (Meteorological Information and Dose/Dispersion Assessment System) software package to predict current and projected isopleths for radiological and/or chemical hazard levels. These plume calculations are performed in real time, and are based on the current wind field snapshot (updated every 15 minutes from four meteorological observation stations located around LANL) as well as current estimates of total release and release rate.

Timely evacuation of threatened areas requires up-to-date display of the plume information in relation to facilities, residences, and transportation routes. ArcView supports the flow of data from the meteorological stations to the MIDAS modeling application through an automated interface. LANL has found ArcView to be a user-friendly application for display and analysis of geographic information.

Through this application, which is part of the Environmental Restoration (ER) project's activities at LANL, managers at the Emergency Operations Center (EOC) will be able to see current and projected at-risk areas. They can then make informed decisions regarding the necessary protective actions for mitigating an atmospheric release, and plan for specific release scenarios.

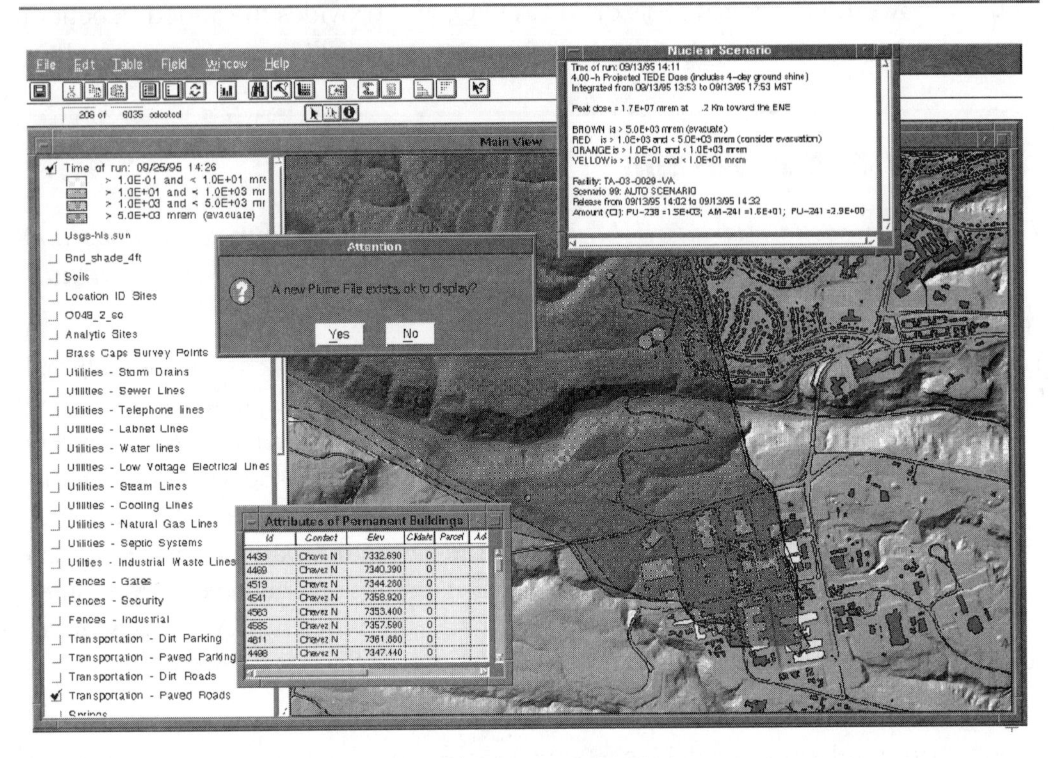


*ArcView facilitating MIDAS data.*

## ArcView and Commercial Products

The ArcView application and database that ViGYAN, Inc., has developed for the U.S. Environmental Protection Agency (EPA) is one example of the commercial products emerging with ArcView. This application, a customization of ArcView using Avenue and Visual Basic, is intended to be a set of analysis tools for novice users within community groups, academic institutions, state and federal agencies, and non-profit organizations.

In one or two steps, several GIS functions can be performed with only a simple knowledge of Windows. This application can be used for analysis such as potential environmental exposure, demographic characterization, and demographic targeting. Data sets have been designed and developed to accompany the tools, including EPA databases that incorporate facility location and release information, demographic information, census geography, political boundaries, transportation networks, and hydrology.

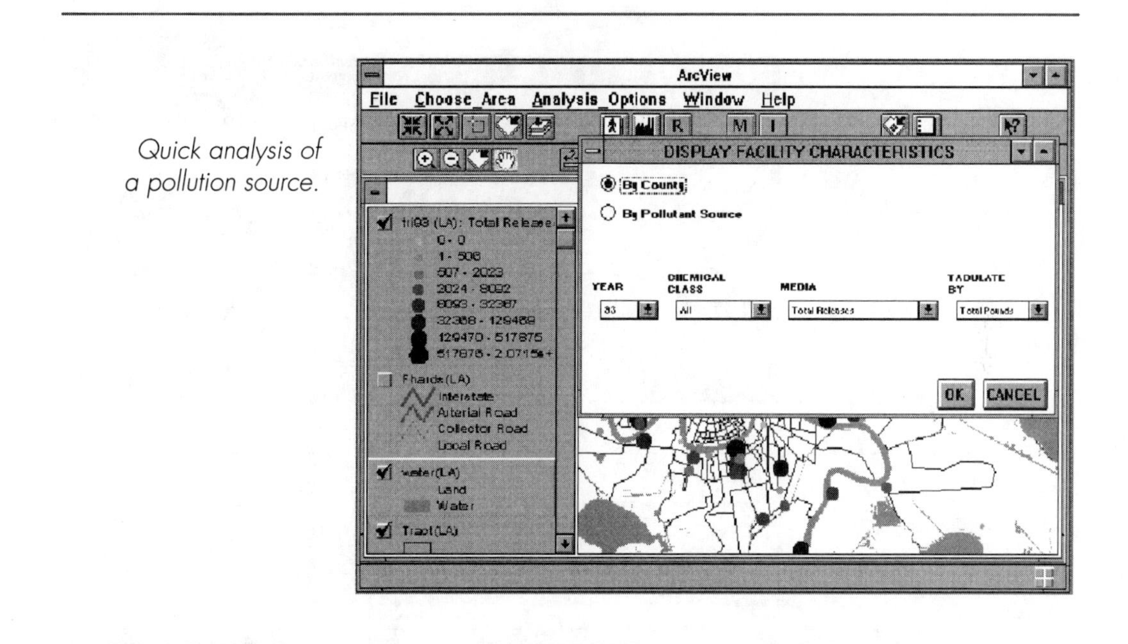


*Quick analysis of a pollution source.*

*Rapid retrieval of demographic information.*

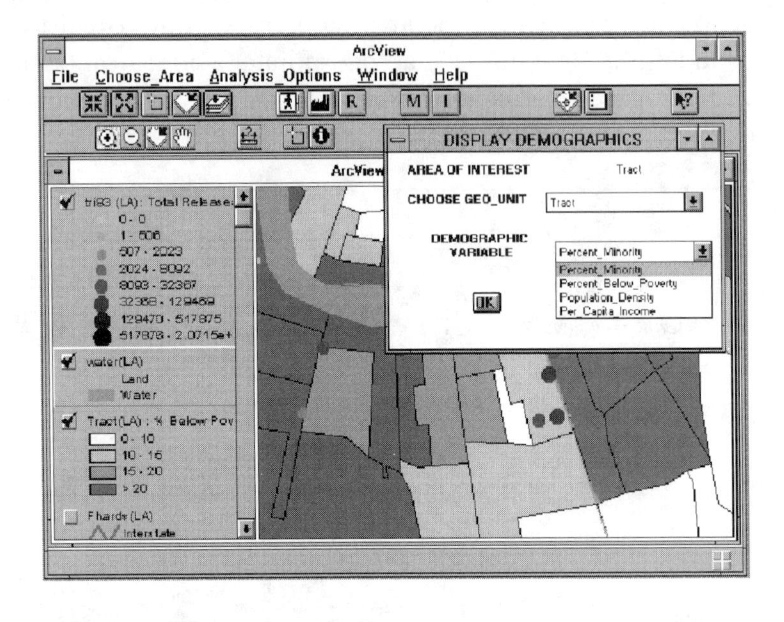


*More demographic analysis.*

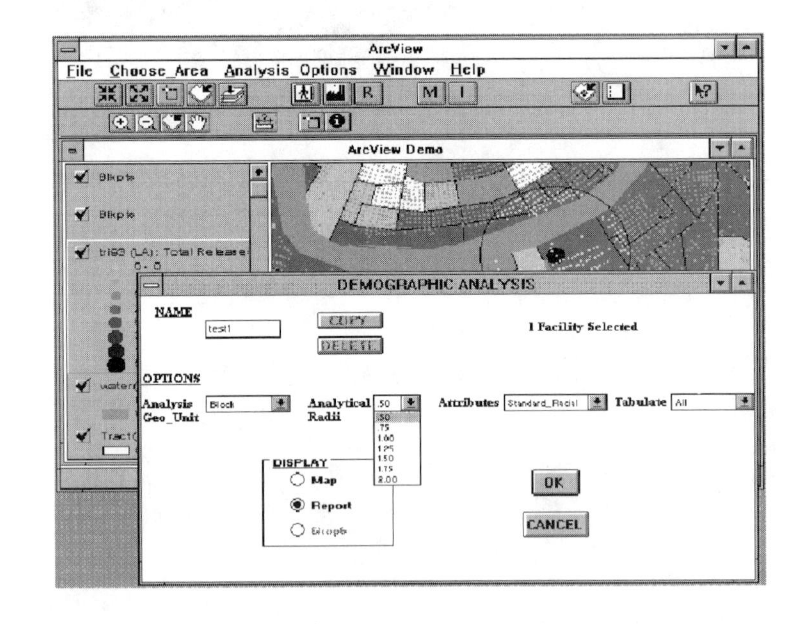

This product (and others in the ArcView format) provides easy-to-use software and readily available data at a low cost, thereby allowing access to data and analysis tools for groups that would not otherwise have the resources.

## Elsewhere

The world of desktop mapping and ArcView seems to be expanding daily. Scan the ESRI User Conference agenda and you will find parties reporting on nearly every aspect of life. From business to agriculture, municipal services to defense, economics to environmental studies, ArcView has something to offer.

We encourage you to attend the annual ESRI User Conference to explore more ideas for using ArcView. You should also consider attending other national conferences that demonstrate the power of desktop mapping. Local user groups can also help you expand your thinking about the range of methods and tasks that can be performed.

Let the rest of the mapping community know about your work. Inquire about presenting at conferences by contacting the editors of pertinent professional journals or by contacting your local ESRI representative.

# Appendix A

# Installation and Configuration

Our purpose in this section is not to provide step-by-step instructions for installing ArcView GIS. Installation is relatively straightforward, and the *ArcView GIS Installation Guide* does a good job of taking you through the process. We do have some general observations to pass along that can improve the stability and performance of ArcView GIS.

## Memory Considerations

### ArcView 2.0/2.1

Memory constraints were a major issue throughout the development of the 2.*x* versions of ArcView. Developers struggled to accommodate the demand for increased functionality while retaining the ability to run ArcView on a "standard" office PC. At that time, the standard could be defined as a PC with a 486-66 processor and 8 Mb of RAM. The final release achieved this goal, but the question remained, Did you *want* to run ArcView 2.*x* on an 8 Mb machine?

As we have gathered from personal observation and contact with other sites, the general consensus was that although ArcView 2.*x* would run on an 8 Mb machine, performance would not be optimal. The money to upgrade to 16 Mb would be well spent. Under Windows, ArcView requires a minimum of 12 Mb swap space on your hard disk, in addition to 8 Mb of RAM. ArcView spent a *lot* of time swapping to disk on an 8 Mb machine, severely degrading overall performance. Users reported that swapping was substantially reduced with 16 Mb, and that performance was entirely acceptable.

## ArcView GIS 3.0

With the release of ArcView GIS 3.0, the system requirements issue becomes more complex. Specifically, you not only need to determine the minimum system resources needed for memory, swap space, and a graphics card, you need to take into account your choice of operating system.

The system and memory requirements for installing and running ArcView GIS on a Windows-based PC remain similar to those required for ArcView 2.*x*; namely, a 486 or higher processor, 12 Mb of memory, and 17 Mb of swap space. This same level of system and memory requirements is achieved, in part, by the modularization of the software through the use of extensions, ensuring that the user need only load those modules that are actively being used. The recommended PC, however, now stands at a Pentium-class processor with a minimum of 16 Mb physical memory. It is our observation that these recommendations should be considered minimum requirements on any system for which ArcView will be the primary application. For ArcView development work, our experience also shows a significant improvement in system stability and performance with an increase in physical memory to 32 Mb.

With regard to operating system and environment, you are strongly advised to run ArcView GIS 3.0 under Windows 95 or Windows NT. While the core modules of ArcView will run under Windows 3.1/Windows for Workgroups 3.11, several of the 3.0 extensions, including SDE Themes, Network Analyst, and Spatial Analyst, will run only under Windows 95 or Windows NT. Additionally, memory management is handled automatically under Windows 95/NT, eliminating the need to specifically allocate swap space for virtual memory and improving system performance in general. ArcView is a 32-bit application, and the consensus among users is that performance and stability is improved when it is run in a 32-bit environment such as Windows 95 or Windows NT.

> **NOTE:** *While it was possible to install and run ArcView 2.*x *under OS/2, this support does not extend to ArcView GIS 3.0.*

## Running ArcView

Managing memory within an ArcView session is another matter. During the operation of ArcView, a certain amount of memory is associated with each object. As you work in a project, the total amount of memory used increases as you open and manipulate different documents. As you switch operations and documents, ArcView flags the memory associated with the old objects as available for reuse, but does *not* immediately free it up. Only when ArcView needs to load new objects is this memory reallocated. The result is that although your ArcView session will continue to run indefinitely, you may have a problem maintaining enough memory to run additional applications. This may become evident when you initialize an additional operation with a high memory requirement, such as printing a view or layout. At this point it may be necessary to either close additional applications before printing or print to a file so that a document can be sent to the printer after the application has been closed.

One word of caution: because ArcView associates a certain amount of memory with each object, operations that involve repeated manipulation of many objects (such as editing shapefiles or manipulating chart markers) may eventually consume all available memory, if they continue over an extended period of time. Our advice is to listen for hard disk activity. If you begin to hear a significant amount of activity during an intensive editing session, this indicates that ArcView is having to swap large amounts of data from memory to disk. Stop and wait for the system to catch up, then save your work. Closing and reopening an ArcView document can also reallocate memory.

Finally, if you are experiencing erratic behavior or intermittent system crashes under either ArcView 2.*x* or ArcView 3.0, the problem may be traced to the installation of a beta or pre-release version of the software on the system. Removing the pre-release version of ArcView and performing a clean installation may help clear up these problems.

# Appendix B

# Functionality Quick Reference

This appendix serves as a reference for the functionality available through the ArcView menu, button, and tool bars. It is designed to be a quick reference to the main functionality available in ArcView, as well as to where these topics are covered in the book.

## Add Event Theme

**Access:** View menu bar—View menu

**Purpose:** Adds a new theme to a view using an event table. An event table contains a locatable field such as an address; X,Y coordinates; or a route location.

**Associated with:** Add Theme, Event Tables, Address Geocoding

**Reference:** Chapters 2, 4

## Add Field

**Access:** Table menu bar—Edit menu

**Purpose:** Adds a new field to the active table. The table must be in dBASE or INFO format.

**Associated with:** Start/Stop Editing, Add Record, Delete Field

**Reference:** Chapter 9

## Add Record

**Access:** Table menu bar—Edit menu

**Purpose:** Adds a new record to the active table, which has been opened to allow editing.

**Associated with:** Start/Stop Editing, Add Field, Delete Record

**Reference:** Chapter 9

## Add Table

**Access:** Project menu bar—Project menu

**Purpose:** Imports a dBASE, INFO, or delimited text file into the active project.

**Associated with:** Import Project, SQL Connect

**Reference:** Chapter 3

## Add Theme

**Access:** View menu bar—View menu; View button bar

**Purpose:** Adds a theme to a view from a spatial data source, such as an ArcView shapefile, ARC/INFO coverage, or supported image source.

**Associated with:** Delete Themes, New Theme

**Reference:** Chapter 3

## Align

**Access:** View menu bar—Graphics menu; Layout menu bar—Graphics menu

**Purpose:** Aligns selected graphics vertically or horizontally in a view or layout.

**Associated with:** Pointer

**Reference:** Chapter 8

## Area Chart Gallery

**Access:** Chart menu bar—Gallery menu; Chart button bar

**Purpose:** Displays format options and changes an active chart to Area chart format.

**Associated with:** Create Chart, Bar Chart Gallery, Column Chart Gallery, Line Chart Gallery, Pie Chart Gallery, XY Scatter Chart Gallery

**Reference:** Chapter 7

## Area of Interest

**Access:** View tool bar

**Purpose:** Sets the Area of Interest for library-based themes in a view.

**Associated with:** Theme Properties

**Reference:** Chapter 4

## Attach Graphics

**Access:** View menu bar—Theme menu

**Purpose:** Attaches the selected graphics to the active theme; attached graphics will display only when the theme is turned on.

**Associated with:** Detach Graphics, Label Features, Auto-label

**Reference:** Chapter 5

## Auto-label

**Access:** View menu bar—Theme menu

**Purpose:** Labels selected features in the active themes using the specified label field. The resultant text graphics are attached to the active themes.

**Associated with:** Label Features, Text

**Reference:** Chapter 5

## Bar Chart Gallery

**Access:** Chart menu bar—Gallery menu; Chart button bar

**Purpose:** Displays format options and changes an active chart to Bar Chart format.

**Associated with:** Create Chart, Area Chart Gallery, Column Chart Gallery, Line Chart Gallery, Pie Chart Gallery, XY Scatter Chart Gallery

**Reference:** Chapter 7

## Bring to Front

**Access:** View menu bar—Graphics menu; Layout menu bar—Graphics menu; Layout button bar

**Purpose:** Brings selected graphics to the front of remaining graphics.

**Associated with:** Send to Back, Pointer

**Reference:** Chapter 8

## Calculate

**Access:** Table menu bar—Field menu; Table button bar

**Purpose:** Performs calculations on all, or selected, records in the active field of a table for which editing has been enabled.

**Associated with:** Start/Stop Editing

**Reference:** Chapter 9

## Chart Color

**Access:** Chart tool bar

**Purpose:** Changes the color of any chart element.

**Associated with:** Show Symbol Palette

**Reference:** Chapter 7

## Chart Element Properties

**Access:** Chart tool bar

**Purpose:** Sets properties of the chart elements of the active chart.

**Associated with:** Chart Axis Properties, Chart Legend Properties, Chart Title Properties

**Reference:** Chapter 7

## Chart Properties

**Access:** Chart menu bar—Chart menu; Chart button bar

**Purpose:** Sets the properties of the active chart by adding/deleting data series or groups, selecting the fields for labeling data series/groups, and setting the order of displaying data series/groups.

**Associated with:** Create Chart, Table Properties

**Reference:** Chapter 7

## Clear All Breakpoints

**Access:** Script menu bar—Script menu

**Purpose:** Clears all breakpoints in the active script.

**Associated with:** Toggle Breakpoint, Compile, Step, Run

**Reference:** Chapter 14

## Clear Selected Features

**Access:** View menu bar—Theme menu; View button bar; View popup menu

**Purpose:** Deselects any selected features in the active themes.

**Associated with:** Select Features, Select Features Using Shape, Switch Selection

**Reference:** Chapter 6

## Close

**Access:** View, Table, Chart, Layout, and Script menu bars—File menu

**Purpose:** Closes the active component of a project.

**Associated with:** Close All

## Close All

**Access:** View, Table, Chart, Layout, and Script menu bars—File menu

**Purpose:** Closes all project components currently open.

**Associated with:** Close

## Close Project

**Access:** Project menu bar—File menu

**Purpose:** Closes the active project and all components.

**Associated with:** Close, Close All, Exit

## Column Chart Gallery

**Access:** Chart menu bar—Gallery menu; Chart button bar

**Purpose:** Displays format options and changes an active chart to Column Chart format.

**Associated with:** Create Chart, Area Chart Gallery, Bar Chart Gallery, Line Chart Gallery, Pie Chart Gallery, XY Scatter Chart Gallery

**Reference:** Chapter 7

## Combine Graphics

**Access:** View menu bar—Edit menu

**Purpose:** Combines selected graphics in a view into one graphic element.

**Associated with:** Select All Graphics, Union Graphics, Subtract Graphics, Intersect Graphics

## Comment

**Access:** Script menu bar—Edit menu

**Purpose:** Comments selected text in a script.

**Associated with:** Remove Comment

**Reference:** Chapter 14

## Compile

**Access:** Script menu bar—Script menu; Script button bar

**Purpose:** Compiles the script in the active script window.

**Associated with:** Run, Step

**Reference:** Chapter 14

## Convert Overlapping Labels

**Access:** View Menu Bar—Theme menu

**Purpose:** Converts labels that overlap as a result of the Auto-label function (drawn in green) to the same text symbol and color as the "good" labels.

**Associated with:** Remove Labels, Remove Overlapping Labels

**Reference:** Chapter 5

## Convert to Shapefile

**Access:** View menu bar—Theme menu

**Purpose:** Converts a theme derived from an ARC/INFO coverage to ArcView shapefile format, either for the entire theme or for a selected set of features.

**Associated with:** New Theme, Add Theme

**Reference:** Chapter 9

## Copy

**Access:** Table, Layout, and Script menu bars—Edit menu; Table, Layout, and Script button bars

**Purpose:** Copies selected features to the clipboard. For tables, the selected feature is the data in the active cell; for layouts, graphics, and scripts, it is text.

**Associated with:** Cut, Paste

## Copy Graphics

**Access:** View menu bar—Edit menu

**Purpose:** Copies the selected graphics in a view to the clipboard.

**Associated with:** Cut, Paste

## Copy Themes

**Access:** View menu bar—Edit menu

**Purpose:** Copies the active themes to the clipboard.

**Associated with:** Cut, Paste

**Reference:** Chapter 4

## Create Chart

**Access:** Table menu bar—Table menu; Table button bar

**Purpose:** Creates a chart from the selected records of the active table, or from the entire table if no records are selected.

**Associated with:** Chart Properties

**Reference:** Chapter 7

## Create/Remove Index

**Access:** Table menu bar—Field menu

**Purpose:** Creates or removes an ArcView index for the active field. If the active field is the Shape field, a spatial index will be created or deleted.

**Associated with:** Open Theme Table

**Reference:** Chapter 10

## Customize

**Access:** Project menu bar—Project menu

**Purpose:** Accesses the dialog window for customizing the ArcView interface; accessible only if Avenue has been installed.

**Associated with:** Control Properties, Project Properties

**Reference:** Chapter 11

## Cut

**Access:** Table, Layout, and Script menu bars—Edit menu; Table, Layout, and Script button bars

**Purpose:** Cuts the selection and places it on the clipboard. For tables, the selection is the data in the active cell; for layouts, selected graphics, and scripts, it is text.

**Associated with:** Copy, Paste

## Cut Graphics

**Access:** View menu bar—Edit menu

**Purpose:** Cuts the selected graphics from a view and places them on the clipboard.

**Associated with:** Cut, Copy, Paste, Delete Graphics

## Cut Themes

**Access:** View menu bar—Edit menu

**Purpose:** Cuts the active themes from a view and places them on the clipboard.

**Associated with:** Copy Themes, Paste, Delete Themes

## Delete

**Access:** Layout menu bar—Edit menu

**Purpose:** Deletes the selected graphics in the active layout.

**Associated with:** Delete Graphics, Cut, Cut Graphics

## Delete Field

**Access:** Table menu bar—Edit menu

**Purpose:** Deletes the active field from a table for which editing has been enabled.

**Associated with:** Add Field, Start/Stop Editing

**Reference:** Chapter 9

## Delete Graphics

**Access:** View menu bar—Graphics menu

**Purpose:** Deletes selected graphics from a view.

**Associated with:** Cut Graphics, Copy Graphics

## Delete Last Point

**Access:** View popup menu

**Purpose:** Deletes the last point added to the line or polygon currently being drawn.

**Associated with:** Undo Edit, Redo Edit

**Reference:** Chapter 10

## Delete Left

**Access:** Script menu bar—Edit menu

**Purpose:** Deletes text from the location cursor to left margin.

**Associated with:** Shift Left, Shift Right, Cut

## Delete Records

**Access:** Table menu bar—Edit menu

**Purpose:** Deletes the selected records for the active table, providing editing has been enabled.

**Associated with:** Cut, Copy, Paste, Add Records

**Reference:** Chapter 9

## Delete Themes

**Access:** View menu bar—Edit menu

**Purpose:** Deletes the active themes from a view.

**Associated with:** Cut Themes, Copy Themes

## Detach Graphics

**Access:** View menu bar—Theme menu

**Purpose:** Detaches graphics from the active themes in the view.

**Associated with:** Attach Graphics

**Reference:** Chapter 5

## Draw

**Access:** View tool bar; Layout tool bar

**Purpose:** Allows graphics (points, lines, polylines, rectangles, circles, and polygons) to be added to a view or layout. Also allows existing point and line features to be split, and polygons to be appended to existing polygons.

**Associated with:** Text, Label, Pointer

**Reference:** Chapter 5

## Edit

**Access:** Table tool bar

**Purpose:** Selects cell data for editing in an active table for which editing has been enabled.

**Associated with:** Start/Stop Editing

**Reference:** Chapter 9

## Edit Legend

**Access:** View menu bar—Theme menu; View button bar

**Purpose:** Accesses the Legend Editor for changing the symbols and/or classification of the active theme.

**Associated with:** Theme Properties

**Reference:** Chapter 5

## Embed Script

**Access:** Script menu bar—Script menu

**Purpose:** Embeds the script in the project.

**Associated with:** Unembed Script

**Reference:** Chapter 14

## Erase

**Access:** Chart tool bar

**Purpose:** Removes data markers from a chart and deselects the records from the associated table.

**Associated with:** Erase with Polygon, Undo Erase

## Erase with Polygon

**Access:** Chart tool bar

**Purpose:** Removes one or more data markers from an XY Scatter chart by defining a polygon around the markers.

**Associated with:** Erase, Undo Erase

## Examine Variables

**Access:** Script menu bar—Script menu; Script button bar

**Purpose:** Displays current values of local and global variables.

**Associated with:** Compile, Step, Run

**Reference:** Chapter 14

## Exit

**Access:** Project, View, Table, Chart, Layout, and Script menu bars—File menu

**Purpose:** Ends the ArcView session.

**Associated with:** Close Project, Save Project, Save Project As

## Export

**Access:** View and Layout menu bars—File menu

**Purpose:** Exports a view or layout to a file. Supported output formats include Encapsulated PostScript, Adobe Illustrator, and CGM on all platforms, as well as Windows Metafile and Windows Bitmap, and Macintosh PICT.

**Associated with:** Print, Print Setup

## Export Table

**Access:** Table menu bar—File menu

**Purpose:** Exports a table to a file. Supported file types include dBASE, INFO, and delimited text.

**Associated with:** Print, Print Setup

**Reference:** Chapter 9

## Extensions

**Access:** Project menu bar—File menu

**Purpose:** Loads or removes available ArcView extensions.

**Associated with:** Open Project

**Reference:** Chapter 9

## Find

**Access:** View menu bar—View menu; Table menu bar—Table menu; Chart menu bar—Chart menu; Script menu bar—Edit menu; View, Table, and Chart button bars

**Purpose:** Finds a particular feature in an active theme, table, or chart, based on an entered text string; or finds selected text in a script.

**Associated with:** Locate, Query, Find Next, Replace

**Reference:** Chapter 6

## Find Next

**Access:** Script menu bar—Edit menu

**Purpose:** Finds next occurrence of selected text in a script.

**Associated with:** Find, Replace

## Frame

**Access:** Layout tool bar

**Purpose:** Adds a view frame, legend frame, scale bar frame, north arrow frame, chart frame, table frame, or picture frame to a layout.

**Associated with:** Draw, Text

**Reference:** Chapter 8

## Full Extent

**Access:** View menu bar—View menu

**Purpose:** Zooms to the full extent of all themes in a view.

**Associated with:** Zoom In, Zoom Out, Zoom to Selected, Zoom to Themes, Zoom Previous

**Reference:** Chapter 3

## General Snapping On/Off

**Access:** View popup menu

**Purpose:** Toggles general snapping on or off during theme editing.

**Associated with:** Interactive Snapping On/Off

**Reference:** Chapter 10

## Geocode Addresses

**Access:** View menu bar—View menu

**Purpose:** Begins geocoding addresses against a theme for which a geocoding index has been built.

**Associated with:** Add Event Theme, Locate, Rematch

**Reference:** Chapters 2, 4

## Group

**Access:** View menu bar and Layout menu bars—Graphics menu; Layout button bar

**Purpose:** Groups selected graphics as a single graphic.

**Associated with:** Ungroup, Pointer

**Reference:** Chapter 8

## Hide/Show Grid

**Access:** Layout menu bar—Layout menu

**Purpose:** Toggles the display of the layout grid on the active layout.

**Associated with:** Layout Properties, Show/Hide Margins

**Reference:** Chapter 8

## Hide/Show Legend

**Access:** View menu bar—Theme menu

**Purpose:** Hides or shows the legend of the active themes in the Table of Contents.

**Associated with:** Edit Legend

## Hide/Show Margins

**Access:** Layout menu bar—Layout menu

**Purpose:** Toggles the display of the layout page margins on the active layout.

**Associated with:** Layout Properties, Show/Hide Grid

**Reference:** Chapter 8

## Hot Link

**Access:** View tool bar

**Purpose:** Invokes the defined hot link for an active theme on a view.

**Associated with:** Open View, Open Project

**Reference:** Chapter 9

## Identify

**Access:** View, Table, and Chart tool bars

**Purpose:** Displays the attributes of a feature in an active theme, table, or chart.

**Associated with:** Select Feature, Select, Select Features Using Shape

**Reference:** Chapter 5

## Import

**Access:** Project menu bar—Project menu

**Purpose:** Imports components of another ArcView project, including ArcView 1.0 projects, into the active project.

**Associated with:** Open Project

## Interactive Snap

**Access:** View tool bar

**Purpose:** Sets the interactive tolerance for snapping vertices on an editable theme.

**Associated with:** Snap, Start/Stop Editing, Draw, Pointer

**Reference:** Chapter 10

## Interactive Snapping On/Off

**Access:** View popup menu

**Purpose:** Toggles interactive snapping on or off during theme editing.

**Associated with:** General Snapping On/Off

**Reference:** Chapter 10

## Intersect Graphics

**Access:** View menu bar—Edit menu

**Purpose:** Creates an output graphic formed from the area held in common by all selected graphics.

**Associated with:** Select All Graphics, Combine Graphics, Union Graphics, Subtract Graphics

**Reference:** Chapter 10

## Join

**Access:** Table menu bar—Table menu; Table button bar

**Purpose:** Joins a table to the active table based on the values of a common field.

**Associated with:** Link, Remove All Joins

**Reference:** Chapters 4, 9

## Label

**Access:** View tool bar

**Purpose:** Labels a feature in the active theme with the attribute from the field specified in that theme's properties.

**Associated with:** Auto-label, Text

**Reference:** Chapter 5

## Layout

**Access:** View menu bar—View menu

**Purpose:** Creates a layout using a specified template.

**Associated with:** Use Template, Store as Template

**Reference:** Chapter 8

## Line Chart Gallery

**Access:** Chart menu bar—Gallery menu; Chart button bar

**Purpose:** Displays format options and changes an active chart to Line Chart format.

**Associated with:** Create Chart, Area Chart Gallery, Bar Chart Gallery, Column Chart Gallery, Pie Chart Gallery, XY Scatter Chart Gallery

**Reference:** Chapter 7

## Link

**Access:** Table menu bar—Table menu

**Purpose:** Establishes a one-to-many relationship between the source table and the active table, based on the values of a common field.

**Associated with:** Join, Remove All Links

**Reference:** Chapter 4

## Load System Script

**Access:** Script menu bar—script menu; Script button bar

**Purpose:** Inserts the source code of a system script.

**Associated with:** Load Text File, Write Text File

**Reference:** Chapter 14

## Load Text File

**Access:** Script menu bar—Script menu; Script button bar

**Purpose:** Inserts the contents of a text file into the active script.

**Associated with:** Load System Script, Write Text File

**Reference:** Chapter 14

## Locate

**Access:** View menu bar—View menu; View button bar

**Purpose:** Locates a specific address on an active theme for which geocoding properties have been set.

**Associated with:** Find

**Reference:** Chapter 4

## Measure

**Access:** View tool bar

**Purpose:** Measures distance on a view.

**Associated with:** Draw

## Merge Graphics

**Access:** View menu bar—Edit menu

**Purpose:** Combines or aggregates selected features from an ArcView shapefile into a single shape.

**Associated with:** Select Feature, Summarize

**Reference:** Chapter 9

## New Project

**Access:** Project menu bar—File menu

**Purpose:** Creates a new ArcView project.

**Associated with:** Save Project As

**Reference:** Chapter 3

## New Theme

**Access:** View menu bar—View menu

**Purpose:** Creates a new theme based on the ArcView shapefile format.

**Associated with:** Edit, Copy Theme

**Reference:** Chapter 9

## Open Project

**Access:** Project menu bar—File menu

**Purpose:** Opens an existing ArcView project.

**Associated with:** New Project, Save Project, Save Project As

**Reference:** Chapter 3

## Open Theme Table

**Access:** View menu bar—Theme menu; View button bar

**Purpose:** Opens the attribute tables for those themes that are active in a particular view.

**Associated with:** Add Table

**Reference:** Chapter 6

## Page Setup

**Access:** Layout menu bar—Layout menu

**Purpose:** Defines the characteristics of the layout page.

**Associated with:** Layout Properties, Use Template, Show/Hide Grid, Show/Hide Margins

**Reference:** Chapter 8

## Pan

**Access:** View and Layout tool bars; View popup menu

**Purpose:** By dragging the mouse, pans across the currently displayed view or layout.

**Associated with:** Zoom In, Zoom Out

## Paste

**Access:** View, Table, Layout, and Script menu bars—Edit menu; Table, Layout, and Script button bars

**Purpose:** Pastes the contents of the clipboard into the active document.

**Associated with:** Copy, Cut

## Pie Chart Gallery

**Access:** Chart menu bar—Gallery menu; Chart button bar

**Purpose:** Displays format options and changes an active chart to Pie Chart format.

**Associated with:** Create Chart, Area Chart Gallery, Bar Chart Gallery, Column Chart Gallery, Line Chart Gallery, XY Scatter Chart Gallery

**Reference:** Chapter 7

## Pointer

**Access:** View and Layout tool bars

**Purpose:** Selects graphics in a view or layout for subsequent editing and manipulation.

**Associated with:** Select All, Select All Graphics

**Reference:** Chapter 5

## Print

**Access:** View, Table, Chart, Layout, and Script menu bars—File menu; Layout button bar

**Purpose:** Prints the active project component.

**Associated with:** Print Setup

## Print Setup

**Access:** View, Table, Chart, Layout, and Script menu bars—File menu

**Purpose:** Controls the output format and printing environment.

**Associated with:** Print

## Promote

**Access:** Table menu bar—Table menu; Table button bar

**Purpose:** Displays selected records at the top of the table.

**Associated with:** Select, Sort Ascending, Sort Descending

**Reference:** Chapter 6

## Properties—Graphic

**Access:** View and Layout menu bars—Graphics menu

**Purpose:** Displays and edits graphics properties for graphic primitives, text, and frames.

**Associated with:** View Properties, Layout Properties, Pointer, Text

**Reference:** Chapter 8

## Properties—Layout

**Access:** Layout menu bar—Layout menu; Layout button bar

**Purpose:** Displays and edits layout properties, including name, grid spacing, and snapping to grid.

**Associated with:** Graphics Properties, View Properties, Table Properties, Chart Properties

**Reference:** Chapter 8

## Properties—Project

**Access:** Project menu bar—Project menu

**Purpose:** Displays and edits project properties, including start-up and shutdown scripts, work directory, name of creator and creation date, and selection color.

**Associated with:** View Properties, Table Properties, Chart Properties, Layout Properties

**Reference:** Chapter 9

## Properties—Script

**Access:** Script menu bar—Script menu

**Purpose:** Displays and edits script properties such as name, creator, creation date, comments, and behavior (whether the script will remain active during execution).

**Associated with:** View Properties, Table Properties, Chart Properties, Layout Properties

**Reference:** Chapter 14

## Properties—Table

**Access:** Table menu bar—Table menu

**Purpose:** Displays and edits properties of the active table, including name, creator, visible fields, and field alias names.

**Associated with:** View Properties, Chart Properties, Layout Properties

**Reference:** Chapter 6

## Properties—View

**Access:** View menu bar—View menu

**Purpose:** Displays and edits the properties of the current view, including name, creation date, creator, map units, distance units, and projection.

**Associated with:** Table Properties, Chart Properties, Layout Properties

**Reference:** Chapter 3

## Query

**Access:** View menu bar—Theme menu; Table menu bar—Table menu; View and Table button bars

**Purpose:** Opens the Query Builder dialog window, which allows feature(s) in a view or records in a table to be selected by a logical expression based on attribute values.

**Associated with:** Find, Locate, Theme Properties

**Reference:** Chapter 6

## Redo Edit

**Access:** Table menu bar—Edit menu; View popup menu

**Purpose:** Restores the last "undo" edit on a shapefile or table.

**Associated with:** Undo Edit, Delete Last Point

**Reference:** Chapter 10

## Refresh

**Access:** Table menu bar—Table menu

**Purpose:** Causes ArcView to reread source data for the active table.

**Associated with:** Open, Add Table, Join, Link

**Reference:** Chapter 7

## Rematch

**Access:** View menu bar—Theme menu

**Purpose:** Opens the Geocoding Editor dialog window, allowing features to be rematched in the geocoded theme.

**Associated with:** Add Event Theme

**Reference:** Chapter 4

## Remove All Joins

**Access:** Table menu bar—Table menu

**Purpose:** Removes all joins from the active table.

**Associated with:** Join, Remove All Links

**Reference:** Chapter 4

## Remove All Links

**Access:** Table menu bar—Table menu

**Purpose:** Removes all links to other tables for the active table.

**Associated with:** Link, Remove All Joins

**Reference:** Chapter 4

## Remove Comment

**Access:** Script menu bar—Edit menu

**Purpose:** Uncomments selected text in a script.

**Associated with:** Comment

**Reference:** Chapter 14

## Remove Labels

**Access:** View menu bar—Theme menu

**Purpose:** Removes all labels attached to the active theme.

**Associated with:** Remove Overlapping Labels, Convert Overlapping Labels

**Reference:** Chapter 5

## Remove Overlapping Labels

**Access:** View menu bar—Theme menu

**Purpose:** Removes all overlapping labels attached to the active theme. Overlapping labels result from the Auto-label function, and they appear in green in the view.

**Associated with:** Remove Labels, Convert Overlapping Labels

## Rename

**Access:** Project menu bar—Project menu

**Purpose:** Renames the selected project component.

**Associated with:** Properties

## Replace

**Access:** Script menu bar—Edit menu

**Purpose:** Replaces the selected string in a script.

**Associated with:** Find, Find Next

## Run

**Access:** Script menu bar—Script menu; Script button bar

**Purpose:** Runs the compiled script.

**Associated with:** Compile, Step

**Reference:** Chapter 14

## Save Edits

**Access:** View menu bar—Theme menu; Table menu bar—Table menu

**Purpose:** Saves all current edits to the theme or table being edited.

**Associated with:** Save Edits As, Start Editing, Stop Editing

**Reference:** Chapters 10, 11

## Save Edits As

**Access:** View menu bar—Theme menu; Table menu bar—Table menu

**Purpose:** Saves the current theme or table, including all edits, to a new shapefile or table.

**Associated with:** Save Edits, Start Editing, Stop Editing

**Reference:** Chapters 10, 11

## Save Project

**Access:** Project, View, Table, Chart, Layout, and Script menu bars—File menu; Project, View, Table, Chart, Layout, and Script button bars

**Purpose:** Saves the active project.

**Associated with:** Save Project As

**Reference:** Chapter 3

## Save Project As

**Access:** Project menu bar—File menu

**Purpose:** Saves the active project to a new name and/or directory.

**Associated with:** Save Project

**Reference:** Chapter 3

## Select

**Access:** Table tool bar

**Purpose:** Selects records in the active table.

**Associated with:** Select All, Select None, Switch Selection

**Reference:** Chapter 6

## Select All

**Access:** Table, Layout, and Script menu bars—Edit menu; Table button bar

**Purpose:** Selects all records in the active table, all graphics drawn in the active layout, or all text in the active script.

**Associated with:** Pointer, Select, Select Name, Switch Selection

**Reference:** Chapter 6

## Select All Graphics

**Access:** View menu bar—Edit menu

**Purpose:** Selects all graphics drawn in the view.

**Associated with:** Pointer

## Select by Theme

**Access:** View menu bar—Theme menu

**Purpose:** Selects features from active themes based on features chosen in the selector theme.

**Associated with:** Select Feature, Select Features Using Shape

**Reference:** Chapter 9

## Select Feature

**Access:** View tool bar

**Purpose:** Selects features in the active theme using the mouse.

**Associated with:** Select Features Using Shape, Select by Theme

**Reference:** Chapter 6

## Select Features Using Shape

**Access:** View button bar

**Purpose:** Selects features in the active theme using selected graphics in the view.

**Associated with:** Select Feature, Select by Theme, Pointer

**Reference:** Chapter 6

## Select None

**Access:** Table menu bar—Edit menu; Table button bar

**Purpose:** Clears the selected set in the active table.

**Associated with:** Select, Select All, Switch Selected

**Reference:** Chapter 6

## Send to Back

**Access:** View and Layout menu bars—Graphics menu; Layout button bar

**Purpose:** Places the selected graphics behind the remaining graphics.

**Associated with:** Bring to Front, Pointer

**Reference:** Chapter 8

## Series from Records/Fields

**Access:** Chart menu bar—Chart menu; Chart button bar

**Purpose:** Toggles between using records and using fields for plotting the data series in a chart.

**Associated with:** Create Chart, Chart Properties

**Reference:** Chapter 7

## Set Work Directory

**Access:** View menu bar—File menu

**Purpose:** Sets or changes the current working directory.

**Associated with:** Project Properties

**Reference:** Chapter 2

## Shift Left

**Access:** Script button bar

**Purpose:** Shifts the currently selected line (or lines) two spaces to the left.

**Associated with:** Shift Right

## Shift Right

**Access:** Script button bar

**Purpose:** Shifts the currently selected line (or lines) two spaces to the right.

**Associated with:** Shift Left

## Show/Hide Grid

**Access:** Layout menu bar—Layout menu

**Purpose:** Toggles the display of the active layout's grid on and off.

**Associated with:** Layout Properties, Show/Hide Margins

**Reference:** Chapter 8

## Show/Hide Legend

**Access:** Chart menu bar—Chart menu

**Purpose:** Toggles the display of the active chart's legend on and off.

**Associated with:** Chart Properties, Show/Hide X Axis, Show/Hide Y Axis, Show/Hide Title

## Show/Hide Margins

**Access:** Layout menu bar—Layout menu

**Purpose:** Toggles the display of the active layout's page margins on and off.

**Associated with:** Layout Properties, Show/Hide Grid

**Reference:** Chapter 8

## Show/Hide Title

**Access:** Chart menu bar—Chart menu

**Purpose:** Toggles the display of the active chart's title on and off.

**Associated with:** Chart Properties, Show/Hide Legend, Show/Hide X Axis, Show/Hide Y Axis

**Reference:** Chapter 7

## Show/Hide X Axis

**Access:** Chart menu bar—Chart menu

**Purpose:** Toggles the display of the active chart's X axis (with tick marks) on and off.

**Associated with:** Chart Properties, Show/Hide Y Axis, Show/Hide Legend, Show/Hide Title

**Reference:** Chapter 7

## Show/Hide Y Axis

**Access:** Chart menu bar—Chart menu

**Purpose:** Toggles the display of the active chart's Y axis (with tick marks) on and off.

**Associated with:** Chart Properties, Show/Hide X Axis, Show/Hide Legend, Show/Hide Title

**Reference:** Chapter 7

## Show Symbol Palette

**Access:** View, Table, Chart, and Layout menu bars—Window menu

**Purpose:** Displays the symbol palette.

**Associated with:** Legend Editor

## Simplify

**Access:** Layout menu bar—Graphics menu

**Purpose:** Explodes the selected graphics, including legends and scale bars, into component graphic elements.

**Associated with:** Frame, Group

**Reference:** Chapter 8

## Size and Position

**Access:** View and Layout menu bars—Graphics menu

**Purpose:** Displays the dialog window for controlling the size and position of the selected graphics.

**Associated with:** Graphics Properties, Pointer

**Reference:** Chapter 8

## Snap

**Access:** View tool bar

**Purpose:** Sets the general tolerance of snapping vertices for an editable theme.

**Associated with:** Snap Feature, Start/Stop Editing, Draw, Pointer

**Reference:** Chapter 10

## Snap to Boundary

**Access:** View popup menu

**Purpose:** Snaps the next entered point to the nearest line segment within the interactive snapping tolerance.

**Associated with:** Snap to Endpoint, Snap to Intersection, Snap to Vertex

**Reference:** Chapter 10

## Snap to Endpoint

**Access:** View popup menu

**Purpose:** Snaps the next entered point to the nearest node held in common by two or more features within the interactive snapping tolerance.

**Associated with:** Snap to Boundary, Snap to Intersection, Snap to Vertex

## Snap to Intersection

**Access:** View popup menu

**Purpose:** Snaps the next entered point to the nearest line endpoint within the interactive snapping tolerance.

**Associated with:** Snap to Boundary, Snap to Endpoint, Snap to Vertex

## Snap to Vertex

**Access:** View popup menu

**Purpose:** Snaps the next entered point to the nearest vertex within the interactive snapping tolerance.

**Associated with:** Snap to Boundary, Snap to Endpoint, Snap to Intersection

**Reference:** Chapter 10

## Sort Ascending/Sort Descending

**Access:** Table menu bar—Field menu; Table button bar

**Purpose:** Sorts all records in the active table on the active field.

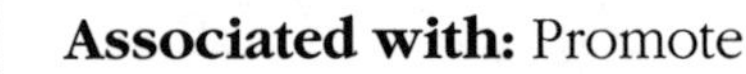

**Associated with:** Promote

**Reference:** Chapter 6

## SQL Connect

**Access:** Project menu bar—Project menu

**Purpose:** Opens the SQL Connect dialog window to enable connection to a database server and subsequent retrieval of records based on an SQL query.

**Associated with:** Export

## Start/Stop Editing

**Access:** Table menu bar—Table menu; View menu bar—Theme menu

**Purpose:** Controls the enabling of editing on a theme or table.

**Associated with:** Edit, Pointer

**Reference:** Chapter 9

## Statistics

**Access:** Table menu bar—Field menu

**Purpose:** Obtains statistics about the currently active numeric field in the active table.

**Associated with:** Summarize, Query

**Reference:** Chapter 9

## Step

**Access:** Script menu bar—Script menu; Script button bar

**Purpose:** Executes one request or object reference in the compiled script.

**Associated with:** Compile, Run, Toggle Breakpoint

**Reference:** Chapter 14

## Store As Template

**Access:** Layout menu bar—Layout menu

**Purpose:** Creates a layout template from the current layout.

**Associated with:** Use Template

**Reference:** Chapter 8

## Store North Arrows

**Access:** Layout menu bar—Layout menu

**Purpose:** Stores each graphics group in the current layout as a north arrow.

**Associated with:** Draw, Use Template

**Reference:** Chapter 8

## Subtract Graphics

**Access:** View menu bar—Edit menu

**Purpose:** Subtracts the shape of one graphic from the shape of another (the second selected one from the first selected one).

**Associated with:** Select All Graphics, Combine Graphics, Union Graphics, Intersect Graphics

## Summarize

**Access:** Table menu bar—Field menu; Table button bar

**Purpose:** Displays the Summary Table Definition dialog window for preparing a summary table based on the active field.

**Associated with:** Statistics, Merge, Query

**Reference:** Chapter 9

## Switch Selection

**Access:** Table menu bar—Edit menu; Table button bar

**Purpose:** Switches the selected set of records in the active table to all records previously unselected.

**Associated with:** Select, Select All, Select None

**Reference:** Chapter 6

## Table

**Access:** View menu bar—Theme menu; View button bar

**Purpose:** Opens the attribute tables for the active themes in a view.

**Associated with:** Add Table

**Reference:** Chapter 6

## Text

**Access:** View and Layout tool bars

**Purpose:** Adds or edits text in the active view or layout.

**Associated with:** Pointer, Draw

**Reference:** Chapter 5

## Theme Properties

**Access:** View menu bar—Theme menu; View button bar

**Purpose:** Reviews and sets properties of the active theme, including name, logical queries, field for feature labeling, range of scales for display, hot link definition, geocoding properties, and snapping.

**Associated with:** View Properties, Edit Legend

**Reference:** Chapter 3

## Themes On/Themes Off

**Access:** View menu bar—View menu

**Purpose:** Toggles all themes in a view on or off.

**Associated with:** Theme Properties

**Reference:** Chapter 3

## Toggle Breakpoint

**Access:** Script menu bar—Script menu; Script button bar

**Purpose:** Toggles a breakpoint on or off at the cursor location.

**Associated with:** Compile, Step, Run, Clear All Breakpoints

**Reference:** Chapter 14

## Undo

**Access:** Script menu bar—Edit menu; Script button bar

**Purpose:** Undo the last change made in a script window.

**Associated with:** Redo

## Undo Edit

**Access:** Table menu bar—Edit menu; View popup menu

**Purpose:** Undo the last edit made to a shapefile or table

**Associated with:** Redo Edit, Delete Last Point

**Reference:** Chapter 10

## Undo Erase

**Access:** Chart menu bar—Edit menu; Chart button bar

**Purpose:** "Undeletes" the last data markers that were erased from the active chart.

**Associated with:** Erase, Erase with Polygon

## Unembed Script

**Access:** Script menu bar—Script menu

**Purpose:** Removes the selected script so that it is no longer embedded in the project.

**Associated with:** Embed Script

**Reference:** Chapter 14

## Ungroup

**Access:** View and Layout menu bars—Graphics menu

**Purpose:** Ungroups a previously grouped graphic into its original individual pieces.

**Associated with:** Group, Pointer

**Reference:** Chapter 8

## Union Graphics

**Access:** View menu bar—Edit menu

**Purpose:** Combines the selected graphics from a view into a single graphic. For polygon graphics, only the exterior boundary is preserved in the output graphic.

**Associated with:** Select All Graphics, Combine Graphics, Subtract Graphics, Intersect Graphics

## Use Template

**Access:** Layout menu bar—Layout menu; View menu bar—View menu

**Purpose:** Using a specified stored template, creates a new layout or updates the current layout.

**Associated with:** Store As Template, Layout Properties

**Reference:** Chapter 8

## Vertex Tool

**Access:** View tool bar; Layout tool bar

**Purpose:** Adds, moves, or deletes vertices from a selected shapefile feature or graphic element.

**Associated with:** Pointer Tool

**Reference:** Chapter 10

## Write Text File

**Access:** Script menu bar—Script menu; Script button bar

**Purpose:** Writes an entire script or selected text to a text file.

**Associated with:** Load Text File, Load System Script

**Reference:** Chapter 14

## XY Scatter Chart Gallery

**Access:** Chart menu bar—Gallery menu; Chart button bar

**Purpose:** Displays format options and changes an active chart to XY Scatter Chart format.

**Associated with:** Create Chart, Area Chart Gallery, Bar Chart Gallery, Column Chart Gallery, Line Chart Gallery

**Reference:** Chapter 7

## Zoom In (button)

**Access:** View menu bar—View menu; Layout menu bar—Layout menu; View and Layout button bars; View popup menu

**Purpose:** Zooms in on the center of the active view or layout by a factor of two.

**Associated with:** Zoom Out, Zoom to Full Extent, Zoom to Selected, Zoom to Themes, Zoom to Page, Zoom to Actual Size

## Zoom In (tool)

**Access:** View and Layout tool bars

**Purpose:** Zooms in on the area you click on or the area you describe on a view or layout.

**Associated with:** Zoom Out, Zoom to Full Extent, Zoom to Selected, Zoom to Themes, Zoom to Page, Zoom to Actual Size

## Zoom Out (button)

**Access:** View menu bar—View menu; Layout menu bar—Layout menu; View and Layout button bars; View popup menu

**Purpose:** Zooms out from the center of the active view or layout by a factor of two.

**Associated with:** Zoom In, Zoom to Full Extent, Zoom to Selected, Zoom to Themes, Zoom to Page, Zoom to Actual Size

## Zoom Out (tool)

**Access:** View and Layout tool bars

**Purpose:** Zooms out from a position you click on or the area you describe on a view or layout.

**Associated with:** Zoom In, Zoom to Full Extent, Zoom to Selected, Zoom to Themes, Zoom to Page, Zoom to Actual Size

## Zoom Previous

**Access:** View menu bar—View menu; View button bar

**Purpose:** Zooms to the previously displayed extent in the view.

**Associated with:** Zoom In, Zoom Out, Zoom to Full Extent, Zoom to Selected, Zoom to Themes

## Zoom to Actual Size

**Access:** Layout menu bar—Layout menu; Layout button bar

**Purpose:** Zooms to the actual size (1:1) of the layout page.

**Associated with:** Zoom In, Zoom Out, Zoom to Page, Zoom to Selected

**Reference:** Chapter 8

## Zoom to Full Extent

**Access:** View menu bar—View menu; View button bar

**Purpose:** Zooms to the full extent of all themes in a view.

**Associated with:** Zoom In, Zoom Out, Zoom to Selected, Zoom to Themes

**Reference:** Chapter 3

## Zoom to Page

**Access:** Layout menu bar—Layout menu; Layout button bar

**Purpose:** Zooms to the full extent of the layout page.

**Associated with:** Zoom In, Zoom Out, Zoom to Actual Size, Zoom to Selected

**Reference:** Chapter 8

## Zoom to Selected

**Access:** View menu bar—View menu; Layout menu bar—Layout menu; View and Layout button bars; View popup menu

**Purpose:** Zooms to selected features of the active themes in a view, or to the selected graphics of a layout.

**Associated with:** Zoom In, Zoom Out, Zoom to Full Extent, Zoom to Themes, Zoom to Page, Zoom to Actual Size

**Reference:** Chapter 6

## Zoom to Themes

**Access:** View menu bar—View menu; View button bar

**Purpose:** Zooms to the extent of the active themes in a view.

**Associated with:** Zoom In, Zoom Out, Zoom to Full Extent, Zoom to Selected

**Reference:** Chapter 3

*Appendix C*

# About the MicroVision Segments

In this book, we have referred to several lifestyle segmentation names marketed by Equifax National Decisions Systems. Below are descriptions of the nine primary Tempe segments for the block groups associated with survey responses.

Note that, for the purposes of demonstration, we elected to use the general purpose MicroVision segments. A total of 50 general purpose segments exist. Specific, industry-tailored versions of these segments are also available from Equifax. In our restaurant exercises, for example, we could have easily employed the MicroVision Restaurant data set.

## *The Primary MicroVision Segments for Tempe*

**Prosperous Metro Mix.** Only 2% of the population is located in rural areas. Ranking fourth (of 50 segments) in average household size, this segment also ranks high on the number of adults in the 30 to 39 age cohort, and above average for children age 0 to 17. This segment earns the fifth highest median household income at 75% above the national average. Over 70% of the families include two or more workers. This segment ranks high in the number of working female adults with children, including the highest number of working mothers with children under age six.

**Movers and Shakers.** This highly urban segment ranks second in the total number of two-person households (37%). Median household income is 55% higher than the national average, and the segment ranks fifth in per capita income. The segment ranks fourth in the number of adults with undergraduate and graduate degrees, fifth in total white collar employment (81%) and third in the number of workers in the professional specialties (27%).

**Home Sweet Home.** With median household income 42% higher than the national average, this segment ranks third among the 50 segments in the proportion of households earning between $35,000 and $50,000. Nearly eight of ten households own their home. This segment ranks fourth in the proportion of properties valued between $100,000 and $150,000.

**A Good Step Forward.** Only 1% of the population resides in rural areas. This segment ranks among the top four for the number of individuals age 22 to 34. Along with an above-average senior population (65 and older), this segment ranks among the lowest for children age 0 to 17. This segment ranks fifth in the number of persons in non-family households (33%), and fourth in the percentage of single-person households (43%).

**Great Beginnings.** Ranking at the top of the 50 segments for adults with either some college or an associate degree, this segment is also above average for higher education degrees attained. Less than half of the segment live in their own homes, at 25% below the national average. Median property value is high and may contribute to rent payments that are nearly 40% above the national norm.

**White Picket Fences.** A segment of average age, the population is evenly distributed among the age ranges in accordance with the national average. Household size is also average, as is the proportion of married individuals and family households. Ranking fifth on the proportion of adults with only a high school diploma, this segment is well below average on the proportion of graduate and undergraduate degrees, and ranks second on the proportion of households earning between $25,000 and $35,000 (19%).

**Books and New Recruits.** With 40% of all individuals between ages 18 and 24, this segment is the lowest ranking segment for median age and near the bottom for the proportion of married and divorced individuals and family households. Over 28% of the population resides in group quarters such as dormitories and barracks, and 67% of students are enrolled in college. This segment has the shortest average commute time, with over 35% of its workers taking less than ten minutes to travel to work.

**University USA.** This is the number one segment for persons living in group quarters (30%), and ranks first for the proportion of non-family households (26%). Only one in ten persons are under age 18. Of all persons identified as students, 89% are presently attending college. Over 75% of housing is renter-occupied, the fourth highest of any segment.

**Urban Single.** With a median age 23% higher than average, this segment has concentrations of individuals age 18 to 29 and 70 and older. This is the number one segment for adults over the age of 84 and also ranks first in the proportion of single-person households. Nearly one third of the adults do not have a high school diploma. Employment is noticeably high in the service occupations. An above-average proportion of workers use public transportation or walk to work. This segment ranks fifth in the proportion of households without vehicles (38%).

## Appendix D

# Moving from ArcView 2.x to 3.0

For those anticipating the need for major training or retooling to make the transition to Version 3.0—relax. ESRI has designed ArcView Version 3.0 (renamed ArcView GIS) such that the transition is almost effortless. For the most part, any project created with ArcView 2.*x* will be fully functional in 3.0 and will not require editing. To the same end, enhancements in functionality are presented within the same GUI used in 2.*x*. Users encountering new or changed functionality in 3.0 should find that most changes are intuitive, and that they flow organically from the overall ArcView functional design.

To the extent that the transition is not 100 percent transparent, we have prepared this appendix.

## ArcView Project Considerations

For the most part, any ArcView project file created under 2.*x* will open and function properly under 3.0. There are, however, some exceptions.

### Geocoding Indexes

Most significant are the changes made to geocoding in Version 3.0. Changes have been made to several address styles. These, in turn, affect the geocoding indexes that are created when the theme is made matchable. The following address styles are affected:

- U.S. Streets
- U.S. Streets with Zone
- U.S. Single Range
- U.S. Single Range with Zone

The geocoding indexes created when a theme was made matchable under 2.*x* will not be properly recognized in 3.0. This will result in incorrect address matching against these themes in 3.0. To create the proper geocoding indexes, the theme should be deleted from the project, then added again to the project and made matchable under 3.0. This will ensure that the proper 3.0 geocoding indexes are created for the theme.

> **NOTE:** *These changes pertain to matchable themes only, and not to any point shapefiles that were created via geocoding under 2.x. Point themes that resulted from geocoding are no longer tied to the geocoding indexes of the street network they were matched against and, as such, do not need to be recreated.*

## CAD Drawing Themes

In ArcView 2.1, the ability to add themes based on CAD drawings was part of ArcView's core functionality. In 3.0, this functionality is enabled via the CadReader extension. If this extension is not already loaded at the time a 2.1 project containing CAD themes is opened, an error will result. To correct this problem, load the CadReader extension before opening or importing the 2.1 project containing CAD drawing themes, then save the project under Version 3.0. This ensures that the CadReader extension is loaded automatically the next time the project is opened.

## Charts

In ArcView 3.0, you can create charts only if you have 100 or fewer total selected records (it could be 100 records for a chart formed from one field, or 50 records for a chart formed from two fields, and so forth). If your chart was created under

2.*x*—and from more than 100 selected records—you will receive an error message when you open the project under Version 3.0. To correct this problem, delete or modify the chart under 3.0 so that the total number of records is 100 or fewer.

# Customization

In ArcView 3.0, several changes have been made that affect customization you may have done under 2.*x*. These changes pertain to customized GUIs, start-up scripts, modification to the *default.apr,* and the Avenue programming language.

## Customized GUIs

In ArcView 2.*x*, the project window contained an icon for each of the five document types: Views, Tables, Charts, Layouts, and Scripts. Documents containing a customized GUI were listed under the appropriate document type—for example, a view document that had a customized GUI would be listed under the Views icon. In Version 3.0, the document type list has been extended so that each customized document is listed as its own separate type in the project window, and each is associated with an icon for this customized GUI.

It is still possible to place customized GUIs in the default groupings—for example, to list all customized view documents when the Views icon is selected—by using the Avenue request SetGroupGUI.

## Start-up Scripts

In ArcView 3.0, the start-up script identified under Project Properties executes at a different point during project initialization than in 2.*x*. Specifically, the start-up script now executes *after* all project documents are opened, not before, as in 2.*x*. If your start-up script performed actions for which the sequence of project initialization was critical, you may have to modify the script to accommodate this change. Conversely, because the start-up script is now executed after all documents are opened, it is possible to perform certain actions via a start-up script that were not possible under 2.*x*.

## Modifications to *default.apr*

Any *default.apr* created under 2.*x* is not compatible with, and cannot be read by, ArcView 3.0. If a customized *default.apr* created under 2.*x* is present in your *$HOME* or *$TEMP* directories, ArcView 3.0 will not start.

To start ArcView 3.0, either remove your old *default.apr* or rename your *default.apr* to *default.old.* Any customization made to your 2.*x* *default.apr* will need to be made to your 3.0 *default.apr* as well.

## Changes to Avenue

While the changes are not extensive, some modifications have been made to Avenue classes and requests. If a script ran correctly under 2.*x*, but now produces an error message (for example, a certain object does not recognize a certain request), your script probably references an object or request that has been modified for Version 3.0.

You need to update your Avenue scripts if you have used any of the requests listed here.

1. When an operation failed in Version 2.1, the following requests returned *nil.* When an operation fails in Version 3.0, the following requests return a null shape. Test the success of an operation with the *IsNull* request instead of *nil.*
   - ❒ ReturnUserCircle
   - ❒ ReturnUserLine
   - ❒ ReturnUserPolygon
   - ❒ ReturnUserPolyLine
   - ❒ ReturnUserRect
2. Change the GetExtent request to ReturnExtent for instances of the following classes:
   - ❒ FTheme
   - ❒ ITheme
   - ❒ Theme

3. For instances of the Display class, change the SetReportUnits request to SetDistanceUnits. Likewise, change GetReportUnits to GetDistanceUnits.
4. For instances of the Application class, change the GetSysDefaults and GetUserDefaults requests to GetSysDefault and GetUserDefault.
5. The RampColors request is no longer available for objects of the Legend class. Use this request with the symbol list of your legend.
6. In ArcView 3.0, ArcView no longer recognizes this enumeration element: *#UNITS_LINEAR_PRJMETERS.*
7. For a project instance, change GetActive and SetActive requests to GetSelectedDocs and SetSelectedDocs.
8. For objects of the Project class, change the GetActiveClass request to GetSelectedGUI.
9. Change the statement *aDoc.GetPluralClassName* to *aDocGUI.GetTitle.*
10. In ArcView 3.0, for objects of the DLL class, the GetFileName request returns a file name object instead of a string.
11. Change the statement *av.LoadLibrary* to *System.LoadLibrary.*
12. For instances of the ITheme class, change the GetISrc request to GetImgSrc.

The ArcView online help contains a detailed list of all changes made to Avenue from 2.*x* to 3.0. (Go to the Index tab in the Help Topics window and type *What's new.* Click on the Display button for a complete list of related topics.)

## Migrating from ArcView 2.x to 3.0

Many sites will find it desirable to run both ArcView 2.*x* and ArcView 3.0 concurrently while migrating to Version 3.0. This can facilitate project customization, allow time for training materials and documentation to be updated to 3.0, and ensure that all users are comfortable with the changes in 3.0 before removing 2.*x* from the system. It should be noted, however, that project files created under 3.0 are not downwardly compatible with 2.*x*. Accordingly, modifications to projects will have to be made under both 2.*x* and 3.0, and separate project files maintained, until the transition is complete.

# Glossary

## Address Events

An address event is a feature located by a unique address. ArcView-supported addresses include street addresses (the most common) as well as polygon and point addresses, such as zip codes or land parcel numbers, respectively. (See also: *Event Table, Polygon Features.*)

## Address Matching

Address matching involves assigning an absolute location through X,Y coordinates to each address in an address event table. This process occurs through interpolating specific address locations against a geocoded street theme coded with address ranges. (See also: *Event Table.*)

## Alias Table

ArcView allows an alias table to be used for address geocoding. In an alias table, you use place name aliases in which a place name, such as Madison County Hospital, is assigned a street address. To facilitate locating addresses in a view, the alias table is associated with an address event table in address geocoding or with a matchable theme. (See also: *Event Table.*)

## Annotation

Annotation is made up of text used to label features on a map. In the context of ARC/INFO data sets, annotation refers to text elements stored as part of an ARC/INFO coverage. Annotation contains not only a text string, but additional properties that define the font, color, size, and angle of the annotation. Annotation from an ARC/INFO coverage can be added as a separate feature class to an ArcView view.

## Attribute Data/Table

Attribute data, also known as tabular data, are linked to themes. ArcView shapefiles (and ARC/INFO coverages) contain spatial data and attribute tables. The attribute tables linked to a particular theme may contain geographic information (e.g., addresses or zip codes). Attribute tables can also contain information associated with features in a theme, such as soil properties or land use descriptions.

A one-to-one relationship exists between features in a shapefile (or coverage) and records in the theme attribute table. At a minimum, the theme attribute table contains a shape field. In addition, theme attribute tables derived from coverages contain an additional field of a unique numeric identifier for each feature. Fields can be added to the theme attribute table to identify additional characteristics of features. For example, a block group number could be associated with each polygon of a census geography theme. Fields can also be used to join additional attribute data to a theme attribute table. (See also: *Field, Join, Record, Table.*)

## CAD (computer-aided design)

CAD software is designed for drafting and for the manipulation of graphic elements. CAD software is commonly used in engineering, surveying, and mapping applications.

## CAD Drawing

A file created by CAD software containing the graphics elements that comprise the drawing. These drawing files include lines, polylines, text, and other elements. In ArcView 3.0, a theme can be created directly from an AutoCAD (.dwg) or MicroStation (.dgn) drawing file.

## Chart

A chart is a graphic representation of attribute or tabular data. In ArcView, a chart references the data from a project table in one of six formats: area chart, bar chart, column chart, line chart, pie chart, and XY scatter chart. An ArcView chart is dynamic in that it represents the current status of the data in the table. Changes to either data values or the selected records in the table are immediately reflected in the corresponding chart. (See also: *Dynamic.*)

## Classification

To classify is to assign features in a theme to classes based on attribute values. Symbols are then assigned to each class so that the distribution of features in each class may be viewed on a map. Within ArcView, classification methods include unique values, natural breaks, quantile, equal area, equal interval, standard deviation, and dot density. The type of classification appropriate for each field depends on the nature of the data.

## Continuous Events

A continuous event is a type of route event whose features are located in a continuous fashion. Assume the route system is a natural gas distribution network. The continuous event might be pipeline age organized into categories such as very old (installed before 1965), old (1966 to 1975), moderate (1976 to 1985), and new (1986 to the present). The gas distribution network could then be coded according to locations where pipeline age category changes. (See also: *Route Events.*)

## Coordinate System

A coordinate system is a map reference system in which precise geographic position can be referred to by means of a rectangular grid. The use of a rectangular grid allows features to be located using X,Y coordinates. This system facilitates the integration of survey data into a larger national grid.

Each coordinate system is derived from a specific map projection. The Universal Transverse Mercator System is derived from the Transverse Mercator projection. This system divides the world into 60 north-south zones and 20 east-west zones, for a total of 1,200 unique grid zones.

The State Plane Coordinate System is used in the United States. In this system, each state is divided into one or more zones extending north-south or east-west. Zones extending north-south are based on the Transverse Mercator projection; zones extending east-west are based on the Lambert Conformal Conic projection. (See also: *Geographic Coordinates, Map Projection.*)

## Coverage

In the context of this book, a coverage refers to an ARC/INFO coverage. An ARC/INFO coverage is a database that stores geographic and tabular data in a set of files. These data files are organized within a common directory. An ARC/INFO coverage can serve as a spatial data source for a theme in ArcView.

## Data Group

A data group is the aggregating unit of a chart. It consists of a set of related elements that describe the same variable. If the data series is formed from records, the data group is aggregated by fields. If the series is formed from fields, the data group is aggregated by record. (See also: *Chart, Field, Record, Table,* and the insert titled, "Clarification of Data Markers, Series, and Groups" in Chapter 7.)

## Data Marker

A data marker is an element of one of these types of charts: column, bar, area, pie slice, or point symbol. This element represents the value of a particular field for a specific record in a table. Data markers are analogous to cells in a spreadsheet. In other words, a data marker in a chart represents the intersection of a field and record in a table, also known as a field value. (See also: *Chart, Field, Record, Table,* and the insert titled, "Clarification of Data Markers, Series, and Groups" in Chapter 7.)

## Data Series

A data series is a set of values compared in a chart. The individual elements of the data series, which may be formed from the records or fields in a table, are displayed in the chart legend. (See also: *Chart, Field, Record, Table* and the insert titled, "Clarification of Data Markers, Series, and Groups" in Chapter 7.)

## Database Theme

A database theme is a theme within a view. In ArcView 3.0, a database theme is based on spatial data stored in an RDBMS, such as Oracle or Informix, using the Spatial Database Engine (SDE). The ability to add a database theme from an SDE database is enabled by loading the SDE Themes extension from the ArcView project window. Database themes allow you to efficiently display and query spatial data comprised of hundreds of thousands of features.

## Datum

A datum is a reference system used to describe the surface of Earth. Coordinate systems that are used when surveying point locations on Earth, such as the State Plane Coordinate System or the Universal Transverse Mercator Grid, are linked to a specific datum. For North America, there are two: the North American datum of 1927, and the North American datum of 1983. (See also: *Geodetic Control.*)

## DOS (MS-DOS) 8.3 Convention

Under MS-DOS, file names are limited to a maximum of eight characters, followed by an extension comprising a maximum of three characters. The field name and extension are separated by a period. Examples of the convention are *highways.dbf,* and *marker.ai.* Directory names in MS-DOS are also restricted to the same naming convention.

## Dynamic

In ArcView, *dynamic* refers to a document (table, chart, or layout) that reflects the current status of the source data it is based on. As the source data changes, the document associated with the source data changes accordingly.

## DXF (Drawing Interchange Format)

A format used for exchanging vector data across CAD software, such as between AutoCAD and MicroStation. DXF files can be in either ASCII or binary in format. ArcView 3.0 lets you create a theme in a view directly from a .dxf file.

## Event Table

An event table contains geographic locations ("events"); however, the table is not in a spatial data format. The geographic locations may be absolute (e.g., latitude and longitude coordinates) or relative (e.g., street addresses). Event tables contain of one of three general types of events: X,Y; route; or address. (See also: *Address Events; Route Events; X,Y Events.*)

## Event Theme

An event theme is a theme based on an event table. Event themes are based on X,Y events, address events, or route events. When the geographic locations are absolute, as in the case of X,Y events, points are created in a theme directly from the X,Y coordinate values. When the geographic locations are relative, feature locations are translated from relative to absolute locations, and the resultant features are stored in the ArcView shapefile format. A theme referencing this shapefile is subsequently added to the active view. (See also: *Event Table.*)

## Export

This term is used in both a general and very specific sense in this book. The general definition is simply the process of moving a file that has been prepared in ArcView to another application. Specifically, Export is a utility provided with PC or UNIX ARC/INFO. This utility creates an ARC/INFO interchange format file. The ARC/INFO interchange format file is given the suffix E*nn* (where *nn* is a number from 00 to 99) for each successive volume created.

## Field

A field is a unique descriptor or characteristic of a record (instance) in a database. Fields are also called columns or items. For a customer database, examples of fields could be name, address, city, and zip code. ArcView supports four field types: Number, String, Boolean, and Date. (See also: *Record, Table.*)

## Geocoding

Geocoding is the process of assigning an absolute location (X,Y coordinates) to a geographic feature referenced by a relative location, such as a street address or zip code.

## Geodetic Control

A geodetic control consists of a correlated network of points for which accurate elevation and position locations have been determined. Local surveys are subsequently adjusted to such a correlated network of points. Control points are tied into existing horizontal and vertical control networks and then adjusted to the appropriate datum. The appropriate datum for North America is either the 1927 datum or the 1983 datum. (See also: *Datum, Map Projection.*)

## Geographic Coordinates (Latitude/Longitude)

Geographic coordinates refer to the geographic reference system of latitude and longitude in which Earth is treated as a sphere and divided into 360 equal parts (degrees). This division is performed along two axes, one running east-west along the equator and the other running north-south along the Greenwich Prime Meridian. Using this coordinate system, any location on Earth can be identified with a unique X,Y coordinate pair. Geographic coordinates are commonly measured in degrees, minutes, and seconds, but can also be formatted as decimal degrees. The format used by ArcView is decimal degrees. (See also: *Coordinate System, Map Projection, Map Units.*)

## GIS (Geographic Information System)

A GIS is a geographic database manager. In other words, a GIS treats all geographic (spatial) features as records in a database, not simply as graphics. Nearly all concepts in traditional relational databases apply to GIS, but with the added dimension of geography.

A GIS builds a bridge between geography and descriptive information through a georelational model. This model provides a one-to-one relationship between a spatial data set and an attribute table. Some database fields are predefined in attribute tables, but you can add any fields you desire. The georelational model also permits you to connect to other tabular databases, whether internal or external to the GIS software.

## Grid

A grid is a spatial data set comprised of a regular array of equally spaced cells. Grid data is also known as raster data. Each grid cell is of uniform size and is referenced by a row and column location that precisely locates the cell within the grid. The grid is referenced to Earth by means of a world file which ties the origin point of the grid to a specific X,Y location on Earth.

## Hot Link

A hot link causes a predefined action to occur when you select the Hot Link tool from the button bar and click on an appropriate feature in a theme. A user defines the action executed at this point through specifying a field value from a theme attribute table and an action to be performed. Examples of hot link actions are displaying an image or opening another view.

## Image Data

Image data are a type of raster data whose features have been converted to a series of celi values by an optical or electronic device. Image data typically refers to satellite imagery or scanned aerial photographs in which each grid cell contains a brightness value representative of the portion of the light spectrum being measured. Imagery may be single band (grayscale) or multiple band (multispectral). ArcView supports the display of both single band and multiple band imagery. (See also: *Raster Data.*)

## Import

This term is used in both a general and very specific sense in this book. The general sense is simply the process of loading or retrieving data into ArcView. Examples include importing dBase or delimited text files into ArcView. Both of these file types are created outside of ArcView and then incorporated (imported) into the program.

Import is also a standalone utility—one that is supplied with ArcView—that allows an ARC/INFO interchange file to be converted to an ARC/INFO coverage or database data file. The Import utility allows ARC/INFO interchange format data to be added to an ArcView project by creating a PC ARC/INFO format coverage (under the Windows platform) or a workstation ARC/INFO coverage (under UNIX). The Import utility also allows project files created under ArcView Version 1 (.av files) to be imported into an ArcView 2.*x* or 3.0 project.

## Join

Joining is a process by which two or more tables are merged into a virtual table through a common field. The resultant table appears as a single table in ArcView, despite being formed from two or more separate data source files.

Join is used when there is a one-to-one or one-to-many relationship between records in a source table and records in a destination table. In the case of a many-to-one relationship between records in a source table and records in a destination table, the Link feature should be used instead of Join. (See also: *Field, Link, Table.*)

## Layout

A layout is a map composition document used to prepare output from ArcView. A layout allows you to define a page and place ArcView documents (views, charts, and tables), imported graphics, and graphics primitives on the page. A layout is dynamic in that graphics linked to ArcView documents can immediately reflect changes made in those documents. Dynamic layouts are live-linked. (See also: *Live Link*.)

## Line Features

Line features are used to represent linear entities, such as highways or streams. Other important uses for lines are to delineate perimeters of polygons and to lay down the positions of route systems and regions. Lines are located and defined through the assignment of a unique series of X,Y coordinate pairs.

Lines (also known as arcs) are made up of connected strings of line segments. Each line segment is delineated by a vertex. The vertices at the endpoints of lines are called nodes. (See also: *Vertex*.)

## Linear Events

A linear event is a type of route event. Linear events are features located along a specific segment of a route. Assume the route system is a road network. A linear event could be a pothole repair segment occurring from milepost 10.5 to 10.8 on County Road 41. (See also: *Route Events*.)

## Link

Linking is used in the case of a many-to-one relationship between records in a source table and records in a destination table. A link defines the relationship between the two tables, but does not bring the tables into a single virtual table. Linking displays all candidate records in the source table that match each unique value in the destination table. For example, a table that identifies apartment complexes by parcel number could be linked to a second (source) table that lists all the tenants in each apartment complex. (See also: *Join*.)

## Live Link

A live link is a dynamic link between a layout frame and the corresponding ArcView document or element. Chart and table frames are always live-linked to respective chart and table documents. View, legend, and scale bar frames can be either live-linked or static. (See also: *Dynamic, Layout, Static Link.*)

## Look-up Table

A look-up table is a secondary table that contains additional information about features identified in a specific field of a primary table. For example, a primary table could contain a list of soil mapping units and a field that holds a symbol for each mapping unit. A look-up table could then be prepared that identifies the soil drainage class for each soil mapping unit.

A one-to-one or many-to-one relationship exists between records in a primary table and records in a secondary table. In ArcView, the Join function can be used to create a virtual table in which the records of the look-up table are associated with records in the primary table, based on values for a common field.

## Map Projection

A map projection is a system by which the curved surface of Earth is represented on the flat surface of a map. The challenge inherent in all map projections is to preserve the properties of area, shape, elevation, distance, and direction present on Earth's surface through the transformation to a map surface. Because it is impossible to preserve all properties simultaneously, you need to select a map projection optimized to preserve the property you most desire.

## Map Units

Map units are the units in which spatial data coordinates are stored. Map units are used in ArcView to set the scale of the view. By default, map units are set to Unknown. The map units for a view are set from the View Properties dialog window, which is accessed from the View menu.

Map units are differentiated in ArcView from distance units. Distance units are used to display measurements and dimensions within a view.

## Matchable Theme

A theme becomes matchable after geocoding indexes have been built on it that conform to an ArcView-supported address style. A matchable theme is used to create an Address Event Theme by address matching from a table that contains address information. (See also: *Event Theme.*)

## Normalize

When applied to thematic classification, normalize refers to expressing the values for one field relative to the values for a second field. Some data, such as those expressed by percentages, may already be normalized.

## Point Events

A point event is a type of route event. Point events are located at a specific point along a route. Assume the route is a highway system coded by route (e.g., Route 41 linking Tijeras Canyon to the town of Milbank) and mile post. Point events could be accident locations identified by a specific mile post on the route system. (See also: *Route Events.*)

## Point Features

Point features represent entities found at discrete locations, such as well sites, transformer sites, or customer locations. Each point feature is located using a single X,Y coordinate pair.

## Polygon Features

Polygon features are used to represent entities of a real extent, such as land parcels, geologic zones, or islands. Polygon features are defined by a series of X,Y coordinate pairs that identify the polygon's perimeters.

Imagine four discrete points (X,Y coordinate pairs) and four lines (arcs) connecting the points in a closed system. The points and lines define the polygon's perimeters. The entity (area) thus defined is a polygon feature.

When polygon features are derived from an ARC/INFO coverage, a special type of point feature known as a *labelpoint* may be associated with each polygon. The labelpoint contains the same attributes as the polygon and enables the polygons to be modeled alternatively by using point features.

## Project

A project is the overall structure used in ArcView to organize component documents such as views, tables, charts, layouts, and scripts. A project maintains the current state of all component ArcView documents, as well as the configuration of the graphical user interface. This information is stored in an ASCII file with an .apr extension.

## Raster Data

Raster data is also referred to as grid cell data. Raster data sets store spatial data as cells within a two-dimensional matrix. This matrix is a gridded area of uniformly spaced rows and columns in which each grid cell contains a value that represents part of the feature being depicted. Grid cell values may be continuous, as in elevation data, or discrete, as in land use data, for which each grid cell value is associated with a specific land use. The resolution of raster data is dependent on the size of the grid cell.

## Record

A record is a specific instance or member of a database. Records are also called rows. In a customer database, records could be individual customers. (See also: *Table.*)

## Relational Join

A relational join is the merging of two attribute tables to produce a single output table based on equivalent values of a common attribute field. For example, the feature attribute table for a theme comprised of land parcels, each identified by an assessor's unqiue parcel number, can be joined to a second table containing additional attributes for each parcel by using the common field of assessor's parcel number in the two tables. In ArcView, the result of a relational join is the creation of a single virtual table within the ArcView project. The underlying source tables are still maintained separately on disk.

## Route Events

Route events are relative locations of features along a route system, such as a road network or power lines. In this case, locations are relative in that they are referenced as distances from a known starting point, such as 1.2 miles from the Salt

Creek substation. Route events come in three forms: point, linear, and continuous. (See also: *Continuous Events, Event Table, Linear Events, Point Events.*)

## Script

An ArcView script is a macro written in Avenue (ArcView's programming language) in order to customize the ArcView environment. Avenue scripts are stored and executed within a project.

## Shapefile

A shapefile is the native ArcView spatial data format. In contrast to an ARC/INFO coverage, a shapefile is a simpler, non-topological format that offers the advantages of faster display and the ability to be created or edited within ArcView. An ArcView shapefile also serves as an effective interchange format for moving data in and out of ARC/INFO or other supporting software.

## Static Link

A static link is a non-dynamic link between a layout frame and the corresponding ArcView document or element. If a view, legend, or scale bar frame is static, the frame represents a snapshot of the view at the time the frame was created. (See also: *Layout, Live Link.*)

## Street Net

A street net (or network) is a spatial data set that represents streets as a connected series of line segments. Each line segment is typically coded with the address range of the street represented, thereby allowing the theme resulting from the street net to be used for geocoding.

## Table

A table is the basic unit of storage in a database management system. It is a two-dimensional matrix of attribute values. ArcView, ARC/INFO, and SQL (Structured Query Language) share this term.

Fields are the vertical components in a table. In SQL, fields are also called columns. Another common term used for a field is *item.* Records are the horizontal components in a table. In SQL, they are called rows. The intersection of a field and a record is a single, discrete entity called a field value or item value. In SQL, the intersection is called a datum.

In ArcView, a table references tabular data from several sources—dBASE, INFO, or delimited text files—in a uniform display format composed of fields and records. ArcView tables are dynamic in that they reference the source data rather than containing the data itself. Consequently, an ArcView table represents the current state of data at the time the project is opened, including all changes made to the data after the last time the project was accessed.

Tables can be displayed, queried, and analyzed. In addition, tables joined to a theme can be queried and analyzed spatially. Tables can also be joined based on sharing equivalent values of a common field, even if the underlying format for the source data differs (such as dBASE and INFO).

## Thematic Classification

Thematic classification refers to the assignment of symbols to a theme's features based on field values in the theme's attribute table, including tables that have been joined to the theme attribute table. A classification may be applied to a theme based on either a range of values or unique values for the field. In ArcView, the maximum number of classes, regardless of the classification type, is 64.

## Thematic Map

A thematic map is a map that displays a set of related geographic features. Typically, a classification has been applied so that the map displays attribute information associated with the geographic features. Examples include maps that display land use or soil drainage classes.

## Theme

A theme is a spatial data set (geography) linked with attribute data that contain a locational component. Imagine a map of a census tract, upon which you superimpose two other maps. The other two maps represent streets and stores in a pizza chain. Each of these maps would be a separate theme.

A theme is a group of similar geographic features in a view. A theme is based on a set of features taken from a specific feature class which is, in turn, derived from a spatial data source. Feature classes include regions, routes, polygons, arcs, points, annotation, labels, and nodes. Certain feature classes, such as regions or routes, are complex classes made up of more basic feature types. The three basic feature types are points, lines (arcs), and polygons.

A theme can be comprised of all features within a specific feature class from a spatial data source, such as all land use polygons, or it can be a subset of features from a spatial data source, such as all commercial land use polygons.

A spatial data source may contain more than one feature class. For example, a census geography may contain both lines representing streets and polygons representing census tracts.

## Vector Data

Vector data sets, also referred to as arc node data, contain features with discrete positions. These positions are stored as X,Y coordinate pairs. Vector data consist of a series of nodes that define line segments, which are in turn joined to form more complex features, such as line networks and polygons. ARC/INFO coverages and ArcView shapefiles are both examples of vector data. (See also: *Line Features, Polygon Features.*)

## Vertex

A vertex is a specific X,Y coordinate pair that forms a line. Vertices are often referred to as shape points. The more vertices or shape points making up the line, the more accurate the representation of the feature. (See also: *Line Features.*)

## View

Essentially, a view is a collection of themes. Assume you are examining the market for home improvement products in a three-county area (Bernalillo, Sandoval, and Valencia) in the state of New Mexico. You want geographic detail, including streets, roads, and zip code areas. You also want lifestyle characteristics of the people who live in specific locales and neighborhoods in the tri-county area.

One of your views could consist of four themes: a map of the tri-county area, the street net, zip code areas, and a market segmentation overlay that defines market segments by zip code area. You could choose to display all themes in the view simultaneously, with the tri-county land map on the bottom and the market segment theme on top, or you could choose to display only two themes in the view simultaneously, and so on.

In ArcView, views organize themes. A view window contains a graphics display area and a table of contents that lists all themes present in the view. You select which theme or themes will be drawn in the view.

More specifically, a view is an interactive map. All themes present in a view share a common geographic coordinate system, and most share a coincident geographic extent (area) as well.

Views are the primary document type in ArcView. Other ArcView documents (tables, charts, and layouts) are typically linked to themes contained in views. A view imposes an organizational structure on themes derived from spatial data, and it provides a means for displaying, querying, and analyzing the data. This is accomplished through the interaction of the three primary components of a view: the map display, the table of contents, and the view's graphical user interface. (See also: *Theme.*)

## X,Y Coordinates

X,Y coordinates are used to refer to the unique geographic location of a spatial feature. All spatial data is maintained in some form of map coordinate system, the most widely used being the geographic coordinates of latitude and longitude.

## X,Y Events

In GIS, X,Y events are the exact locations of features that have been pinpointed on a map by X,Y coordinates. Commonly used map coordinate systems are latitude/longitude, UTM, and State Plane Coordinates. (See also: *Event Table.*)

## World File

A world file is a text file used to tie the location of a raster data set, such as an image or a grid, to a location on Earth. A world file contains the X and Y coordinates for the origin of the grid and, optionally, the grid cell size and rotation of the grid.

# Index

## B



## C

# D

# E

## G



## H

## M

## N



## O



## P

## Q



## R

## T

## U



## V



## W

## X



## Z

*MicroStation Exercise Book 5.X*
$34.95 Includes Disk
Optional Instructor's Guide $14.95

*MicroStation Reference Guide 5.X*
$18.95

*Build Cell for 5.X*
Software $69.95

*101 MDL Commands (5.X and 95)*
Executable Disk $101.00
Source Disks (6) $259.95

## Pro/ENGINEER and Pro/JR.

*Automating Design in Pro/ENGINEER with Pro/PROGRAM*
$59.95 Includes CD-ROM

*INSIDE Pro/ENGINEER, 3E*
$49.95 Includes Disk

*Pro/ENGINEER Exercise Book, 2E*
$39.95 Includes Disk

*Pro/ENGINEER Quick Reference, 2E*
$24.95

*Thinking Pro/ENGINEER*
$49.95

*Pro/ENGINEER Tips and Techniques*
$59.95

*INSIDE Pro/JR.*
$49.95

## Softdesk

*INSIDE Softdesk Architectural*
$49.95 Includes Disk

*Softdesk Architecture 1 Certified Courseware*
$34.95 Includes CD-ROM

*Softdesk Architecture 2 Certified Courseware*
$34.95 Includes CD-ROM

*INSIDE Softdesk Civil*
$49.95 Includes Disk

*Softdesk Civil 1 Certified Courseware*
$34.95 Includes CD-ROM

*Softdesk Civil 2 Certified Courseware*
$34.95 Includes CD-ROM

## Interleaf

*INSIDE Interleaf (v. 6)*
$49.95 Includes Disk

*Interleaf Quick Reference (v. 6)*
$24.95

*Interleaf Exercise Book (v. 5)*
$39.95 Includes Disk

*Interleaf Tips and Tricks (v. 5)*
$49.95 Includes Disk

*Adventurer's Guide to Interleaf LISP*
$49.95 Includes Disk

## Other CAD

*Manager's Guide to Computer-Aided Engineering*
$49.95

*Fallingwater in 3D Studio*
$39.95 Includes Disk

## Windows NT

*Windows NT for the Technical Professional*
$39.95

## SunSoft Solaris

*SunSoft Solaris 2.* for Managers and Administrators*
$34.95

*SunSoft Solaris 2.* User's Guide*
$29.95 Includes Disk

*SunSoft Solaris 2.* Quick Reference*
$18.95

*Five Steps to SunSoft Solaris 2.**
$24.95 Includes Disk

*SunSoft Solaris 2.* for Windows Users*
$24.95

## HP-UX

*HP-UX User's Guide*
$29.95

*HP-UX Quick Reference*
$18.95

*Five Steps to HP-UX*
$24.95 Includes Disk

# OnWord Press Distribution

## End Users/User Groups/Corporate Sales

OnWord Press books are available worldwide to end users, user groups, and corporate accounts from local booksellers or from Softstore/CADNEWS Bookstore: call 1-800-CADNEWS (1-800-223-6397) or 505-474-5120; fax 505-474-5020; write to SoftStore, Inc., 2530 Camino Entrada, Santa Fe, NM 87505-4835, USA or e-mail orders@hmp.com. SoftStore, Inc., is a High Mountain Press Company.

## Wholesale, Including Overseas Distribution

High Mountain Press distributes OnWord Press books internationally. For terms call 1-800-4-ONWORD (1-800-466-9673) or 505-474-5130; fax to 505-474-5030; e-mail orders@hmp.com; or write to High Mountain Press, 2530 Camino Entrada, Santa Fe, NM 87505-4835, USA.

**On the Internet: http://www.hmp.com**

OnWord Press, 2530 Camino Entrada, Santa Fe, NM 87505-4835 USA

# CD-ROM Credits and Copyrights

The companion CD-ROM contains data and project files for use with the exercises in the accompanying book.